Ecopoetics of Reenchantment

ECOCRITICAL THEORY AND PRACTICE

Series Editor: Douglas A. Vakoch, METI

Advisory Board

Sinan Akilli, Cappadocia University, Turkey; Bruce Allen, Seisen University, Japan; Zélia Bora, Federal University of Paraíba, Brazil; Izabel Brandão, Federal University of Alagoas, Brazil; Byron Caminero-Santangelo, University of Kansas, USA; Chia-ju Chang, Brooklyn College, The City College of New York, USA; H. Louise Davis, Miami University, USA; Simão Farias Almeida, Federal University of Roraima, Brazil; George Handley, Brigham Young University, USA; Steven Hartman, Mälardalen University, Sweden; Isabel Hoving, Leiden University, The Netherlands; Idom Thomas Inyabri, University of Calabar, Nigeria; Serenella Iovino, University of Turin, Italy; Daniela Kato, Kyoto Institute of Technology, Japan; Petr Kopecký, University of Ostrava, Czech Republic; Julia Kuznetski, Tallinn University, Estonia; Bei Liu, Shandong Normal University, People's Republic of China; Serpil Oppermann, Cappadocia University, Turkey; John Ryan, University of New England, Australia; Christian Schmitt-Kilb, University of Rostock, Germany; Joshua Schuster, Western University, Canada; Heike Schwarz, University of Augsburg, Germany; Murali Sivaramakrishnan, Pondicherry University, India; Scott Slovic, University of Idaho, USA; Heather Sullivan, Trinity University, USA; David Taylor, Stony Brook University, USA; J. Etienne Terblanche, North-West University, South Africa; Cheng Xiangzhan, Shandong University, China; Hubert Zapf, University of Augsburg, Germany

Ecocritical Theory and Practice highlights innovative scholarship at the interface of literary/cultural studies and the environment, seeking to foster an ongoing dialogue between academics and environmental activists.

Recent Titles

Ecopoetics of Reenchantment: Liminal Realism and Poetic Echoes of the Earth, by Bénédicte Meillon
Indian Feminist Ecocriticism, edited by Douglas A. Vakoch and Nicole Anae
Interrogating Boundaries of the Nonhuman: Literature, Climate Change, and Environmental Crises, edited by Sune Borkfelt and Matthias Stephan

Ecopoetics of Reenchantment

Liminal Realism and Poetic Echoes of the Earth

Bénédicte Meillon

LEXINGTON BOOKS
Lanham • Boulder • New York • London

Published by Lexington Books
An imprint of The Rowman & Littlefield Publishing Group, Inc.
4501 Forbes Boulevard, Suite 200, Lanham, Maryland 20706
www.rowman.com

86–90 Paul Street, London EC2A 4NE

Copyright © 2023 by The Rowman & Littlefield Publishing Group, Inc.

All rights reserved. No part of this book may be reproduced in any form or by any electronic or mechanical means, including information storage and retrieval systems, without written permission from the publisher, except by a reviewer who may quote passages in a review.

British Library Cataloguing in Publication Information Available

Library of Congress Cataloging-in-Publication Data

Names: Meillon, Bénédicte, 1977- author.
Title: Ecopoetics of reenchantment : liminal realism and poetic echoes of the earth / Bénédicte Meillon.
Description: Lanham : Lexington Books, [2022] | Series: Ecocritical theory and practice | Includes bibliographical references and index.
Identifiers: LCCN 2022037120 (print) | LCCN 2022037121 (ebook) | ISBN 9781666910421 (cloth) | ISBN 9781666910445 (pbk.) | ISBN 9781666910438 (epub)
Subjects: LCSH: Ecofiction--History and criticism. | Magic realism (Literature) | Liminality in literature. | Human ecology in literature. | Ecology in literature. | Nature in literature. | LCGFT: Literary criticism.
Classification: LCC PN3448.E36 M45 2022 (print) | LCC PN3448.E36 (ebook) | DDC 813/.54093553--dc23/eng/20220910
LC record available at https://lccn.loc.gov/2022037120
LC ebook record available at https://lccn.loc.gov/2022037121

PERMISSIONS

Ann Pancake, excerpts from *Strange as This Weather Has Been*. Copyright © 2007 by Ann Pancake. Reprinted with the permission of The Permissions Company, LLC on behalf of Counterpoint Press, counterpointpress.com.

From *Prodigal Summer* by Barbara Kingsolver. Copyright © 2000 by Barbara Kingsolver. Courtesy of HarperCollins Publishers.

Extract from *Prodigal Summer* by Barbara Kingsolver (c) Barbara Kingsolver. Reprinted by permission of Faber and Faber Ltd.

From *Animal Dreams* by Barbara Kingsolver. Copyright © 1990 by Barbara Kingsolver. Courtesy of HarperCollins Publishers.

From *Animal Dreams*. Copyright © 1990 by Barbara Kingsolver. All rights reserved. Reproduced with permission of Little Brown Book Group Limited through PLSclear.

Jean Giono, excerpts from *Joy of Man's Desiring*, translated by Katherine Allen Clark. Copyright © 1980 by Aline Giono. Reprinted with the permission of The Permissions Company, LLC on behalf of Counterpoint Press, counterpointpress.com.

From *The Overstory: A Novel* by Richard Powers. Copyright © 2018 by Richard Powers. Used by permission of W. W. Norton & Company, Inc.

Excerpt(s) from *The Overstory: A Novel* by Richard Powers, copyright © 2018 Richard Powers. Reprinted by permission of Vintage Canada, a division of Penguin Random House Canada Limited. All rights reserved.

From *The Overstory* by Richard Powers. Copyright © Richard Powers 2018, published in the United States by W. W. Norton & Company, 2018, in Great Britain by William Heinemann, 2018, Vintage, 2019, Vintage Classics, 2022. Reprinted by permission of The Random House Group Limited

Robin Wall Kimmerer, excerpts from Braiding Sweetgrass: Indigenous Wisdom, Scientific Knowledge and the Teachings of Plants. Copyright © 2013, 2015 by Robin Wall Kimmerer. Reprinted with the permission of The Permissions Company, LLC on behalf of Milkweed Editions, www.milkweed.org.

From *Braiding Sweetgrass* by Robin Wall Kimmerer. Copyright © Robin Wall Kimmerer, 2013, published by Milkweed Editions 2013, Penguin Books 2020. Reprinted by permission of Penguin Books Limited.

From *The Sacred Hoop* by Paula Gunn Allen Copyright © 1986, 1992 by Paula Gunn Allen Reprinted by permission of Beacon Press, Boston.

From *Power* by Linda Hogan. Copyright © 1988 by Linda Hogan. Used by permission of W. W. Norton & Company, Inc.

From *Power* by Linda Hogan. Copyright © 1988 by Linda Hogan. Used by permission of Beth Vesel Literary Agency.

Excerpt(s) from *Becoming Animal: An Earthly Cosmology* by David Abram, copyright © 2010 by David Abram. Used by permission of Pantheon Books, an imprint of the Knopf Doubleday Publishing Group, a division of Penguin Random House LLC. All rights reserved.

Excerpt(s) from *The Fifth Sacred Thing* by Starhawk, copyright © 1993 by Miriam Simos. Used by permission of Bantam Books, an imprint of Random House, a division of Penguin Random House LLC. All rights reserved.

From *Symbiotic Planet* by Lynn Margulis, copyright © 1998. Reprinted by permission of Basic Books, an imprint of Hachette Book Group, Inc.
From *Symbiotic Planet* by Lynn Margulis, copyright © 1998. Reproduced with permission of Orion Publishing Group Limited through PLSclear.

Contents

Acknowledgments ix

Introduction 1

PART I: FROM DISENCHANTMENT TO AN ECOPOETICS OF REENCHANTMENT 19

Chapter One: Disenchanted, Enchanted, and Reenchanted Worldviews 21

Chapter Two: Toward an Ecofeminist Project of a Rational Reenchantment 41

Chapter Three: An Ecofeminist Remystification of Narrative: The Many Faces of Gaia in the Anthrop-o(bs)cene 51

Chapter Four: Sowing the Seeds of an Ecopoet(h)ics of Wonder and Enchantment: Reincorporating Language and the Human into the Flesh and Song of the World 109

PART II: ECOPOETIC REENCHANTMENT VIA LIMINAL REALISM 125

Chapter Five: Why Liminal, Rather than "Magical," "Spiritual," "Mystical," "Ontological," or "Epistemological" Realism? 127

Chapter Six: Postcolonial Liminality and (Re)initiation into a Multispecies World: Moving betwixt and between Human and Other-than-Human Realms in Linda Hogan's *Power* and Leslie Marmon Silko's *Ceremony* 143

Chapter Seven: Post-Pastoral Thought-Experiments with Totemic and Animistic Liminality 169

Chapter Eight: Liminal Realism and Interspecies
  Thought-Experiments in Contemporary Fiction   195

**PART III: WRITING AND DWELLING
  ECOPOETICALLY**   241

Chapter Nine: Ecopoets and the Art of Anamorphosis   243

Chapter Ten: Postmodern Shamanism: Making Headway Toward
  Other-than-Human Perspectives   263

Chapter Eleven: Reweaving Word to World: Ecopoets as
  Instruments of the Sympoietic Song of the Earth   283

Chapter Twelve: Restor(y)ing and Rewor(l)ding: Writing in a
  Grounded Middle Voice   307

Chapter Thirteen: Translating the Song of the Earth: Reen-*chanting*
  Earthly Harmonies   319

Conclusion   333

Bibliography   341

Index   359

About the Author   375

Acknowledgments

Because writing springs from watershed flows and processes, this book owes to more people than I could possibly name. The bibliography and notes point to most of its irrigation system.

I will start by expressing my deepest gratitude to Charlie Grandjeat, for guiding me throughout the long-winded French academic process called *Habilitation à Diriger des Recherches* and through the writing of this book, always providing encouragement as well as exceptionally sound and caring advice. His most thoughtful suggestions for revision, his illuminating reading advice, comments, and questions, and his own pioneering work in ecopoetics have long been a great influence over my work.

I am thankful toward the other members of my HDR committee, Angela Biancofiore, Nathalie Cochoy, Claire Omhovère, Jonathan Pollock, and Frédérique Spill who have labored over a first version of this book and whose comments, questions, and feedback have been of tremendous help in the subsequent revision process. Special thanks go to Frédérique Spill, for bringing Ron Rash's writing to my attention, and to Nathalie Cochoy, whose fine love of literature and arts is of a wonderfully contagious kind.

I am very grateful to the various institutions that have supported my research in the past two years allowing this book to come to completion: first, the CRESEM and the Université de Perpignan via Domitia, for a precious six-month sabbatical, and then the LARCA-CNRS at the Université Paris Cité, for a one-year fellowship freeing me of half of my teaching duties so that I could finish my HDR and complete a few research projects including the revision and publication of this book. I am also thankful for the local support of my English Studies department team, who have generously compensated for the classes that temporarily had to be filled during the time I have devoted to this work.

My colleagues working with me in the OIKOS ecopoetics workshop have also been of great support and inspiration, especially Pascale Amiot, Marie-Pierre Ramouche, Caroline Durand-Rous, Margot Lauwers, Jonathan

Pollock. Special thanks go out to all the colleagues whose dedication, hard work, and even friendship have one way or another trickled down into the work presented in this book. I am thinking particularly of Joni Adamson, Athane Adrahane, Aline Bergé, Françoise Besson, Nathalie Blanc, Anne-Laure Bonvalot, Rachel Bouvet, Claire Cazajous-Augé, Joanne Clavel, Fanny Fournié, François Gavillon, Caroline Granger, Wendy Harding, André-Marie Harmat, Terry Harpold, Serenella Iovino, Olivier Panaud, Anne Simon, Scott Slovic, Frédérique Spill, and Davide Vago. As I cannot name them all, I would like to thank all the members of the AFEA, the SAES, EASLCE, and ASLE, who do such important work and make our conferences, events, research projects, and publications happen. I am especially indebted to all the colleagues taking part over the years in the various conferences, events, and publications on reenchantment organized at UPVD since 2016, all of which have enriched my thinking. And of course, many of my most astute students have raised questions and shared thoughts in illuminating ways I am thankful for.

My deep gratitude also goes out to the writers and artists that have inspired my work, many of whom I have been very fortunate to meet and exchange with, and most particularly here Belinda Cannone, Linda Hogan, Nathanael Johnson, Barbara Kingsolver, Anne Sibran, Starhawk, and Ron Rash.

Amélie Chevalier has been of precious help in obtaining permissions to quote, with the Index, and with various aspects of the layout work. The team at Rowman & Littlefield have also been of great assistance in bringing this book into fruition.

A million *mercis* to my *sœurcières* for their powerful magic, and to my dear friends for their sustaining love. Thank you to the octopus, the hedgehog, and the turtle who have come my way to share some of their great wisdom. Last but not least, my deepest thanks to Jesse, my sweetest, ever supportive and energizing partner. This work is dedicated to our amazing daughters, Zoë and Leah, who make everyday life full of wonder, song, laughter, and dance. May their future be joyful and enchanting.

Introduction

As I launched into the writing of this book, France was timidly preparing to emerge from two months of confinement at the beginning of an unprecedented global pandemic. Due to the halt of much human bustling, planes, car engines, and various other machines and motors had gone silent. Many people rejoiced at the return of wildlife in suddenly hushed cities. Through an astonishing springtime—and potentially to cope with the dismal picture emerging with the sanitary crisis—some people waxed lyrical about how much quieter urban milieus had suddenly become, and, simultaneously, how the abating of noise pollution had let some of the more-than-human voices of the earth break through. A great number of observers reported feeling enchanted at the increase in birdsong they perceived. We may wonder to what extent such enhanced polyphony corresponded to an actual phenomenon—and bioacoustics has proved that birds were indeed moving back into acoustic niches temporarily vacated by humans—or whether this heightened perception might also be accounted for by the unusually calm moment allowing us to listen out for the sounds that are ordinarily muffled, drowned out by the din of our daily activities. In any case, that era of dire, multifaceted disenchantment has paradoxically brought about a certain dose of reenchantment. As the French were invited to return their children to school—while picking up professional and shopping activities—I wondered how this temporary widely shared feeling might play out as we resumed our pre-Covid lives, falling back into the hustle and bustle of crowded routines.

Could this reawakening to the wealth and intriguing beauty of the earth's "soundscapes"[1] have enduring effects over the choices we make as humans sharing our habitat with so many other-than-human lifeforms? Might we have experienced firsthand that, through the voices of our many co-dwellers, the Earth is loquacious in its own, en-*chant*-ing manner? How deeply might the human species be moved by such extraordinary planetary trouble occurring in the wake of a virus that likely has emerged because of our trafficking with wildlife in ways that we have long known to be irresponsible?[2] Might we seriously start taking account of how animals, plants, and the entire biosphere respond to outrageous anthropic activity, and, possibly, to its partial ceasing? Would most humans quickly revert to the blind mutilating and muting of the Earth in the name of "progress," boosting the economy, in spite of everything that we should have learned by now—both from the science on climate change and from our more-fully felt connections to other species through these times of crisis? Or would we finally start paying serious heed to our responsibility as earth-dwellers, and come *to our senses* as a species?

Before running into this unforeseen planetary crisis, many scientists and thinkers have long been ringing the alarm, insisting on the urgency of finding ways to curb global warming and the ongoing, speedy loss of biodiversity. Working in different spheres and on various levels, many are striving to figure out what might prompt the most powerful and destructive of us to change priorities and ways of life so that our inhabitation of the planet might make sense in the long run and from a biocentric perspective.[3] In this context, it seems relevant to ponder what *sense* can be made of the notion of a "song of the Earth"—that time-old, universal myth tied to the notion of a *harmonia mundi*—resurfacing at the turn of the century,[4] as humans around the globe grow more and more distressed by the unfurling ecocide.

In his paramount study on soundscapes, Murray Schafer approaches the world as an immense musical composition. He studies the thick acoustic environment where humans evolve in direct contact with the sonorous textures of the world which are emitted by humans as they hunt, farm, wage war, run trains and industries, or else by the sea, the wind, trees, animals, birds, insects, and so forth. Schafer moreover broaches aspects of human language such as onomatopoeia that evidently arose in direct response to our acoustic experience of the world. He thus paves the way for an approach to *ecopoiesis* as a form of poetic echoes created by humans in contact with the sonorous flesh of the world, relaying "the voices of the sea," "the voices of the wind," "the symphony of the world's birds."[5] For him, much of our languages and music are an expression of how the sonorous world impresses itself onto our hearing selves.[6]

A chunk of contemporary literature has returned to a *poiesis* relaying the polyphonic "song of the world"—as French writer Jean Giono once put it in his eponymous novel—thus calling attention to that invisible, yet concrete dimension of our lives that we are constantly immersed in, whether we are being mindful or not. Hence my choice to include Giono in my discussion of ecopoetics, whose influence over contemporary literature of reenchantment must be ascertained. As musicologist Joachim Berendt or bioacousticians such as Murray Schafer and Bernie Krause have established, recent sciences have lent veracity to the notion of a song of the earth. Going way beyond an ancient myth or a mere topos for lyrical poets to muse on, the concept of the song of the world does correspond to a biophysical reality.[7] While ethologists have been paying greater attention to bird language or cetacean singing, and how these interact with the larger soundscapes they emerge from, string theory and quantum field physics have besides proposed a theoretical model of the universe based on laws of resonance and vibration. As I have been investigating the humming of vibrant matter and the songs of the world which are mediated through postmodernist ecopoetic prose, I have come to wonder to what extent this widespread recent rehabilitation of such an ancient metaphor might tie in with bioacoustic explorations—both scientific and creative—of the world as music. Bernie Krause's writings on the "biophony," "geophony," and "anthropophony" constituting the soundscapes of the world have proved of particular interest. Krause distinguishes the sounds emanating from geological forces and elements ("geophony"), from those produced by the nonhuman organisms living in a specific biome ("biophony"), from those produced by humans ("anthropophony"). Together with other scientists in his field, Krause has revealed how biomes orchestrate the world's wild music, each organism finding her own acoustic niche in the sound territory she inhabits. As research has shown, human encroachment onto wild habitats induces biodiversity losses that are related to the disorientation and communication problems generated by the introduction of noise in harmonious soundscapes.[8] Mulling this over, I have formulated the hypothesis that in paying attention to and reproducing the world's music, the ecopoetic prose I am looking at might reveal something of the actual capacities of humans, animals, and plants to hear and emit sounds, songs, vibration, and rhythms in harmony. Within a world made of many complex multispecies enmeshments, might some sort of cross-species communication then take place via literature and the arts?

A branch of ecocriticism has developed in the past two decades that is referred to as "ecopoetics"[9]—a field of study delving into the poetic echoes of the earth, with its "sympoietic"[10] soundscapes and multisensorial textures that writers attune to and translate.[11] Picking up Scott Knickerbocker's title in his rich exploration of the field, one might define ecopoetics as the study of the relationships between "the language of nature and the nature of language." In other words, ecopoetics analyzes "the foregrounding of poetic artifice as a manifestation of our interrelation with the rest of nature."[12] Ecopoetics thus reveals how language is "infused with the natural world."[13] It focuses on the poetic, sonorous, and rhythmic aspects of writing to find out how "sound is one way nature in its broadest sense pushes through the poem at the level of form."[14] As made clear from the title chosen for his seminal monograph, *The Song of the Earth* (2000), Jonathan Bate eloquently envisions poetry as both a "response" to and an "echoing of the song of the earth itself."[15] In my own work, I have encouraged ecopoetic readings of prose fiction as well as of artistic media other than literature—such as dance, photography, or cinema—and I am particularly interested here in a kind of prose fiction that exhibits lyrical aspects while resorting to an ecopoetic form of liminal, magical realism—that is, an enchanting realism that challenges the separation between human and other-than-human realms.

Together with the de-divinization and eradication of magic from our world which, according to Max Weber, resulted from adherence to modern science and rationalization, Western minds have cultivated an ontology divesting altogether the more-than-human world of the capacity to feel, sense, reason, or express itself. Associated with the Moderns' notion of personhood, such interior qualities were upheld as the privilege of humans (when not restricted further and perceived as a White Man's prerogative). As ecofeminists decried early on, in pitting nature against culture, the dualistic ontology of Moderns induced both "the death of nature" and our alienation from the latter,[16] proclaiming a radical discontinuity between humans and the rest of nature.[17] Until very recently, those who insisted that they could hear the voices of the Earth, that they could communicate with animals and plants, were typically dismissed as superstitious mystics, new-age morons, or, more seriously, as primitivistic animists. Exceptionally, they were celebrated as great poets or inspired philosophers. If on top of claiming to be sensitive to other-than-human existences, these same people happened to be female or from indigenous backgrounds, they were—and in many places still are—treated with disdain, as backward people, wannabee shamans or as preposterous pseudo-witches whose refusal to relinquish spirituality—that deep feeling of being profoundly connected to the world beyond oneself in a way that provides existence with meaning—evidenced their inferior academic education, deficient intellectual faculties, or at least some overall faulty frame of mind.[18] According to the almighty reductionist science that had disanimated the world, one should abide by the *fact* that "nature"[19] had been decreed a mindless machine, to be broken down and controlled by humans. Matter had been proved inert and speechless. Along with silly ideas of an enchanted world, some rejoiced, myths and metaphors of a *harmonia mundi* had overall long been tossed away in the disenchantment bin. This bleak picture of a world ruled by meaninglessness and disorder was reinforced by scientific propositions such as the second law of thermodynamics. Prevailing throughout most of the nineteenth and twentieth century, the notion of universal entropy—the idea that matter and energy naturally tend toward degradation, loss, and chaos—introduced more suspicion than ever as to the belief that there could exist any sort of cosmic order or earthly harmony.[20]

And yet, in the past few decades, the environmental sciences and humanities have radically shifted to a vision that reconsiders "agencies" and "intra-actions" on both macroscopic and microscopic scales, producing an enchanting, kaleidoscopic vision of a highly complex, ever-emerging world, where matter, bodies, minds, and meanings cannot be disentangled from one another.[21] In the process, there has been a salient tendency toward reenchantment in many places, countering the modern tale and attitude of disenchantment.[22] With discoveries brought about by chaos theory, quantum physics, molecular biology, ethology, bioacoustics, cognitive neuroscience, biosemiotics, ecoanthropology, multispecies ethnology, or ecopsychology—to name just those—many of today's thinkers have been toiling across disciplinary fields to create new paradigms susceptible of bettering our transactions with more-than-human nature.

The ignorance of many nostalgic, self-proclaimed "Cartesian" thinkers notwithstanding—should those be reminded that Descartes actually invoked a divine telos to explain causation?—postmodern science has in fact long broken away from the purely mechanistic and dualistic paradigms of the so-called Enlightenment.[23] In addition to the budding new materialist or vitalist trends in academia, many scientists have themselves been calling for a reenchantment of our rapport with the world.[24] In so doing, these scientists have collectively been working toward a non-dualistic mindset, a reunification of matter and consciousness, and of the "humanities" and the "sciences." Some, including Starhawk, Joni Adamson, and myself, have been looking further at ways of interweaving the fabrics of poetry, myth, and science. Many of us have hypothesized that therein might lie some of the keys for those striving for more wholesome, less destructive ways of inhabiting the planet. It seems that ultimately, a part of academia might be moving toward the inclusive, transdisciplinary approaches to the biosphere that early ecopsychologists and ecofeminists initiated in the late 1970s and 1980s.[25]

Following in the steps of Isabelle Stengers, even a white male philosopher in the greatest position of authority, Bruno Latour, has come to rehabilitate the figure of Gaia in French academia—and has been taken seriously. Thanks to the research carried out by anthropologists such as Eduardo Viveiros de Castro or Philippe Descola, "perspectivism"—the stance according to which all human and nonhuman sentient beings see themselves as persons, although each kind of being perceives the world differently—and "animism"—recognizing a form of agency, if not sentience, inherent in nonhumans—have meanwhile gained recognition as valid worldviews.[26] It is accepted today that forests, for instance, are animate, and that they can think, albeit in their own ways. In formulating how it can be argued that forests do think, ecoanthropologist Eduardo Kohn studies "that complex web of relations that [he calls] an 'ecology of selves,'" and consequently claims that the world is indeed "enchanted."[27] Turning the tables on enlightenment and naivety, Bruno Latour goes so far as to argue that "[one] of the great enigmas of Western history is not that 'there might still exist people who are naïve enough to believe in animism,' but rather, lies in the still rampant, naïve belief in a 'material world' that would supposedly be disanimated."[28]

Such worldviews existed long before white European men "discovered" them. They were embedded in the stories, rituals, places, and practices of many indigenous peoples—groups whose traditional knowledges based on empirical observation are just now starting to be recognized by the dominant culture. There is a field of anthropology now focusing on TEK—an acronym for Traditional Ecological Knowledge, or Traditional Environmental Knowledge. Indigenous worldviews have long permeated postcolonial literature, which has of necessity been mostly written in the colonizers' tongues. Benefiting from liminal positions and hybrid cultures, postcolonial writers often cross-pollinate different languages and visions.[29] Yet overall, it has been assumed that their wonderful stories of shapeshifters, of spirits, enchanted forests, painted or storied characters coming alive, of Spider Woman or Corn Goddess, animate drums, grass-dancers, and shamans, together with their orality-influenced poetics, need not be granted too much serious attention from an ontological standpoint. Indeed, indigenous stories of transformational beings, of material metalepses (a character jumping from a story or painting into the "real" world),[30] and of interspecies communication were reduced to myth, fiction, or poetry—the stuff that is investigated in cultural and literary studies, but not by truly serious scholars of philosophy or science. Such mythopoeic tinkering with folklore was seen as culturally meaningful in terms of collective identity and cultural heritage. Indigenous ecopoetics might be unpacked in the light of anthropological findings, but because on the face of it, it could not be reconciled with rational science and modern epistemologies, few recognized such stories as ontologically insightful.

Meanwhile, within a small circle of literary critics, much of this literature has been safely labelled "magic(al) realism." To most, this means fiction obeying "supernatural" laws and intrinsically disconnected from the real world. By and large, magical realism is read as political or historical allegory, if not pure flights of fancy. Not often has magical realism been considered as skillfully crafted ecopoetic thought-experiments that might actually affect the way we humans perceive our *oikos*, with its many inhabitants, thereby making us better at reading some of the invisible dimensions of the extradiegetic, very real and concrete world we live in.

Only recently has indigenous *ecopoiesis* started being reinterpreted as creative, self-reflexive, and intelligent braiding of various epistemologies. Much influenced by the anthropological groundwork achieved by Eduardo Viveiros de Castro and Marisol de la Cadena, postcolonial ecocriticism and environmental justice scholars have since the turn of the century produced considerable work rehabilitating the value of indigenous ecological knowledges relayed in their stories.[31] These pioneering studies cast a different light onto enchantment and "cosmovisions." According to Joni Adamson's seminal work on the matter, cosmovisions are "seeing instruments," drawing our attention to long-observed entanglements between human and nonhuman agencies.[32] These scholars have been inspired by cogent nonfiction essays where indigenous scientists and thinkers themselves bridge the different ontologies and forms of discourse which they have learned to negotiate.[33] Thus, various takes on reality are being reconfigured by the reclaiming of indigenous epistemologies and ecological practices that were long denigrated as irrational, ignorant, and grounded in superstitious beliefs in the "supernatural." It is now being recognized that, albeit tied up with mythology and poetry, these stories were in fact very much rooted in careful observation of nature, of the land, and of actual places, with their material features, dwellers, intra-actions, and behaviors. Simultaneously, the value of storytelling itself is being redeemed. As we are starting to gain insight into the puzzling "magic" at the heart of many indigenous stories interweaving science, myth, poetry, and cosmopolitics, the need for new paradigms, for new stories of ecological interconnection and new forms of down-to-earth spirituality is being heralded everywhere. Logically, this has much affected contemporary art and literature, constantly at work in creating new stories and myths.

In parallel to postcolonial literatures that cunningly use and abuse the standards of European novelistic discourse the better to deconstruct and challenge European ideologies, hierarchies, and epistemologies, there has also emerged a nonindigenous ecoliterature of reenchantment—and I am here using the term "reen-*chant*-ment" as signaling both an ontological shift and a reawakening to the world's "wild music."[34] Moving away from the (mostly white and male) American Nature Writing tradition that constituted much of the corpus first explored by leading ecocritics such as Lawrence Buell, Scott Slovic, Mark Tredinnick, or Yves-Charles Grandjeat, to name just a few, the contemporary texts I have been sounding can for the most part be labeled "ecofeminist."[35] Indeed, the writing I am looking at follows in the tracks of ecofeminists thinkers, activists, and artists who have long "connected the devastation of the Earth with the exploitation of women" and of indigenous peoples, while creating languages that "[reach] across and beyond the boundaries of previously defined categories. These languages [recognize] the *lived* connections between reason and emotion, thought and experience. They [embrace] not only women and men of different races, but all forms of life—other animals, plants, and the living Earth itself." As Irene Diamond and Gloria Orenstein go on, "[the] diverse strands of this retelling and reframing [lead] to a new, more complicated experiential ethic of ecological interconnectedness."[36]

As a matter of fact, most of the writers in my corpus at one point or another craft relationships between their human characters and the more-than-human world that might be considered as "totemic" or "animistic."[37] Mindful of the pitfalls of cultural appropriation, we might consider using the term "neopagan" to characterize such postmodernist, cross-fertilizing *ecopoiesis* by both white and other-than-white authors engaged in reinhabiting the world. One area this book seeks to investigate is indeed a growing trend in contemporary literature bridging animistic, totemic, and naturalistic worldviews. Moreover, as I will show, such texts draw away from the transcendentalism that infuses much nature writing. Instead, they capture the immanent, material vibrancy of the world and the many invisible and multispecies entanglements that we, humans, cannot do away with in our terrestrial co-becoming.

In their approach of place, the texts in my corpus further differ from most of those traditionally associated with nature writing in that they restore the many interlaced dimensions of what Kent C. Ryden has called "the invisible landscape." In his landmark study, Ryden builds on geographer Yi-Fu Tuan's definition of "place" as "a center of meaning constructed by experience"[38] for both the individuals and the groups familiar with it. Ryden first calls attention to the sense of place acquired "through the senses and through movement: color, texture, slope, quality of light, the feel of wind, the sounds and scents carried by that wind."[39] These sensuous textures of the world make for what Tuan defines as "a unique blend of sights, sounds, and smells, a unique harmony of natural and artificial rhythms," allowing for "a feel of place" that is "registered in one's muscles and bones."[40] The ecopoets I am looking at likewise foreground the synesthetic meshwork of our somatic responses to the world. They give ample credence to smell, touch, taste, and sound, whereas much earlier nature writing predominantly relayed a scopic take influenced by naturalistic descriptions and the picturesque.[41] As a result, the landscapes extolled in my corpus are more thickly layered as they are extensively interlaced with soundscapes, odorscapes, and feelscapes.

Besides, and most importantly, the writers in my corpus promote a sense of place that is not formed simply by the experience of one single individual going out into the wild. Their characters' reading of land rather depends on a regional palimpsest inscribed with the experiences and stories of an entire community, itself grounded in a specific place. Drawing from Edward Relph's work, Ryden stresses the connection between one's sense of self and one's sense of place, since many tightly identify with the place they come from and inhabit. Consequently, according to Ryden, "the sense of place—the sense of dwelling in the invisible landscape—is in large part a creation of folklore and is expressed most eloquently in folklore."[42] The relations to place in my corpus are similarly deeply rooted in those of families, communities, and ancestors, which shape the characters' enchanted perception of their dwelling places—a perception that is formed and transmitted through the generations, in great part via stories, art, practices, and myths. Consequently, the *ecopoiesis* under study is nurtured by rituals and stories, by craft and tradition, as much as by one single character's experience, which often constitutes just the latest layer of nature's palimpsest. To a certain extent, the latter thus perpetuates the enchanting qualities of specific places, thereby revitalizing them as storied places.

Unlike the "culture of enchantment" endorsed by sociologist James William Gibson, which, he claims, "is nothing less than the reinvestment of nature with spirit"[43]—and although he and I do share important references and viewpoints—the reenchantment dealt with in this book radically diverges from Gibson's in that I propose an ecofeminist, nondualist approach of the agency and biosemiotic expressivity of matter. To put it in a nutshell, rather than reintroducing 'spirit' in ways adhering to classic, religious worldviews hinging on *"transcendence* [. . .], glimpses of a numinous world beyond our own,"[44] what I am interested in is an ecopoetics that sheds light on the entanglements between matter, mind, and discourse. Instead of seeking "a numinous world *beyond our own*," the ecopoetics I mobilize explores texts that reveal *immanence* in a world that is very much *here and now*. As we shall see, the writers I take on use the potential and freedom of ecopoetic fiction the better to provide insight into the world's glory in a way that sticks overall to what ecophenomenologist David Abram terms "matter realism."[45] Briefly put, my proposition here is that we take seriously the stories and ecopoetic thought-experiments that could be labeled as *"magical* realism" or "environmental *fiction*," the better to sift their ontological and scientific validity as well as their epistemological, affective, and ethical performativity.

Reading prose fiction through an ecopoetic lens, I have formed a primary corpus for this study that includes works sharing the main features of what has goaded me to coin the umbrella term "an ecopoetics of reenchantment." First, most of these books resort at some point to an enchanting, magical realist mode—an artistic and literary mode that has provoked much controversy and brilliant studies, and part of which I have come to reconceptualize as "liminal realism." Suffice it to say for now that the works I have been exploring produce hybrid onto-epistemologies of in-betweenness, spurring us to reconsider our multispecies ways of being, of knowing and becoming. While some are crafted by indigenous and/or ecofeminist writers, and a few by white males of European descent, all these works reconcile to a certain extent "magic" with "realism," poetry with science, as well as forms of animism or totemism with postmodern epistemologies. Second, the books in my corpus translate the songs, the humming, and rhythms of the more-than-human world. In that sense, they can be said to *chant* and dance with the Earth. Although penned in novelistic prose, they nevertheless exhibit a poetic fiber justifying my approach of these texts from the standpoint of ecopoetics. And third, many of the authors under study have been trained as scientists, many also as musicians, and all of them display in-depth knowledge of scientific ecology, biology, botany, entomology, ethology, geography, and history. Therefore, the mysticism or spirituality at the heart of their writing is finely balanced with potential adherence to current scientific worldviews. In addition, they are compatible with theories of the human psyche and perception drawing on philosophy, psychology, and anthropology. In many ways that I will go into, such writing takes part in an Anthropocene ecopoet(h)ics,[46] exhibiting a tight grasp on the scientific and philosophical stakes of our current era and ecological crisis.

Now that I have given a rough overview of the subject of this book, I will outline the three parts constituting this monograph. To finely situate my take on the notion of reenchantment, I will first launch into a detailed exposition of my theoretical and conceptual framework when positing the existence of an "ecopoetics of reenchantment." To clarify how I intend to contribute to current discussions in the large overlapping fields of ecological studies and ecopoetics, my first part covers the ground from the disenchantment era to a postmodern epoch, with its apparent propensity toward reenchantment. I first synthesize previous theories and studies on enchantment, disenchantment, and reenchantment. Then I focus on the many faces of Gaia as they are self-reflexively rehabilitated in ecofeminist and indigenous writing. Through an ecofeminist perspective, I will propose the term "Anthrop-o(bs)cene" in relationship with the reclaiming of Mother-Earth mythologies. Tricky notions such as the "sacred" and "spirituality" will be disentangled from restrictive associations with religion and the divine. Using some of the texts in my corpus, I will argue for a reinvention of a terrestrial spirituality and bodymindfulness, based on ecological and biophysical knowledge of the material living world. Finally, I situate my take on the notion of "a voice" or "a song of the Earth" in relationship with an ecopoetics of reenchantment, much of which at that point stems from a close reading of some of Aldo Leopold's and Rachel Carson's work.

Having clearly situated and defined my overall theoretical and conceptual approach, I argue in my second part for a shift from the concept of "magical realism" to that of "liminal realism." This will allow me to rigorously problematize a liminal ecopoetics of reenchantment. Starting from an overview of the history of magical realism as a literary mode, my discussion also draws on Victor Turner's anthropological accounts of initiation rites and his seminal work on liminality. Casting characters who experience an initiation back into the flesh of the world by becoming "liminal *personae*" or "threshold people," the texts in my corpus are indeed best tackled in light of Turner's concepts of "liminality" and "communitas." According to Turner—himself drawing from Arnold van Gennep's anthropological study of rites of passage in indigenous oral cultures—liminal entities "elude or slip through the network of classifications that normally locate states and positions in cultural space." They "are betwixt and between the positions assigned and arrayed by law, custom, convention, and ceremonial."[47] With Turner's work in mind, I therefore scrutinize those characters who can communicate with plants, animals, insects, and even the elements. I linger on characters endowed with such *extra-ordinary*—often synesthetic and greatly empathetic—capacity to feel that they are cast in the roles of shamans, healers, psychics, witches, and shapeshifters.

In a long chapter, I have devoted elsewhere to Linda Hogan's *People of the Whale* and *Solar Storms*, I have demonstrated how the Chickasaw writer explores the liminal potential of magical realism to refine our sense of belonging to a multispecies, biotic communitas.[48] I have examined how Hogan taps into the totemism of her elders to immerse her readers in threshold states via her characters' multisensorial experiences of the world, staging their initiation back into an organic multispecies world. Hogan indeed frequently casts liminal characters moving in-between human and other-than-human worlds, with sensory capacities challenging the boundaries between the sensate and the subliminal, between the physical and the imaginary. In modern psychology, the concept of a subliminal threshold evokes the "limit below which a certain sensation ceases to be perceptible."[49] In other words, the frontier between the sensate and the subliminal is located at a similar threshold as the one between consciousness and the unconscious, as messages processed by our bodyminds are not always interpreted consciously. Consequently, liminal realism frequently explores that sensory subliminal threshold.

Throughout this study, it will appear that a motif common to many of the books in my corpus has to do with initiation rites entailing experiences of "liminality":

> Van Gennep has shown that all rites of passage or 'transition' are marked by three phases: separation, margin [. . .], and aggregation. The first phase (of separation) comprises symbolic behavior signifying the detachment of the individual or group either from an earlier fixed point in the social structure, from a set of cultural conditions (a "state"), or from both. During the intervening "liminal" period, the characteristics of the ritual subject (the "passenger") are ambiguous; he passes through a cultural realm that has few or none of the attributes of the past or coming state. In the third phase (reaggregation or reincorporation), the passage is consummated.[50]

Following Van Gennep, Turner defines society and "communitas" in anthropocentric ways. For the sake of adapting these anthropological concepts to the ecopoetic corpus under study, they shall be expanded here to include nonhuman earth others: "[Society is] a structured, differentiated, and often hierarchical system of politico-legal-economic positions with many types of evaluation, separating men [, women, and nonhuman lifeforms] in terms of 'more' or 'less.'" Considering a second model for human interrelatedness which, as Turner contends, alternates with the society model, the anthropologist focuses on "communitas." Turner differentiates "communitas" from "community," which he regards as a mere "area of common living." According to Turner, communitas "emerges recognizably in the liminal period [. . . It] is of society as an unstructured or rudimentarily structured and relatively undifferentiated *comitatus*, community, or even communion of equal individuals."[51] One aspect common to most of the books in my study is the staging of moments when the human world merges with more-than-human worlds so that plants, animals, and elements can be apprehended for a moment as equals in a biocentric communitas.

I intend to make the case for liminal realism by looking at episodes in ecopoetic fiction whereby characters respond to multisensorial stimuli emitted by other-than-human beings in ways that disrupt our habitual perception of the world. Endowed as they are with psychic and somatic powers superior to those commonly accepted as ordinary or valid in the dominant version of Western rationalism, such characters feel and act upon senses that may read according to what neurocognitive scientist Antonio Damasio has formulated as his "somatic marker hypothesis."[52] Studying how affective and various sensorimotor perceptions work simultaneously to influence our conscious reasoning, Damasio showed that "emotion was in the loop of reason, and that emotion could assist the reasoning process rather than necessarily disturb it."[53] Underscoring how emotional and somatic responses can occur both above and below the radar of awareness, Damasio demonstrated that what we sometimes wrongly perceive as pure reasoning is in fact always intertwined with complex neuronal and sensory activity, engaging many different pathways in our brains and our whole bodies at once. As biologist E. O. Wilson recalls in his study on the origins of creativity, even though "[in] our daily lives we imagine ourselves to be fully aware of everything in the immediate environment," it turns out that "we sense fewer than one thousandth of one percent of the diversity of molecules and energy waves that constantly sweep around and through us."[54] I am thus interested in these highly sensitive characters whose learned and intuitive ways of relating with their environment may corroborate Damasio's statement that "[intuition] is simply rapid cognition with the required knowledge swept under the carpet, all courtesy of emotion and much past practice."[55]

It is fundamental to keep in mind that, where some might be tempted to start waving their red flag against irrational belief in some sort of paranormal or supernatural phenomena, what Damasio scientifically studies as "intuition"—and the marginal, somatic perceptions relayed in the texts I am examining—basically corresponds to unconsciously acquired habits of sensorimotor perception leading to rewarding decision-making. Before he can conclude anything about how even unconscious, somatic knowledge affects our intuition, Damasio's studies rest heavily on brain imagery, on research on human behavior after severe head injury, as well as on studies of animal behavior. He goes as far as to propose, after Jonas Salk, that even in the scientific field, new insights and breakthroughs are "not logically derived from preexisting knowledge." Owing much to intuition, Salk has argued, even the "creative processes on which the progress of science is based operate on the level of the subconscious."[56]

Bringing in characters evolving both within and off the grids of modern thinking and science, I mean to verify to what extent ecopoetic experiments may corroborate Salk's and Damasio's conclusion that "creativity rests on a 'merging of intuition and reason.'"[57] Besides, it is in this sense too that my concept of a liminal realism operates, straddling those realms of conscious feeling, of subliminal somatic responses, and of reason—realms that the Cartesian legacy has erroneously taught us to perceive as separate. Throughout, I unravel different strands of "liminality," "magic," and "realism" interlaced in the books in my corpus, relying mainly on stories written by Jean Giono, Linda Hogan, Leslie Marmon Silko, Anne Pancake, Richard Powers, and Starhawk.

Finally, my third part explores rewritings of a *harmonia mundi* in North American and French fiction. Paying attention mostly to texts that have emerged in indigenous as well as in nonindigenous cultural contexts in the past thirty years or so, I analyze how these immerse their readers into sympoietic soundscapes and how they thereby translate and reentangle the anthropophony, the geophony, and biophony that are constantly at play in the world. Throughout, I investigate the different textures of the world that are diffracted in the texts via sensuous, poetic prose. In this last part, I scrutinize passages from books by Jean Giono, Robin Wall Kimmerer, Barbara Kingsolver, Ann Pancake, Susan Power, Richard Powers, Ron Rash, Starhawk, and Terry Tempest Williams.

NOTES

1. My take on "soundscapes" owes much to Raymond Murray Schafer's groundbreaking work in *Tuning*.

2. See Marie-Monique Robin's extensive study on the emergence of zoonotic viruses in *Fabrique*.

3. The term "biocentric" designates a stance where humans are no longer considered at the center of the world, standing as the one great and meaningful perceiving mind at the heart of an environment deemed only secondary in value. From a biocentric perspective, humans are granted no more intrinsic value than any other

being. Paul W. Taylor first advanced the word "biocentrism" as a proposition that "all organisms are teleological centers of life in the sense that each is a unique individual pursuing its own good in its own way" (*Respect*, 100). According to Christopher Manes, "the biocentric stance of deep ecology may be understood as focusing evolutionary theory and the science of ecology onto the idiom of humanism to expose and overcome the unwarranted claim that humans are unique subjects and speakers" ("Nature," 22). Manes goes on: "From the language of humanism one could easily get the impression that *Homo sapiens* is the only species on the planet worthy of being a topic of discourse. Ecology paints a quite different, humbling, picture. If fungus, one of the 'lowliest' of forms on a humanistic scale of values, were to go extinct tomorrow, the effect on the rest of the biosphere would be catastrophic, since the health of forests depend on *Mychorryzal* [sic] fungus, and the disappearance of forests would upset the hydrology, atmosphere, and temperature of the entire globe. In contrast, if *Homo sapiens* disappeared, the event would go virtually unnoticed by the vast majority of Earth's life forms. As hominids, we dwell at the outermost fringes of important ecological processes such as photosynthesis and the conversion of biomass into stable nutrients" ("Nature," 24).

4. My take on the notion of reenchantment owes much to Raymond Murray Schafer's *The Tuning of the World*: *The World as Music*, Joachim-Ernest Berendt's *The World is Sound*, Jonathan Bate's *The Song of the* Earth, and Tredinnick's, *The Land's Wild* Music, together with contributions by bioacoustician Bernie Krause. I have for my part been discussing the notion of a song of the earth in relationship to an ecopoetics of reenchantment since 2015. See my contribution "Le chant de la matière pour désensorceler les modernes: vers une écopoétique du réenchantement à travers quelques romans des Appalaches, my *Ecozon@* paper "Measured Chaos: EcoPoet(h)ics of the Wild in Barbara Kingsolver's *Prodigal Summer*," or again my introduction to the edited volume *Dwellings of Enchantment*: *Writing and Reenchanting the Earth*, which forms a spinoff of the 2016 Perpignan international ecopoetics conference. French scholar Aline Bergé has also developed an approach of French poetry closely tied to the notion of the world's soundscapes; see Bergé "Geste." The same topos underlies Michel Collot's 2019 monograph on contemporary French poetry, *Le chant du monde*. Surprisingly, his book does not address Jean Giono's influence on French ecopoets, while Jonathan Bate's pioneering work receives only a cursory mention.

5. Schafer, *Paysage,* 39, 48, 62.

6. Ibid, 73–76.

7. See Berendt, *Nada*. As illuminating as Berendt's study may be in terms of our capacity to respond to the world's wild music, in no way can the present book endorse his circumvoluted reintroduction of a divine telos. Not only am I interested in a secular, new materialist form of reenchantment, one that is coherent with science and not based on faith, it must be pointed out that Berendt's somewhat hidden agenda to prove the existence of God fails to convince at all. While relying on famous physicists, he mostly grounds his thinking in texts and songs penned by people of religious creed. Moreover, there might be detected throughout his book an increasingly common temptation to instrumentalize the high level of abstraction required by physics and musicology to carry out a nebulous, scientifically undefendable reintroduction of God

in his conception of the universe. Berendt problematically never clearly defines what he might have had in mind when referring to "God." As I happen to have experienced, many New-Age cults in France and elsewhere have been surfing on widespread ecological fear and disenchantment to enroll people via a pernicious oversimplification and distortion of complex science and philosophy—something which ecopoeticians should beware of as we broach reenchantment.

8. Krause, *Chansons*, and *Orchestre*.

9. Bate, *Song*; Knickerbocker, *Ecopoetics*; Pughe, "Réinventer." More recently, Belgian scholar Pierre Schoentjes has written an essay on ecopoetics. Yet, his approach may more rightly pertain to ecocriticism, considering the little attention finally paid to the material texture of language.

10. I am taking up the concept of *sympoiesis* put forward by Donna Haraway in *Staying*.

11. See also for instance Pinson, *Habiter;* Bergé, "Geste"; Meillon, "Ecopoétique," "Agony," and "Measured."

12. Knickerbocker, *Ecopoetics*, 159.

13. Ibid, 185.

14. Ibid, 162.

15. Bate, *Song*, 76, 282.

16. See Merchant's historical survey, *The Death of Nature*.

17. This has long been at the heart of ecofeminist thinking such as Val Plumwood's (*Feminism and the Mastery of Nature*, 1993), or Starhawk's (e.g., "Power," *Magic, and Earth*).

18. In using the term "spirituality," I am referring to the life of the bodymind when engaged in an existential quest for meaning, itself derived from a sense of connection with the more-than-human world.

19. In *Ecology Without Nature,* Timothy Morton has circumscribed the difficulties and pitfalls for ecocritics discussing the slippery concept of "nature." For a good synthesis of the history of this tricky notion, see Ducarme and Couvet's "What does 'nature' mean?" The transdisciplinary collection of essays *Rethinking Nature*, co-edited by Choné, Hajek and Hamman is also of great interest.

20. See James Gleick's enlightening exploration of chaos theory, where he tackles the evolution in scientific studies from the notion of entropy and disorder to those of chaos and universal laws of form creation. See also Prigogine and Stengers's *Order*. Gleick addresses the delicate balance between forces of stability and instability that constantly interact at both microcosmic and macrocosmic levels. Principles of non-linearity and feedback loops together with the existence of strange attractors behind pattern formation have helped explain why "chaos" and "randomness" were first misconceived. Studying the spontaneous formation of snowflakes in the atmosphere, the iterative, fractal processes determining the design and structure of world forms in their emergence and evolution, as well as atmospheric turbulence and the behavior of other complex systems, scientists in the 1970s have built what is often referred to with the umbrella term "chaos theory," establishing the fact that there is always some amount of order being recreated within disorder itself, and certain universal "harmonies" to be discovered (197). Much can be said too about what information theory

and research in bioacoustics have revealed regarding what we perceive as "noise" as opposed to what we perceive as "music." See Berendt's chapters on "Before We Make Music, the Music Makes Us," retracing the evolution of the concept of a *harmonia mundi*, from Plato to Max Planck (57–75). See also Berendt's musicological take on universal harmony in "Harmony as the Goal of the Universe," *Nada*, 112–27.

21. See Meillon, ed., *Dwellings of Enchantment: Writing and Reenchanting the Earth*, containing many earlier contributions to an ecopoetics of reenchantment. I am here implicitly referring, among others, to Bruno Latour in *Nous n'avons jamais été modernes* and *Face a Gaïa*, and to Karen Barad, in *Meeting the Universe Halfway: Quantum Physics and the Entanglement of Matter and Meaning*. I will get back to those further and clarify my take on their work.

22. This is the case, for instance, in work carried out by Morris Berman, David Ray Griffin, Jane Bennett, Serenella Iovino, and Serpil Oppermann, to name but a few. Reenchantment has been at the heart of my own work for over twenty years. Part 1 will clarify the various moves toward reenchantment which this book stems from.

23. Bohm, "Postmodern," 61.

24. See *The Reenchantment of Science*, edited by Ray Griffin, which I take up in Part 1.

25. See for instance Theodore Roszak, *The Voice of the Earth*, or Starhawk in "Power," and *Magic*. I will elaborate on their contributions later.

26. Viveiros de Castro, *Amerindian*, 478.

27. Kohn, *Forests*, 16.

28. Latour, *Gaia*, 94–95, translation mine.

29. I am implicitly referring to Homi Bhabha's work on the concept of hybridity in postcolonial contexts. See *The Location of Culture*.

30. This takes place for instance in Susan Power's *Sacred Wilderness*, or in her short story "Roofwalker."

31. See Viveiros de Castro, "Cosmological," and "Exchanging"; de la Cadena "Indigenous Cosmopolitics"; Huggan and Tiffin, *Postcolonial Ecocriticism*; Nixon, *Slow Violence*; DeLoughrey and Handley, *Postcolonial Ecologies*, and Adamson and Monani, "Cosmovisions."

32. See Adamson's *Middle Place*, *"Avatar,"* and the introduction to *Ecocriticism and Indigenous Studies*, by Adamson and Monani, "Cosmovisions."

33. See for instance Jack Forbes, *Columbus and Other Cannibals*, Linda Hogan, *Dwellings: A Spiritual History of the Living World*, or, more recently, Robin Wall Kimmerer, *Braiding Sweetgrass: Indigenous Wisdom, Scientific Knowledge, and the Teachings of Plants* (2013).

34. I am drawing from Mark Tredinnick's work in *The Land's Wild Music*. I am indebted to French scholar Yves-Charles Grandjeat who initially brought my attention to Mark Tredinnick's work. See Grandjeat, "Retreating," and "Poetic."

35. The influence of ecofeminist thinkers and writers on my research is humongous. Throughout this book, I will constantly situate some of the concepts, theory, and ecopoet(h)ics at the heart of my work within ecofeminist frames and studies. For now, let me quote Irene Diamond and Gloria Feman Orenstein in their "Introduction" to the seminal anthology, *Reweaving the Wounds: The Emergence of Ecofeminism*:

"Ecofeminist politics does not stop short at the phase of dismantling the androcentric and anthropocentric biases of Western civilization. Once the critique of such dualities as culture and nature, reason and emotion, human and animal has been posed, ecofeminism seeks to reweave new stories that acknowledge and value the biological and cultural diversity that sustains all life. These new stories honor, rather than fear, women's biological particularity while simultaneously affirming women as subjects and makers of history. This understanding that biological particularity need not be antithetical to historical agency is crucial to the transformation of feminism" (xi). In addition, ecofeminism "[affirms] and [celebrates] the embeddedness of all the Earth's peoples in the multiple webs and cycles of life" (x-xi).

36. Diamond and Orenstein, "Introduction," *Reweaving*, xi.

37. In Part 2, I clarify my take on these terms, as totemism and animism are tied to my concept of "liminal realism." I am greatly indebted to Caroline Durand-Rous' pioneering work on the reinvention of totemism in contemporary North American literature. Thanks to her, I have learned to pay more attention to the (re)surfacing of totemic bonds in both indigenous and non-indigenous contemporary environmental literature. See Durand-Rous, "Totem, transmission" and "Totem, exploration."

38. Tuan, quoted in Ryden, *Mapping*, 37.

39. Ryden, *Mapping*, 38.

40. Tuan, quoted in Ryden, *Mapping*, 38.

41. Despite passages by Muir or Thoreau for instance that do draw attention to the music produced by the wind in the trees, or that rely on senses other than sight, the awareness to the land which they seek nevertheless remains mediated by sight. Multisensorial mindfulness to nature is already more present in later writings by Wendell Berry and Barry Lopez, as evidenced by Scott Slovic's seminal study, *Seeking Awareness in American Nature Writing*. Moreover, and relating to my next point, as opposed to those in my corpus, the singular experiences of place related by most canonic nature writers hardly ever present themselves as overtly indebted to any folk patrimony inherited through the generations, passed on and constantly revitalized by old and new stories and rituals.

42. Ryden, *Mapping*, 45.

43. Gibson, *Reenchanted*, 11.

44. Ibid, 11.

45. Abram, *Becoming*, 10.

46. I often use the parenthetical h when writing ecopoet(h)ics to highlight the *ethics* at work in ecopoetics. French philosopher Jean-Claude Pinson has paved the way with the French neologism "*poéthique*." Pinson does not use the prefix eco-, and yet his research is clearly that of an ecopoetician. In his book on contemporary French poetry, *Une autothéorie*, he writes, "I have come to foreground the word 'poet(h)ics' to underline that poetry is not simply an art of language (that which is of interest to poetics). Poetry seems to me the vehicle of a greater ambition. I believe that, on the level of existence, of an ethos—of one's customary way of being in the world—, poetry hankers for a different light, a different language that might give meaning to our stay, to our inhabitation of the earth" (Qtd. in Brun, *Habiter*, 326, translation mine). Pinson formulated elsewhere that "[a] poem is more than a verbal object offered for aesthetic

pleasure or analysis. It is also a proposition of a world—a proposition as to a potential modality of inhabitation. To use the portmanteau word 'poethics' is to designate the relationship between a poet's speech and their specific 'habit of inhabitation' (according to the double etymology of the word *ethos*)" (*Habiter*, 135, translation mine). As for potential problems concerning the much-debated term "Anthropocene," I will get back to the issue in my first part, as I deal with Gaian ecopoetics.

47. Turner, *Ritual*, 94–95. These aspects of my research I have been working on for a while; see Meillon, "Entrelacs," and "Writing."

48. See Meillon, "Entrelacs," and "Writing," where I rely on Wendy Faris's definition of magical realism on the one hand, and, on the other hand, on Victor Turner's exploration of liminality in initiation rites (*Ritual* and "Betwixt.") and Caroline Durand Rous's aforementioned work on literary reinventions of totemism.

49. Ashcroft et al. *Postcolonial*, 145.

50. Turner, *Ritual*, 94–95.

51. Ibid, 96.

52. Damasio, *Descartes*, x-xi.

53. Ibid.

54. Wilson, *Origins*, 59.

55. Damasio, *Descartes*, xiii.

56. Quoted in Damasio, *Descartes*, 189.

57. Ibid.

PART I

From Disenchantment to an Ecopoetics of Reenchantment

Chapter One

Disenchanted, Enchanted, and Reenchanted Worldviews

DISENCHANTMENT IN THE WAKE OF MODERN SCIENCE

The most popular tale of disenchantment associated with modernity is that formulated by Max Weber at the end of World War I.[1] The German term Weber used in his narrative of the rationalization and secularization of the modern world was *Entzauberung*, which can be translated as "taking out the magic," or "demagification."[2] The objectifying, mechanistic ontology established during the modern period prompted thinkers to shun the possibility that lifeforms possessed any invisible powers to act upon other things or beings of the world. Any causality that could not be explained by the laws of physics then known was deemed occult, dismissed as irrational. In some cases, unexplainable cause-effect links were read as evidence that the world occasionally evolved in random, meaningless patterns. The dominant view then held was that the entire world could be dissected via linear mathematics, using tools, methods, and paradigms from reductionist physics. However, Weber warned that "increased intellectualization and rationalization do *not* [. . .] indicate an increased and general knowledge of the conditions under which one lives, [but] the knowledge or belief that [. . .] one can, in principle, master all things by calculation. This means that the world is disenchanted."[3] The disenchantment of nature thus served the Enlightenment ideology of *Man's*—not women's—possession and mastery of nature under the reign of the human Logos, as Descartes famously had envisioned in his *Discourse on Method* (1636).[4]

As philosophy of religion scholar David Ray Griffin observes, the "dedivinization of nature" often associated with Weber's disenchantment diagnosis actually goes back to Romanticism, as first formulated by Friedrich Schiller.[5] Tackling the disenchantment of nature a century before Weber, the German poet and thinker chose the phrase "*die Entgötterung der Natur*," which, according to cultural historian Morris Berman, literally translates as "the disgodding of nature."[6] As Berman puts it, this corresponds to the "progressive removal of mind, or spirit, from phenomenal appearances."[7] Thus, Griffin argues, deity could not be immanent to the world; on the contrary, "it was a being wholly external to the world who imposed motion and laws upon it from without." Consequently, "nature was bereft of all qualities with which the human spirit could feel a sense of kinship and of anything from which it could derive its norms. Human life was rendered both alien and autonomous."[8] Hence the rejection of animistic viewpoints, of the divine as immanent, and hence the desacralization of the world.

For all its positive aspects, the flip side of the Scientific Revolution lay, first, in a world rendered meaningless, and second, in the sterilization of humans' capacity for wonder—a condition Weber bewailed as "the fate of our time."[9] Weber decried "the last stage of this cultural development." Quoting Goethe, he lamented: "'Specialists without spirit, sensualists without heart; this nullity imagines that it has attained a level of civilization never before achieved.'"[10] In stripping the world of magic, the modern mechanistic worldview simultaneously rendered the world fragmentary and soulless. Whereas people used to conceive of themselves as belonging to a larger harmonious cosmos giving sense to their own human existence, they suddenly could no longer claim any kinship with the more-than-human world. Nor could they intuit any meaning to their terrestrial sojourn other than that of serving "progress" as it had been defined by the Enlightenment project. In Carl Jung's terms, "Man feels himself isolated in the cosmos. He is no longer involved in nature and has lost his emotional participation in natural events, which hitherto had symbolic meaning for him."[11] Drawing on Jung's concept of "participation," Berman formulated the notion of "participating consciousness," which "bespeaks a psychic wholeness" that used to be derived from pre-scientific beliefs and attitudes:

> The view of nature which predominated in the West down to the eve of the Scientific Revolution was that of an enchanted world. Rocks, trees, rivers, and clouds were all seen as wondrous, alive, and human beings felt at home in this environment. The cosmos, in short, was a place of belonging. A member of this cosmos was not an alienated observer but a direct participant in its drama. His personal destiny was bound up with its destiny, and this relationship gave meaning to his life.[12]

The dark side of disenchantment then rests in the tragic disintegration, for many, of an existential meaning derived from a sense of belonging to, and an ability to emotionally bond with, the broader more-than-human world. Instilling a sense of nonparticipation and insisting that the observer maintain herself separate from the observed, modern science has spawned, in Berman's terms, "a sickness in the soul": "Scientific consciousness is alienated consciousness. [. . .] Subject and object are always seen in opposition to each other. [. . .] The logical end point of this worldview is a feeling of total reification: everything is an object, alien, not-me; and I am ultimately an object too, an alienated 'thing' in a world of other, equally meaningless things."[13]

In his thorough history of the disenchantment of the world, Berman traces the early beginnings of the "disgodding of nature" all the way back to Judaic and Greek cultures. On the one hand, Judaism started rooting out animistic and totemic beliefs, which were gradually dismissed and replaced by monotheism—with its belief in an almighty sky God, existing prior to and outside of His terrestrial creation. This is also an aspect of Christianity that many ecofeminists blame for rendering the earth soulless. It is central to Lynn White Jr's famous lecture on "The Historical Roots of our Ecological Crisis," where he deems that the "victory of Christianity over paganism was the greatest psychic revolution in the history of our culture."[14] White Jr. underscored that "in its Western form, Christianity is the most anthropocentric religion the world has ever seen."[15] Whereas in Antiquity, "every tree, every spring, every stream, every hill had its own *genius loci*, its guardian spirit," making it necessary for humans to "placate the spirit in charge" of a place before cutting down a tree, damming a river, or mining a mountain, by "destroying pagan animism, Christianity made it possible to exploit nature in a mood of indifference to the feelings of natural objects."[16] Moreover, Socrates and Plato widely contributed to the conception of a separation between object and subject. Turning away from pre-Homeric culture and its poetic tradition, where the spell of performed poetry was extolled as bringing about a powerful sensuous experience conducive to intuitive learning, Plato regarded active emotional identification and participating consciousness as pathological. As Berman notes, "[in] the poetic tradition, the basic learning process was a sensual experience. In contrast, the Socratic dictum, 'know thyself' posited a deliberately non-sensual type of knowing."[17]

Deconstructing concepts of nature through the different ages and civilizations of the West, ecofeminist historian and philosopher Carolyn Merchant analyzes how "the image of nature that became important in the modern period was that of a disorderly and chaotic realm to be subdued and controlled."[18] Central to Merchant's findings is the association of both women and indigenous peoples with the concept of "nature as wilderness," justifying the consolidation of a European colonial and patriarchal ideology. Just as the ruling white men of the time conveniently claimed that the "savages" of the New World exhibited a wildness necessitating their civilizing influence, women were likewise metonymically perceived as being on the side of an unruly animal nature.[19] As Merchant analyzes, underlying the persecution of "witches" throughout the early modern period was the fear of a form of "personal animism" associated with witchcraft practices.[20] With "the death of the old order of nature, [. . .] the spirits were removed from nature in coincidence with the waning of witch trials"[21]—an intricate, historical process that may account for the resurfacing of magic, of animistic practices, and liminal ecopoetics in much contemporary ecofeminist art, praxis, politics, and theory.

There was of course earlier resistance to the disenchantment of the world both during and after the Enlightenment period. Turning away from controlled reason and calculus, many nineteenth-century artists and intellectuals glorified the personal revelations one might access via deep-seated knowledge derived from intuition and the senses. The propension of Romantic poets, in particular, to revert to pathetic fallacy and personification reveals that many held on to a vision of nature as animate. Much of their writing encouraged full participation in the more-than-human world via a poetic stance. The lyricism characterizing a large part of their writing furthermore translates their rebuttal of the established, disenchanted picture of their times. In the English-speaking world, the poetry and prose of transcendentalist poets and philosophers such as David Henry Thoreau, Ralph Waldo Emmerson, or of the cosmic poet Walt Whitman, to name just those few, like the writings of Modernist authors such as T. S. Eliot, James Joyce, D. H. Lawrence, Ezra Pound, or later, of Francis Scott Fitzgerald and John Steinbeck, clearly show that an entire fringe of thinkers did not adhere to the vision of the world encapsulated by the disenchantment tale, which they often execrated and actively worked against. Yet, despite many intellectuals' and artists' defiance, nothing could slow down the expansion of the European capitalistic, industrial revolution which was carried out at the expense of healthy ecosystems and mindsets, and which depended on people's overall acceptance of a disenchanted worldview.

The disenchanted vision propagating through the centuries has primarily affected the ontology of those claiming to be "modern" in a Western, white, and male-dominated society. Meanwhile, indigenous alter-tales have consistently worked against the hegemony of what French anthropologist Philippe Descola, in his classification of ontologies, labels "Naturalism."[22] According to this latter ontology, which can be associated with dualistic Cartesian thinking, the forms of the world exist in a certain physical continuity, but are separated on the level of their interiorities. Within that system, consciousness, soul, or spirit, are perceived as pertaining to humans only, and are viewed as singular phenomena contained within each, separate individual. To put it in a nutshell, the modern, naturalistic worldview grants subjectivity, or sentience, to humans only and holds consciousness to exist in the world only in discontinuity. By contrast, first peoples—as well as witches, shamans, poets, or medicine men and women of all times for that matter—have never ceased to perceive the world as filled with consciousness. Still today, it remains widely believed and increasingly attested by science that various practices—and I would include *ecopoiesis* as one of them—may allow for forms of communication across individuals and across realms, starting with the worlds of animal, vegetal, and elemental life. Shifting perceptions and tools, and revising the narratives through which we construct and apprehend the world, we might indeed regain some access to invisible sensuous dimensions of existence and rediscover forms of continuity that are both intra- and interspecific, and which may indissolubly involve both matter and mind.

INDIGENOUS KNOWLEDGE, ECOFEMINISM, AND POSTMODERN SCIENCE

In her book on the recovery of feminine Native American traditions, Paula Gunn Allen taps into her mixed Laguna Pueblo and Sioux ancestry. She attempts to translate into English the indigenous worldview of interrelatedness encapsulated in the eponymous image of "The Sacred Hoop."[23] In this non-dualistic vision, all individuals, whether human or other-than-human, take part in the physical and spiritual fabric of the whole: "The American Indian sees all creatures as relatives (and in tribal systems relationship is central), as offspring of the Great Mystery, as cocreators, as children of our mother, and as necessary parts of an ordered, balanced, and living whole. This concept," Allen explains, "applies to what non-Indian Americans think of as the supernatural, and it applies as well to the more tangible (phenomenal) aspects of the universe. American Indian thought makes no such dualistic division, nor does it draw a hard and fast line between what is material and what is spiritual, for it regards the two as different expressions of the same reality [. . .]."[24]

In her effort to translate this animistic ontology, Allen proposes that the "closest analogy in Western thought is the Einsteinian understanding of matter as a special state or condition of energy." Yet, she warns—and she was writing a few decades before Karen Barad coined her quantum-physics-based theory of agential realism[25]—"even this concept falls short of the American Indian understanding, for Einsteinian energy is believed to be unintelligent, while energy according to the Indian view is intelligence manifested in yet another way." From a Native standpoint, "human intelligence," or awareness, "[rises] out of the very nature of being, which is of necessity intelligent in and of itself, as an attribute of being." It follows that "those attributes possessed by human beings are natural attributes of *all* being."[26] Because in indigenous ontologies, there is no separation between the "natural" and the "supernatural," and because humans are not considered alienated from either, Native individuals moving within their own worldviews do not suffer from disenchantment: "disease is a condition of division and separation from the harmony of the whole."[27] Whether in spirit or in mind, individuals are healthy only in wholeness, as they experience belonging and "enwholement":[28] "each creature is part of a living whole," and "all parts of that whole are related to one another by virtue of their participation in the whole of being."[29]

The turn of the twentieth century has been marked by exacerbated wariness of the alienating ideology of a disenchanted world. Published in 1981, Morris Berman's cogent history of disenchantment and subsequent plea for reenchantment received notable attention across the sciences and humanities. Carefully tracking the evolution of Western epistemologies over time, and seeking to reinstate mind within the phenomenal world, Berman makes a fine point for "the reenchantment of the world"—as goes the title of his book. Relying much on the epistemological framework laid out by anthropologist and cyberneticist Gregory Bateson, Berman's argument for reenchanting the world does not condone a return to a prescientific version of animism, nor to any traditional notion of God, but, rather, to a new animism that would rely on "participant observation," a "revival of the unconscious," and the "development of relational or holistic perception."[30] As will become clear, this ties in closely with the concept of "liminal realism" at the heart of this book. Although Berman hardly refers to French philosopher Maurice Merleau-Ponty's phenomenological exploration of the intermeshing of perceiving subject and perceived object,[31] his call for "the end of ego-consciousness" nevertheless draws from a similar understanding, here derived from quantum physics:

> It turns out that the subject/object distinction of modern sciences, the mind/body dichotomy of Descartes, and the conscious/unconscious distinction made by Freud, are all aspects of the same paradigm; they all involve the attempt to know what cannot, in principle, be known. The subject/object merger intrinsic to quantum mechanics, on the other hand, is part of a very different paradigm that involves a new mind/body, conscious/unconscious relationship. This mental framework [. . .] conceives of the relationship between the mind and the body as a field, alternately diaphanous and solid. In Wolfgang Pauli's terms, it "would be the more satisfactory solution if mind and body could be interpreted as complementary aspects of the same reality."[32]

Rejecting the dichotomous Cartesian paradigm, Berman thus embraces a vision that resonates with Paula Gunn Allen's translations into English of concepts inherent to other-than-modern ontologies.

Berman's world picture, like Bateson's, presents us with a reenchanted world that is filled with neither spirits nor a transcendent God. Instead, Berman envisions a relational ontology: "There are no spirits out there within the rocks or trees, but neither is my relationship to those 'objects' one of a disembodied intellect confronting inert items. My relationship to those 'objects' is systemic, ecological in the broadest sense. The reality lies in my relationship with them."[33] It is a kind of new animism, or "agential realism" as Barad would say, that moves away from anthropocentric paradigms as it makes room for a new type of materialism, allowing for the immanence of mind within matter—itself always entangled into a multidimensional, enfolded, and dynamic universe. As Bateson put it:

> The individual mind is immanent but not only in the body. It is immanent also in the pathways and messages outside the body; and there is a larger Mind of which the individual mind is only a subsystem. This larger Mind is comparable to God and is perhaps what some people mean by "God," but it is still immanent in the total interconnected social system and planetary ecology.[34]

Again, it is hard not to hear echoes from Paula Gunn Allen's elaboration on metaphors of humans as "offspring of the Great Mystery" and of the "sacred hoop" quoted earlier.

In the above quotation, Bateson brings to the fore the trickiness of words that might prove much more slippery than at first sight. This is the case for words and concepts such as "God," that are generally held to have clear, well-identified meanings, and which may yet prove to be rather elusive or complex depending on the context. As I will tackle throughout this book, this is also true of concepts such as "spirituality," "magic," "Mother Earth," or "the sacred," the subtleties of which need to be pondered and clarified beyond common sense.[35] This is a problem that Laguna Pueblo writer Leslie Marmon Silko has carefully exposed in her 1977 novel *Ceremony*. As with much postcolonial writing, Silko's dialogic narrative fabric self-reflexively interweaves history and fiction, poetry, science, and myth, or again "magic" and "realism," the "natural" and the "supernatural," while in many respects it contains a metafictional reflection on language itself, pointing to the creation of meaningful discourse and stories.[36] At one point in Silko's story, the old medicine man named Ku'oosh performs a healing ceremony meant to heal Tayo from trauma and renew his connection to the world. Like many of the soldiers surviving the horrors of war, Tayo has come back from World War Two showing all the symptoms of post-traumatic stress disorder. Standing in between two different cultures, old Ku'oosh speaks from a position similar to the postcolonial writers.' Both must word and perform the traditional worldviews, stories, chants, and practices which they call upon via a different language—that of the colonizers who eradicated *in the first place* most of their culture, knowledge, myths, and practices of enchantment:

"But you know, grandson, this world is fragile."

> The word he chose to express "fragile" was filled with the intricacies of a continuing process, and with a strength inherent in spider webs woven across paths through sand hills where early in the morning the sun becomes entangled in each filament of web. It took a long time to explain the fragility and intricacy because no word exists alone, and the reason for choosing each word had to be explained with a story about why it must be this certain way. That was the responsibility that went with being human, old Ku'oosh said, the story behind each word must be told so there could be no mistake in the meaning of what had been said; and this demanded great patience and love.[37]

In many ways, this crucial passage from *Ceremony* provides a guiding principle as to how to conduct research in ecofeminist, ecopoetic, and postcolonial reenchantment.

Like Berman's, my own reenchantment project has been influenced by Carolyn Merchant's substantial contribution to ecofeminist theory, with the publication in 1980 of *The Death of Nature: Women, Ecology and the Scientific Revolution*. In her detailed study of how evolving European epistemologies and societies have shaped different yet interrelated constructs of nature and women over time, the environmental historian and philosopher wrote down a cornerstone analysis of the mechanistic framework that legitimated the Industrial Revolution. While underscoring the connections underlying the subordination and management of both women and nature within the Western frameworks associated with the ideology of "progress," Merchant makes the case for more holistic paradigms that could concur with the discoveries of scientific ecology, of quantum physics, and chaos theory better than does the modern view of nature. She insists that to deal with the ecological crisis, humans need to be reintegrated within natural ecosystems. Her authoritative appeal for the development of a non-mechanistic science and an ecological ethic proved seminal for both ecofeminism and postmodern science, which were emerging conjointly at the turn of the century.[38]

One of Carolyn Merchant's greatest inputs has been to shed light on the spiritual and ethical value of Mother Earth schemas. "The image of the earth as a living and nurturing organism had served as a cultural constraint restricting the actions of human beings," she explains. Merchant insists that the images and myths humans create as they strive to represent the world are ultimately endowed with normative power: "As long as the earth was considered to be alive and sensitive, it could be considered a breach of human ethical behavior to carry out destructive acts against it."[39] What is true about myth is also true about science, which, in its own way, forms a kind of performative mythology: "Descriptive statements about the world can presuppose the normative; they are ethic-laden. [...] The norms may be tacit assumptions hidden within the descriptions in such a way as to act as invisible restraints or moral ought-nots. [...] To be aware of the interconnectedness of descriptive and normative statements is to be able to evaluate changes in the latter by observing changes in the former."[40]

Max Weber had already intimated the loss of ethics inseparable from the erasure of magical and religious forces from the world. These forces, Weber pinpointed, had erected "spiritual obstacles" to the inconsiderate overexploitation of nature. In *The Protestant Ethic and the Spirit of Capitalism*, Weber argued that the earlier "magical and religious forces" that were perceived to act via nature entailed "ethical ideas of duty based upon them." In turn, the ethical ideas borne from the notion that spirits roamed the world served to oppose "serious inner resistance" to the profit-driven ideology behind the Industrial Revolution and the establishment of Capitalism.[41] As the world was turned into a crude causal mechanism, Weber implied, the use of nature became tied to cold, rational calculation of whatever economic benefit could be derived from the human endeavor to exploit natural resources.

Pioneer ecofeminist theorist Riane Eisler sensibly warned however against the pitfalls that might lie in a wholesale critique of Modernity or in trucking with indigenous worldviews: "While there is much we can learn today from tribal cultures," she underscores that it is crucial "not to indiscriminately idealize all non-Western cultures and/or blame all our trouble on our secular scientific age."[42] Recalling that the practices of torture, female excision, and cannibalism were integral to some premodern tribal groups, she insisted that there have always existed indigenous ways that might be said "as barbarous as the most 'civilized' Roman emperors or the most 'spiritual' Christian inquisitors."[43] Anticipating the "ecological Indian" myth and debates that would appear later,[44] Eisler encouraged a careful examination of past and present peoples said to be living 'close to nature' and who may yet be "blindly destructive of their environment." She reminds us for instance that overgrazing, overcultivation of land, and deforestation were not entirely absent in ancient cultures both Western and non-Western.[45] If ecofeminism often reclaims and regenerates pagan and indigenous lore and practices, Eisler thus nuances, it is not to promote idealistic visions of the past, but rather, to recuperate and recycle holistic and enchanted worldviews that are direly needed today.

In addition, Eisler specifies a crucial point: "the basic issue is not one of technology versus spirituality or nature versus culture. The fundamental issue is how we define nature, culture, technology, and spirituality—which in turn hinges on whether we orient to a dominator or a partnership model of society."[46] Clarifying some of the assumptions underlying ecofeminism, Eisler writes: "It is not science and technology, but the numbing of our innate human sensibilities that makes it possible for men to dominate, oppress, exploit, and kill. What passes for 'scientific objectivity' in a dominator society is the substitution of detached measuring for an inquiry designed to enhance and advance human evolution." Continuing with her disambiguation of slippery terms, Eisler goes on: "Even beyond this, what this system requires is that spirituality be equated with a detachment that often condones and encourages indifference to avoidable human suffering—as in most Eastern religions. Or it leads to the Western dualism that justifies the domination of culture over nature, of man over woman, of technology over life, and of high priests and other so-called spiritual leaders over 'common' women and men."[47]

The loss of a spiritual dimension warranting ethical values and moral restraints in our dealings with nonhuman nature has remained at the heart of contemporary calls for reenchantment. Physicist David Bohm for instance insists on the ethical implications of quantum physics' findings, encouraging a holistic worldview:

> The general way we think of this world will thus be a crucially important factor of our consciousness, and thus of our whole being. If we think of the world as separate from us, and constituted of disjoint parts to be manipulated with the aid of calculations, we will tend to try to become separate people, whose main motivation with regard to each other and to nature is also manipulation and calculation. But if we can obtain an intuitive and imaginative feeling of the whole world as constituting an implicate order that is also enfolded in us, we will sense ourselves to be one with this world.[48]

Taking up both Weber and Merchant, a group of scientists including Griffin and Swimme put together a collection of essays proposing a reenchantment project based on nondualist, postmodern sciences. While Morris Berman had deplored that the latest advances in quantum physics and ecology had "not made any significant dent in the dominant mode of thinking,"[49] many scientists who praised his work consequently joined his effort to reinstill a sense of wonder and wholeness within existence.

A few years after the publication of Berman's book, David Ray Griffin invited scientists and philosophers to contribute to a book defending no less than "The Reenchantment of Science" through a "postmodern science," a "postmodern spirituality," and a "postmodern medicine."[50] In *The*

Reenchantment of Science: Postmodern Proposals, scientists and thinkers such as David Bohm, Frederick Ferré, Stanley Krippner, or Rupert Sheldrake pave the way for "a reenchanted, liberated science [. . . and] a postmodern spirituality" that are to be "relational, ecological, planetary, postpatriarchal."[51] Building also on the foundations of ecofeminism that were being laid down around the same period, these advocates of a "constructive" or "revisionary" postmodernism "[reject] not science as such but only that scientism in which the data of the modern, natural sciences are alone allowed to contribute to the construction of our worldview."[52] Their project requires a paradigm shift "transcending [the] individualism, anthropocentrism, patriarchy, mechanization, economism, consumerism, nationalism, and militarism" propelled by the modern period. In the introduction to the book, David Ray Griffin aligns constructive postmodern thought with support of "the ecology, peace, feminist, and other emancipatory movements of our times, while stressing that the inclusive emancipation must come from Modernity itself."[53] Preventing the charges often pressed against any attempt to criticize modernity, Griffin is careful to distinguish their joint venture from a misguided return to premodern thought: "the modern world," it goes without saying, "has produced unparalleled advances that must not be lost in a general revulsion against its negative features."[54] Nevertheless, Griffin underlines the irony in the fact that "modern science, in disenchanting nature, began a trajectory that ended by disenchanting science itself. [. . .] For some time," he goes on, "many held that science at least gives us the truth, even if a bleak one. Much recent thought, however, has concluded that science does not even give us that. The disenchantment is complete."[55]

In his introduction, Griffin retraces the complex history of disenchantment, highlighting the biases and vested interests of those elaborating mechanistic, dualistic frameworks over the centuries. In addition, he goes over contested notions that form the pinnacle of modern epistemology. Among those, the denial of "downward causation" from personal to impersonal processes,[56] the dualistic separation between mind and body, or between matter and mind— the latter which might sometimes be referred to as 'agency' in more recent new materialistic and agential realistic lingo. Griffin moreover underscores the impossibility of drawing any clear, reliable line between sentient and insentient things and beings. Besides, he addresses the unsolved problem of emergence, as well as the puzzling principle of nonlocal causation. Relying on the advances of quantum physics, molecular biology, and ethology, Griffin pinpoints the "internal relations [. . .] not only of living beings but also of the most elementary physical units."[57] While scientists have established that elementary forms of memory, decision, and organization exist all the way down to bacteria, to DNA and RNA micromolecules, and even to atoms, on the macroscopic level, James E. Lovelock and Lynn Margulis have offered evidence of "the influence of the planet as a living organism, and of the universe as a whole, on their parts."[58] In fact, Griffin and his colleagues cogently argue, postmodern sciences such as physics and biology corroborate a postmodern organicism much inspired by the organic process philosophy of scientist and philosopher Alfred North Whitehead, and which, in the end, does offer an organistic, reenchanted vision of the world.[59]

Writing in the 1980s, Griffin and his co-authors relied mostly on the then most recent scientific findings and philosophical theories to justify their reenchantment project. To achieve this paradigm shift, they insisted, our worldview must be a multifaceted one, where metaphysics and poetry may help complete the picture that can only partly be drawn out by science. These are points that Whitehead had examined early on. Elaborating his theory on "the bifurcation of nature," the British philosopher decried the illusory severing in modern science of "subject" from "object," as well as the separation of scientific disciplines into specialized, non-porous fields of knowledge.[60] According to Griffin, logic, aesthetics, and ethics should actually be implicated in the realm of science.[61] As previously pointed out by scientists Ilya Prigogine and Isabelle Stengers, some of the dead ends of modern epistemology might be overcome by reconciling the arts and humanities with the sciences.[62] Many have expressed concern with the negative consequences of this epistemological cleavage—both a cause and a consequence of humankind's misguided faith in its capacity to extract itself from nature. The "two cultures" split originally pinpointed by Charles Percy Snow indeed implies a quintessential separation between the human world of meaning, mind, and society, or "culture," from that of nonhuman nature and matter.[63]

Even more problematically, as Snow emphasized, the dichotomy at the heart of Western education has now massively produced very limited minds. Having acquired only a partial capacity to form an intelligent picture of the world, most modern minds are effectively handicapped. Hindered by the two-culture divide, most of us turn out incapable of assembling into a coherent worldview the vastly different discourses, reading grids, and forms of knowledge that are available across the boards of the humanities and sciences, allowing us to apprehend various aspects of reality and to inhabit the world meaningfully. As a reaction, cosmologist Brian Swimme proposes that "we learn to interpret the data provided by the fragmented scientific mind within the holistic poetic vision alive in ecofeminism." Swimme embraces the ecofeminist critique of modern ideology and compares "the patriarchal mind-set of our culture" with "a frontal lobotomy." He argues that the typical scientifically-trained mind, with its emphasis on controlling, distancing, calculating, and objectifying, "has been shut down in its fundamental cognitive and sentient powers."[64] With Prigogine and Stengers, Swimme was one of the scientists in the early eighties who incriminated the supremacy of rationality, advocating instead a rehabilitation of the sensitive and defending the value of storytelling. "Without the benefit of a cosmic story that provided meaning to our existence as earthlings," Swimme asserted, "we were stranded in an abstract world and left to invent nuclear weapons and chemical biocides and ruinous exploitations and waste. [. . .] We pursued 'scientific law,' relegating 'story' and 'myth' to the nurseries and tribes."[65]

Swimme's radical claim is that "scientists have entered a new enchantment."[66] Understanding that the universe is a dynamic, partly unpredictable, chaotic system and, renouncing the old, fixed and immutable laws of classical Newtonian physics, scientists have come to accept that the world is governed by orders and disorders always in the making. Owing to twentieth-century advances in physics, scientists have attained a greater level of humility, acknowledging, on the one hand, the limits of linear mathematics, and on the other hand, humans' incapacity to objectively study and analyze phenomena from without. As evidenced by quantum physics, no experiment can be conducted without eventually getting entangled with its object and results. Faced with the yet inexplicable, and with the possibility of more or less visible "intra-actions"—to take up Karen Barad's coinage—always at play in the world, scientists have come to appreciate the "world's effervescence," its "exuberant creativeness."[67] To put it in Swimme's words, the world is acknowledged as "an engendering reality." Thus, scientists have regained an attitude of wonder at "this creative interlacing energy that envelops us entirely."[68] What has been happening with quantum physics is no less than a "reenchantment *with* the universe": "a new love affair between humans and the universe happens. [. . .] Further surprises always occur, for to be in love is to be in awe of the infinite depths of things. [. . . Scientists] are suddenly astonished and fascinated in an altogether new way by the infinite elegance which gathers us into life and existence."[69] It is interesting that Swimme the scientist should envision wonder as an attitude harnessing love and consideration for the world—a tenet of ecocriticism that, as I will get back to later, was already present in Aldo Leopold and Rachel Carson's writings and is also central to Jane Bennett's theory of enchantment.

Going further, it turns out that metaphors at the heart of many traditional worldviews now crop up in scientific discourse. Where Paula Gunn Allen speaks of one's "enwholement" within a universe of interrelated physical and metaphysical realities that "moves and breathes continuously,"[70] physicist David Bohm conceptualizes the "unbroken wholeness," or "flowing wholeness" of a cosmos that is always enfolding and unfolding.[71] Trying to explain the "Sacred Hoop of Be-ing"[72]—the image at the core of her indigenous onto-epistemology, Allen writes: "the concept is one of singular unity that is dynamic and encompassing, including all that is contained in its most essential aspect, that of life."[73] Formulating the worldview theorized by quantum physics, Bohm proposes on his end that "the whole universe is *actively* enfolded to some degree in each of its parts. Because the whole is enfolded in each part, so are all the other parts, in some way and to some degree."[74] It is striking that Allen's effort to translate into English her traditional understanding of worldly entanglements chimes with Bohm's theory of an implicate and explicate order. Going back to her Native understanding of matter and spirit as different expressions of the world, Paula Gunn Allen, ventures that it is "as though life has twin manifestations that are mutually interchangeable and, in many instances, virtually identical aspects of a reality that is essentially more spirit than matter or, more correctly, that manifests its spirit in a tangible way.[75]

To explicit the idea that the universe may be characterized by "enfoldment" Bohm elaborates:

> [The] mechanistic picture, according to which the parts are only externally related to each other, is denied. That is, denied to be the primary truth; external relatedness is a secondary order of things, which I call the explicate order. This is, of course, the order on which modern science has focused. The more fundamental truth is the truth of internal relatedness, because it is true of the more fundamental order, which I call the implicate order, because in this order the whole and hence all the other parts are enfolded in each part.[76]

Bohm aspires to "a very different approach of science," one that is "more consistent and plausible and that fits better the actual development of modern physics than does the current approach."[77] Paving the way for Karen Barad's elaboration of a posthumanist "*ethico-onto-epistem-ology,*"[78] which draws on quantum physics to study the entanglements between matter and meaning, between the material and the discursive, Bohm defends that "[a] postmodern science should not separate matter and consciousness and should therefore not separate facts, meaning and value."[79] To put it in Barad's terms, "knowing is a matter of part of the world making itself intelligible to another part. Practices of knowing and being are not isolable; they are mutually implicated."[80] Of course, Maurice Merleau-Ponty's phenomenology of perception may read as a precursor here. Reincorporating human subjectivity and perception within what he calls the "flesh of the world"—its corporeal existence as we can apprehend it through our senses—Merleau-Ponty's legacy for a postmodern, new material onto-epistem-ology is immense.[81]

In any case, one essential conclusion that may be drawn from quantum physics, with major implications for the arts and humanities, is that the modern ontology, based as it is on the idea that human nature and mind might cut themselves off from the rest of the world, was founded on a delusory ideology of human exceptionalism. In Bohm's words, "[it] is a mistake to think that the world has a totally defined existence separate from our own and that there is merely an external 'interaction' between us and the world."[82] Barad has coined the term "intra-action" to describe the constant flow of matter and energy in a dynamic, complex movement of co-becoming: "The neologism 'intra-action' *signifies the mutual constitution of entangled agencies.* That is, in contrast to the usual 'interaction,' which assumes that there are separate individual agencies that precede their interaction, the notion of intra-action recognizes that distinct agencies do not precede, but rather emerge through, their intra-action."[83] As Barad stresses, "*the notion of intra-action constitutes a radical reworking of the traditional notion of causality.* [. . .] A lively new ontology emerges: the world's radical aliveness comes to light in an entirely nontraditional way that reworks the nature of both relationality and aliveness (vitality, dynamism, agency)." This paradigm shift from interactivity to intra-activity furthermore induces in its wake an essential rethinking of "core philosophical concepts such as space, time, matter, dynamics, agency, structure, subjectivity, objectivity, knowing, intentionality, discursivity, performativity, entanglement, and ethical engagement."[84]

Barad's proposition undergirds Merleau-Ponty's groundbreaking work, which has laid down the foundations of ecophenomenology. This is visible in David Abram's elaboration on Merleau-Ponty's phenomenology of perception, which is all about the inextricable enmeshment between the human and more-than-human worlds. Trained as a theoretical particle physicist, Barad for her part relies on quantum physics to establish how what we consider as phenomena results from an interweaving between the observed and the observer—a notion that comes close to Merleau-Ponty's interlacing of the perceiving and the perceived, of an "intercorporeity" entangling the human bodymind into the world: "[The flesh of the world] is the coiling of the visible around the seeing body, of the tangible around the touching body."[85] Going further, Barad emphasizes that subject and object get creatively intermeshed into an enfolding of co-constitutive material *and discursive* practices. "The separation of epistemology from ontology," Barad points out, "is a reverberation of a metaphysics that assumes an inherent difference between human and nonhuman, subject and object, mind and body, matter and discourse." As she maintains, "what we need is something like an *ethico-onto-epistem-ology*— an appreciation of the intertwining of ethics, knowing, and being—since each intra-action matters, since the possibilities for what the world may become call out in the pause that precedes each breath before a moment comes into being and the world is remade again, because the coming of the world is a deeply ethical matter."[86]

Seeking an image that might best encapsulate her complex worldview of a dynamic "Sacred Hoop of Be-ing," Paula Gunn Allen puts forward the following comparison: "The nature of the cosmos, of the human, the creaturely, and the supernatural universe is like water. It takes numerous forms; it evaporates and it gathers."[87] Physicists have indeed demonstrated the dual nature of matter and energy, which, to take up Bohm again, can "manifest either like a wave or like a particle according to how they [are] treated in an experiment."[88] Proving that the very nature of reality is context-dependent, quantum physics thus undermines the tenets of mechanistic science which had upheld that a particle would objectively remain itself no matter the context and whether or not it was being observed. Added to the principle of nonlocality also brought to the fore by quantum physics—the fact that "things [can] apparently be connected with other things any distance away without apparent force to carry the connection"[89]—scientists have proved that the human mind itself could not be abstracted from matter. In Bohm's phrasing, "evidently the whole world, both society and nature, is internally related to our thinking process through enfoldment in our consciousness."[90]

Bohm was already gesturing toward the deterministic role of the stories and meanings that we, as humans, co-compose with the world: "Because we are enfolded inseparably in the world, with no ultimate division between matter and consciousness, *meaning and value are as much integral aspects of the world as they are of us*. If science is carried out with an amoral attitude, the world will ultimately respond to science in a destructive way."[91] These are the very deep intra-actions entangling the world's becoming with storytelling which were elsewhere identified by Cheryll Glotfelty in the early stages of ecocriticism. As Glotfelty emphasizes, "literature does not float above the material world in some aesthetic ether, but, rather, plays a part in an immensely complex global system in which energy, matter, and ideas interact."[92]

NOTES

1. In speaking of a "tale of disenchantment," I am expressing resistance to a cultural construct while alluding to Jane Bennett's work. In *The Enchantment of Modern Life: Attachments, Crossings, and Ethics*, Bennett proposes a philosophical and new materialist approach to enchantment, which I get back to in the next chapter.

2. See Griffin, "Reenchantment," 2, and Bennett, *Enchantment*, 57.

3. Weber, "Science," 139.

4. I am using the capitalized 'Man' here with much critical ecofeminist distance, to refer to the dualistic, patriarchal, and androcentric mindset of the time that excluded women from the world of reason and control—women being relegated onto the side of wild nature, as needing to be controlled and civilized by (white) men. See Carolyn Merchant's *Death of Nature*, Susan Griffin's *Woman and Nature*, or Val Plumwood's

Feminisms and the Mastery of Nature. As for Descartes' famous saying, this is how the full passage goes: "As soon as I had acquired a few general notions in physics, and as [. . .] I noticed how far they could lead, and how much they differed from the principles that had been available heretofore, I thought I could not hide them without committing a great sin against the law that obliges us to provide, as much as possible, all men with general welfare. Indeed, physics has taught me that it is possible to attain knowledge that is of much utility for life, and that instead of the speculative philosophy that is taught in school, it can lead to practical thinking, by means of which, knowing the force and actions of fire, water, air, stars, the heavens and all the other bodies that surround us [. . .] we could use them [. . .] for all adequate purposes, and thus become like masters and possessors of nature. This is desirable not only for the infinite number of artifices that would grant us the opportunity to easily enjoy the fruits of the earth and all the commodities that it provides, but principally too for the conservation of our health, which is undoubtedly the first good and the foundation of all other goods in life." (*Discours*, 168, translation mine).

5. Griffin, "Reenchantment," 2.
6. Berman, *Reenchantment*, 57.
7. Ibid, 57.
8. Griffin, "Reenchantment," 2–3.
9. Weber, "Science,"155.
10. Weber, *Protestant*, 182.
11. Quoted in Gibson, *Reenchanted*, 10.
12. Berman, *Reenchantment, 2.*
13. *Ibid*, 3.
14. White Jr., "Historical," 43.
15. Ibid, 43. On top of the way in which the destruction of pagan animism made "the old inhibitions to the exploitation of nature [crumble]," White Jr. moreover blames the fact that "for animism the Church substituted the cult of saints" whose eternal dwelling place is not in natural places but in an unearthly heaven.
16. Ibid, 43.
17. Berman, *Reenchantment, 58–60.*
18. Merchant, *Death*, 127.
19. Ibid, 123–32.
20. Ibid, 140.
21. Ibid, 148.
22. Descola, *Nature*, 302–4.
23. Allen, *Sacred.*
24. Ibid 50–60.
25. Barad, *Meeting*, 332–52.
26. Allen, *Sacred*, 60.
27. Ibid, 127.
28. Ibid, 127.
29. Ibid, 60. As I will get back to later, it may be possible to interpret this view of consciousness and the universe following hypotheses laid out on the one hand by anthropologist and cyberneticist Gregory Bateson, and, on the other hand, by

theoretical physicists such as David Bohm—with his concepts of an "implicate order" and "unbroken, flowing wholeness"—or in light of Karen Barad's work, who has revolutionized contemporary philosophical understanding of "the entanglements of matter and meaning" via quantum physics.

30. Berman, *Reenchantment*, 141, 301–2.
31. Merleau-Ponty, *Nature* and *Visible*.
32. Berman, *Reenchantment*, 141.
33. Ibid, 140.
34. Bateson, *Ecology*, 436.
35. As explained earlier, my own take on "spirituality" is not linked to a specific religion, nor to any notion of a God or gods. It refers to the cultivation of a sense of connection tying the individual soul, or psyche, to the energies and principles flowing through the larger, material world, both human and other-than-human. Ecofeminist spiritualities are rooted in non-dualistic frameworks and cannot therefore conceive of a life of the mind totally detached from the feeling body. To the contrary, ecofeminist spiritualities explore both the visible and invisible connections to the larger world which can manifest themselves through the senses, through affects, and intuition.
36. Teaching *Ceremony* as part of a postcolonial studies course, I usually refer to Linda Hutcheon's useful concept of "historiographic metafiction." See Hutcheon, *Poetics*. As for the dialogism of prose, I am relying on Mikhail Bakhtin's study of novelistic heteroglossia in *The Dialogic Imagination*.
37. Silko, *Ceremony*, 35–36.
38. In using the term "postmodern science" I am referring to the project of reenchantment undertaken collectively by philosophers and scientists David Ray Griffin, David Bohm, Rupert Sheldrake, Frederick Ferré, and others in *The Reenchantment of Science*, which I will get back to shortly.
39. Merchant, *Death*, 3.
40. Ibid, 4–5.
41. Weber, *Protestant*, 26.
42. Eisler, "Ecofeminist Manifesto," 32.
43. Ibid, 32.
44. Krech III, *Ecological*. See also, Machet et al.'s introduction to the special issue of *Elohi: The Invention of the Ecological Indian*.
45. Eisler, "Ecofeminist Manifesto," 32.
46. Ibid, 33.
47. Ibid, 33.
48. Bohm, "Postmodern," 67.
49. Berman, 2.
50. See for instance the preface and table of contents, where these notions are salient. They are taken up in detail in the individual chapters.
51. Ray Griffin, *Reenchantment*, (xiii).
52. Ibid, x.
53. Ray Griffin, *Reenchantment*, x–xi.
54. Ibid, xi.
55. Ibid, 3.

56. Ibid, 3.
57. Ibid, 15.
58. Ibid, 15. I get back to the "Gaia hypothesis" as formulated by Lovelock and Margulis in chapter 3.
59. Griffin, ed., *Reenchantment*, 151–57.
60. See Alfred North Whitehead's chapter, "Theories of the bifurcation of Nature" in *The Concept of Nature,* 26–48.
61. Griffin, 27.
62. Prigogine and Stengers, *Order*, xxvii.
63. See Snow, *Two Cultures*.
64. Swimme, "Lobotomy," 15–16.
65. Swimme, "Cosmic," 49.
66. Ibid, 51.
67. Barad, *Meeting,* 177.
68. Swimme, "Lobotomy," 20.
69. Swimme, "Cosmic," 51.
70. Allen, *Sacred*, 59.
71. Bohm, "Postmodern," 65–66.
72. Allen, *Sacred*, 11.
73. Ibid, 56.
74. Bohm, "Postmodern," 66.
75. Allen, *Sacred*, 60.
76. Bohm, "Postmodern," 66.
77. Ibid, 60
78. Barad, *Meeting*, 185.
79. Bohm, "Postmodern," 60.
80. Barad, *Meeting, 185.*
81. Merleau-Ponty, *Nature*, 279–80. I will also take up Merleau-Ponty to shed light on various aspects of liminal realism throughout this study.
82. Bohm, "Postmodern," 67.
83. Barad, *Meeting, 33.*
84. Ibid, 33.
85. Merleau-Ponty, *Visible*, 189, translation mine.
86. Barad, *Meeting, 185.*
87. Allen, *Sacred*, 101.
88. Bohm, "Postmodern," 63.
89. Ibid, 64.
90. Ibid, 67.
91. Ibid, 67.
92. Glotfelty, "Literary Studies," xix.

Chapter Two

Toward an Ecofeminist Project of a Rational Reenchantment

While ecopoetics is first and foremost concerned with poetry, it can nonetheless provide insight into other creative forms of storytelling and artmaking—of *poiesis*—emerging as a direct response to the *oikos*, our earthly home. As such, *ecopoeisis* performs a becoming-with that paves the way for new ways of inhabiting the Earth. In his defense of storytelling, Brian Swimme contends that "we will only have a common story for the human community when poets tell us the story. For until artists, poets, mystics, nature lovers tell us the story [. . .] we have only facts and theories."[1]

Swimme further argues for the reinvention of stories and ceremonies enabling us to "celebrate the cosmic story" via "poetry, chant, dance, painting, music."[2] Referring to traditional communities, Swimme points to one of the aspects central to the ecopoetics of reen-*chant*-ment here explored: "Unless the story is sung and danced the universe suffers from decay and fatigue. Everything depends on telling the story—the health of the people, the health of the soil, the health of the sun, the health of the soul, the health of the sky." The problem with modernity, Swimme argues, is that "|instead] of poets, we had one-eyed scientists and theologians. Neither of these high priests nor any of the rest of us was capable of telling the cosmic story."[3] Although the congruence between postmodern science and ecofeminist or indigenous *poiesis* often gets overlooked—such texts are often misinterpreted by thinkers incapable of conceptualizing anything outside of the dualistic box, some of whom moreover seem thrown off by the deliberate interweaving of poetics, spirituality, science, politics, and philosophy that is characteristic of much indigenous and ecofeminist storytelling—Swimme has on his end stated that "only when the scientific facts are interpreted by an ecofeminist consciousness will we begin to see where we are, who we are, and what we are about."[4]

Insisting that there exists a possibility for a "rational reenchantment" that women can contribute to, Ynestra King has explained:

> We thoughtful human beings must use the fullness of our intelligence to push ourselves intentionally to another stage of evolution. One where we will fuse a new way of being human on this planet with a sense of the sacred, informed by all ways of knowing—intuitive *and* scientific, mystical *and* rational. It is the moment where women recognize ourselves as agents of history—yes, even as unique agents—and knowingly bridge the classic dualisms between spirit and matter, art and politics, reason and intuition. This is the potentiality of a *rational reenchantment*. This is the project of ecofeminism.[5]

Tying in with much ecofeminist *and* scientific visions, the literary thought-experiments and material ecopoetics that I explore throughout this survey also coalesce with Jane Bennett's "enchanted materialism," offering a "picture of the world as a lively and endless flow of molecular events, where matter is animate without necessarily being animated by divine will or intent."[6] Working against the various disenchantment tales associated with modernity, Jane Bennett redeems "a contemporary world sprinkled with natural and cultural sites that have the power to 'enchant.'"[7] Her "alter-tale" is not so much one of "reenchantment"; rather it scrutinizes natural, cultural, and technological "magical sites already there."[8] Cognizant of the performative power of culturally constructed stories, Bennett warns that "the very characterization of the world as disenchanted ignores and discourages affective attachment to that world."[9] Arguing that "the mood of enchantment may be valuable for ethical life,"[10] her main point is that enchantment is characterized by certain positive, somatic responses to the world, producing an affective force that may be harnessed to propel ethical generosity.

In her call to reclaim partnership traditions, Riane Eisler refers to ancient "pagan" traditions. "In ancient times," she says, "the world itself was one. The beating of drums was the heartbeat of the earth—in all its mystery, enchantment, wonder, and terror. What was later broken asunder into prayer and music, ritual and dance, play and work, was originally one. [. . .] There was no splintering of culture and nature, spirituality, science, and technology."[11] Awakening from the nightmare set in an uninhabitable fragmented world, ecofeminism advocates an integrative reinvention of holistic stories, sciences, practices, and ontology that is *not* antithetical with Modernity *per se*, but that yet calls for a redirecting of modernity's basic epistemological and ethical assumptions. Thus, "the reclamation of these [ancient] traditions can be the basis for the restructuring of society: the completion of the modern transformation from a dominator to a partnership world."[12] Ecofeminism's audacious reclaiming of the feminine, the sensitive, of our corporeal attachment to the

Earth, and of spirituality thus emerged from the awareness that modern patriarchal models of "power over" needed to be deconstructed and reweaved into healthier models of "power to" that might "illuminate and transform human consciousness (and with it reality)," making for a more sustainable future.[13] "Poised on the brink of ecocatastrophe," Eisler wrote, "let us gain the courage to look at the world anew, to reverse custom, to transcend our limitations, to break free from the conventional constraints, the conventional views of what are knowledge and truth."[14]

One of the ways in which ecofeminism and indigenous worldviews and *poiesis* often overlap has to do with a relational ontology. Redeeming principles of interconnectedness and interdependence which have culturally been interpreted as "feminine," and which, in a Western culture promoting total autonomy and control, are mostly devalued, ecofeminists and ecopsychologists have been calling for narratives and practices that might restore our sense of bonding with the more than-human-world. In their chapter "The Rape of the Well-Maidens," Mary E. Gomes and Allen D. Kanner make the case for the cultivation of the "self-in-relation" as opposed to the "separative self" that we are mostly taught to develop in the Western world.[15] Hence the books in my corpus, which I have chosen for the relational ontology they diffract in ways aligning with ecofeminism and ecopsychology, and with older animistic and totemic traditions. In these works, patriarchal ideologies and behaviors that, in shunning a part of our most vital connections, deprive us of essential experiences of joy, meaning, and empathy, are offset by ecofeminist, indigenous, and bioregional stances celebrating our interdependence with the more-than-human world. All the books in my corpus thus openly take issue with the myth of the separate self. They push forward a vision of the world where interconnection is empowering, while bringing about feelings of gratitude, bliss, humility, and reciprocity.

As I will attempt to demonstrate, artworks, practices, and theories of reenchantment may restore the "biophilia" that, according to biologist Edward O. Wilson and many ecopsychologists, is genetically ingrained in the human species, leading us to spontaneously seek connection with other-than-human nature because it makes us feel good and at home.[16] Reawakening our "innate love of contact with other living organisms,"[17] an ecopoetics of reenchantment may therefore contribute to the tempering of the ecophobic inclinations and attitudes in the Western world that ecocritic Simon Estok has called attention to. Ecopsychology and ecofeminism can help break down the mechanisms of "ecophobia," which Estok defines as a "uniquely human psychological condition that prompts antipathy toward nature."[18] According to Estok, ecophobia could, like biophilia, find its roots in the evolution of the human species in somewhat hostile environments and, following Wilson's claims about the genetic inscription of biophilia, it may also partly account for the antagonistic relationship between humans and nature. As Estok points out, "[ecophobia] emanates from anxieties about control."[19] As a matter of fact, I would be tempted to read ecophobia as a major consequence of the complex disenchantment process brought about by patriarchal Modernity, rather than as an evolutionary condition applying to humankind as a species. From an ecofeminist and ecopsychological perspective, it can be ventured that, when forced to acknowledge the many ways in which a human self cannot separate from and control his or her 'environment,' our fundamental connections to and interdependence with the more-than-human world which the Moderns have come to perceive as constituting 'nature' "are experienced as threats due to the insistence that we *should* be separate" and able to master both ourselves and the nature outside of us.[20]

As becomes clear from the preface to the second edition of *The Death of Nature*, Carolyn Merchant both helped and was influenced by ecofeminism's renewal of earthly spiritualities. As she calls for "the restoration of sustainable ecosystems that fulfill the basic human physical and spiritual needs," Merchant ventures, that "[perhaps] Gaia will then be healed."[21] In a similar way, some of the contributions in the book edited by Griffin on the reenchantment of science contend that quantum physics is indeed compatible with a form of mysticism, especially with the worldviews expressed in Taoism and Buddhism. Meanwhile, ecofeminists have developed various forms of ecospiritualities—many of them hybrid and syncretic, mythopoeic and self-reflexive—ranging from indigenous practices and beliefs (Paula Gunn Allen, Gloria Anzaldùa), Buddhism (Greta Gaard), to neopagan forms of Earth and Goddess religions (Starhawk, Eisler) including ecofeminist rereadings of Christianity (Charlene Spretnak, Sally Abbott, Sallie McFague).[22] As John Baird Callicot and others have established, even certain strands of Christianism have fostered revisionary interpretations of the Bible.[23] Greener understandings of the Old and New Testaments promote Earth stewardship rather than dominion—a trend that has culminated recently with Pope Francis's *Laudato Si*, exhorting the world to no less than an "ecological conversion" based on the imperative to care for the common *oikos*.[24]

Albeit often underestimated or ignored, the role of ecofeminist thinkers has been fundamental in bringing about the paradigm shift occurring in the twenty-first century—a shift whereby humans no longer perceive themselves outside and above nature, and whereby nonhuman nature, matter, women, and indigenous groups are conjointly reempowered. As touched upon earlier, in the seventies and eighties, many ecopsychologists and ecofeminists alloyed interest in the human psyche and spirituality with a thorough understanding of the latest sciences. While some strands of ecofeminism deliberately and reflexively reinstill a sense of the sacred into their worldview—and again, what various writers may mean when they use the word 'sacred' must be clarified, especially when, from a nondualist, ecofeminist point of view, the sacred does not stand in opposition with the profane—others favor a new materialist approach. As I hope to show, these two tendencies are not as irreconcilable as might first seem. In the early eighties, ecofeminist activist and writer Starhawk advanced "another form of consciousness" that many ecofeminists and ecopsychologists envisioned, holding a promise of healing. Following Starhawk, this form of consciousness can be called "immanence—the awareness of the world and everything in it as alive, dynamic, interdependent, interacting, and infused with moving energies: a living being, a weaving dance."[25] Stressing how ecofeminism reinvents older practices and worldviews that have survived in spite of European patriarchal expansion, Starhawk argues that "it has existed from earlier times, underlies other cultures, and has survived even in the West in hidden streams."[26]

If narrow interpretations of Starhawk's identification as a "witch" and her open claims to be dealing with "neopaganism" and "magic" can be misleading to some, her ecopoetic, self-conscious enterprise is nevertheless driven by her deep awareness that "we [. . .] need to reconcile science and spirituality." Explaining her own take on the sacred, Starhawk contends that "[when] our sense of the sacred is based not upon dogma but upon observations and wonder at what is, no contradiction exists between the theories of science and those of faith."[27] Starhawk relies on Connie Barlow, who has argued for the bridging of scientific and religious worldviews:

> The more we learn about Earth and life processes, the more we are in awe and the deeper the urge to revere the evolutionary forces that give time a direction and the ecological forces that sustain our planetary home. Evolutionary biology delivers an extraordinary gift: a myth of creation and continuity appropriate for our time. [. . .] Finally, geophysiology, including Gaia theory, has reworked the biosphere into the most ancient and powerful of all living forms—something so much greater than the human that it can evoke a religious response.[28]

Defending an "earth-based spirituality [. . .] rooted in three basic concepts"—"immanence, interconnection, and community"[29]—and which seeks to "[ground one's] spirit in the rhythms of nature,"[30] it may well be that Starhawk's ecopoetic vision and thought-experiments embrace a form of spirituality that is, indeed, compatible with scientific ecology, biology, geophysics, quantum physics, and new materialism. "When you understand the universe as a living being," she reflects, "then the split between religion and science disappears because religion no longer becomes a set of dogmas and beliefs we have to accept even though they don't make sense, and science is no longer restricted to a type of analysis that picks the world apart."[31] "Science itself," she reminds us, "has moved beyond the mechanistic model of the universe. Today science is likely to describe the world in terms of networks and probabilities and complexities, as interlocking processes and relationships. Yet," she deplores, "our thinking and understanding as a culture does not often reflect this greater sophistication."[32]

As I set out to demonstrate, Starhawk's ecopoetic thought-experiments (such as those crafted in her novel *The Fifth Sacred Thing*, which I will turn to later) might turn out avant-garde in terms of shifting our ontologies toward a reenchanted, liminal realism. For now, suffice it to recall that in her provocative use of the term 'magic,' Starhawk defines her practice as "the art of changing consciousness at will," which she ties to the vision and activism required for bringing about political consciousness and change.[33] Foregrounding the ethical and political implications of an earth-based spirituality, Starhawk writes: "when we start to understand that the Earth is alive, she calls us to act to preserve her life. When we understand that everything is interconnected, we are called to a politics and set of actions that come from compassion, from the ability to literally feel with all the living beings on the Earth."[34] Moving toward a biocentric vision of community that adheres with Aldo Leopold's plea for an ecological land ethic based on love and respect, Starhawk concludes, "[that] feeling is the ground upon which we can build community and come together and take action and find direction."[35]

Building on Freudian, Jungian, and *Gestalt* theory, as well as on the Gaia hypothesis postulated by Lovelock and Margulis, Theodore Roszak, the father of ecopsychology, formulated the hypothesis of an "ecological unconscious": "Gaia, taken simply as a dramatic image of ecological interdependence, might be seen as the evolutionary heritage that bonds all living things genetically and behaviorally to the biosphere." With Gaia, Roszak claims, "we have the possibility that the self-regulating biosphere 'speaks' through the human unconscious, making its voice heard even within the framework of modern urban human culture."[36] In *The Voice of the Earth,* Roszak calls for a revision of modern natural philosophy: "[The] deeper modern science delves into the nature of things, the more it finds hints and traces of the primordial animist world: *mind in the cosmos*, as we will refer to it. In their own halting, roundabout way, scientists are fashioning a picture of the world as alive, intentional, creative [. . .]."[37] Roszak again ventures the notion of an ecological unconscious, which he defines as "the 'savage' remnant within us, that rises up subjectively to meet the environmental need of the time, [. . .] offering the chance once again to *connect*."[38] Thinking in lines of flight that fall in formation with other thinkers—such as Gregory Bateson, Morris Berman, Gary Snyder, or Alan Watts—of what an ecological mind might mean, Roszak goes on: "Mind, far from being a belated and aberrant development in a universe of dead matter, *connects* with that universe as the latest emergent stage on its unfolding frontier." Relying on psychologist William

James's seminal exploration of religious experience, Roszak extolls the information we may access through "the subliminal door": "Drawing upon the most recent findings of science, we might now say that Gaia [...] gains access to us through the door of the id."[39]

Deploying the image of the "voice of the earth," Roszak exhibited awareness that in reenchantment must lay our salvation: "The voice of the Earth is that close by. If we are, as the Romantic poets believed, born with the gift of hearing that voice, then turning a deaf ear to her appeal must be a wrenching effort, and a painful one to maintain, just as all the efforts to hide from the truth or our identity must be painful. Repression hurts. We call that pain 'neurosis.'"[40] Rebounding on Roszak, it could be that ecopoetic liminal realism expresses those connections with the more than-human world which Moderns have learned to repress. Despite centuries of repression, the deep-down knowledge that humans are made of the same stuff as the rest of the living world still palpitates, alive in a subliminal realm, somewhere in-between the full consciousness of our somatic perceptions and our deep, unconscious responses and drives. Because it may provide access to our ecological unconscious, ecopoetics may well be key in helping us heal from the current state of generalized alienation from the natural world. As Roszak claims, "[the] collective unconscious, at its deepest levels"—and I would add as it gets expressed in *ecopoiesis*—"shelters the compacted ecological intelligence of our species, the source from which culture finally unfolds as the self-conscious reflection of nature's own steadily emergent mindlikeness."[41] Moreover, since humans have co-evolved with their natural milieux and from animal, vegetal, bacterial, and elemental lifeforms, it may well be that our genetic heritage holds an ecological unconscious, containing the memory of long processes of co-becoming. Encoded in our genes then, and following the rules established in the recent field of epigenetics, our ecological unconscious could explain the biophilia ingrained in humans, the firing up of certain sensitivities, capacities, and intuition when immersed in nature, as well as the resurgence of ecopoetic matter in our dreams and art. As Edward O. Wilson has it, "there is a lot of Mother Nature still in our genes."[42]

As should be becoming clear, one of the aims of this book is to analyze contemporary ecoliterature that manifests a reenchanting liminal realism compatible with indigenous worldviews, ecofeminist practices and theory, the ecophenomenology of Merleau-Ponty and David Abram, as well as with Karen Barad's agential realism. This offers a worldview recognizing the agency of storytelling and poetics, perceived as material practices with incidence over the world's mattering. My work draws from the theoretical framework of new materialism elaborated by scholars such as Serpil Opperman and Serenella Iovino. As a matter of fact, their work on narrative agency and material ecocriticism is itself inscribed in a larger reenchantment project:

Material ecocriticism invites us into a polyphonic story of the world that includes the vital materiality of life, experiences of nonhuman entities, and our bodily intra-actions with all forms of material agency as effective actors. [. . .] In this "alter-tale" the new narrative agents are things, nonhuman organisms, places, and forces, as well as human actors and their words. Together, they anticipate an alternative vision of a future where narratives and discourses have the power to change, re-enchant, and create the world that comes to our attention only in participatory perceptions.[43]

The intuition that I have followed in the past twenty years is that the polyphonic, multitextured, and multispecies song of the world translated in my corpus reveals how the worldviews now theorized by material ecocriticism can harmonize with ecofeminism, with traditional, ecological understanding and storytelling, and with ecological science, biosemiotics, and quantum physics. Ecopoetic liminal realism may specifically coincide with Paula Gunn Allen's attempt to convey how, "through the sacred power of utterance [American Indian people] seek to shape and mold, to direct and determine, the forces that surround and govern human life and the related lives of all things."[44]

Although some scientists and philosophers, mired in the dualistic framework still dominant in much of our culture, have failed to acknowledge the ethico-onto-epistemo-logical potential that lies in the ecofeminist interweaving of the many facets of the Gaia metaphor across different domains, it seems that the ecopoetic resurgence of Gaia across the boards of the arts, sciences, humanities, politics, and practices *speaks worlds, signaling* a radical rewor(l)ding under way.

NOTES

1. Swimme, "Cosmic," 52.
2. Ibid, 53.
3. Swimme, "Cosmic," 53.
4. Swimme, "Lobotomy," 17.
5. King, "Healing," 121.
6. Bennett, *Enchantment*, 14.
7. Ibid, 3.
8. Ibid, 8.
9. Ibid, 3.
10. Ibid, 3.
11. Ibid, 33–34.
12. Ibid, 34.
13. Ibid, 30.

14. Ibid, 34.

15. Gomes and Kanner are drawing from important work carried out by psychologists at the Stone Center at Wellesley College.

16. See Wilson, *Biophilia*, and *Origins*.

17. Wilson, *Origins*, 132.

18. Estok, *Ecophobia*, 1.

19. Ibid, 92.

20. Gomes and Kanner, "Rape,"114.

21. Merchant, *Death*, xviii.

22. See Starhawk, *The Spiral Dance: A Rebirth of The Ancient Religion of the Great Goddess*, and *The Earth Path: Grounding Your Spirit in the Rhythms of Nature*. Many great essays on ecofeminist spiritualties are available in the anthologies edited by Judith Plaskow and Carol P. Christ, *Weaving the Visions: Patterns in Feminist Spirituality*, by Judith Plant, *Healing the Wounds: The Promise of Ecofeminism*, or by Irene Diamond and Gloria Feman Orenstein, *Reweaving the World: The Emergence of Ecofeminism*. Readers interested in exploring some of the rich spiritual potential of ecology outside of ecofeminist contributions may turn to Bron Taylor's *Dark Green Religion*, and Leslie E. Sponsel's *Spiritual Ecology: A Quiet Revolution*. See also essays by Kate Rigby, "Spirits that Matter: Pathways toward a Rematerialization of Religion and Spirituality," and Greta Gaard, "Mindful New Materialisms: Buddhist Roots for Material Ecocriticism's Flourishing," edited by Serenella Iovino and Serpil Oppermann.

23. Of major importance in that area is Callicot's extensive work on ecology and religion, which includes a redeeming reading of Genesis challenging Lynn White Jr's position on the matter (see "Genesis and John Muir," *Beyond*, 187–219). See also Gibson's chapter on the "greening of religion" (*Reenchantment*, 93–123).

24. Pope Francis, *Laudato*, # 217.

25. Starhawk, *Dreaming*, 9. I take up the weaving metaphor again in the next chapter.

26. Ibid.

27. Starhawk, *Earth*, 11.

28. Quoted in Starhawk, Ibid, 11.

29. Starhawk, "Power," 73.

30. See the subheading of her book *The Earth Path: Grounding Your Spirit in the Rhythms of Nature*.

31. Starhawk, "Power," 73.

32. Starhawk, *Earth*, 10.

33. Starhawk, "Power," 76. See also *Dreaming*: "The answers I propose involve magic, which I define as *the art of changing consciousness at will*. According to that definition, magic encompasses political action, which is aimed at changing consciousness and thereby causing change," 13.

34. Starhawk, "Power," 74.

35. Ibid.

36. Roszak, "Psyche," 14.

37. Roszak, *Voice*, 96.

38. Ibid, 96.
39. Ibid, 303–04.
40. Ibid, 304.
41. Ibid, 304.
42. Wilson, *Origins*, 130.
43. Iovino and Oppermann "Material Ecocriticism," 88.
44. Allen, *Sacred*, 56.

Chapter Three

An Ecofeminist Remystification of Narrative

The Many Faces of Gaia in the Anthrop-o(bs)cene

In her book dedicated to magical realism, Wendy Faris remarks that the mode "[constitutes] a latent tendency to include a spirit-based element within contemporary literature—a possible remystification of narrative in the West."[1] According to Faris, "[by] incorporating a mysterious dimension into the discourse of literary realism, magical realism both questions and replenishes it."[2] So far, these observations could apply to the ecopoetics of reenchantment that is the focus of this book. However, where Faris is looking at a magical realism that "would then represent a moment of cultural retrospection that is the reverse image of the moment of 'desacralization' that Frederic Jameson investigates," and that "might even begin to reverse [. . .] progressive secularization," I would be extremely guarded in interpreting the liminal realism at the heart of my corpus as simply a form of regressive desecularization of the world.

Rather than reinstating literal beliefs in nonmaterial spirits, gods, and goddesses, the "liminal realism" I am looking at tends to exhibit here and there a certain awareness of its own, enchanting processes as it deliberately situates us in-between scientific and ecopoetic worldviews of the material agencies at work in the world. At the end of the day, the principal way in which nature might potentially seem resacralized by such ecopoetic fiction lies in its foregrounding of the inviolable laws of interconnection and symbiosis ruling all life on Earth, laws which should force humans to renounce anthropocentrism and to adopt a biocentric perspective instead. By restoring our sense of wonder, what I propose to call "liminal realism" then positively emphasizes the inherent value, beauty, and complexity of all lifeforms in a way that can confer a secular sense of the sacred.

That said, it cannot be denied that the texts at hand exhibit a tantalizing, mythopoeic dimension. Yet, if liminal realism tends to reproduce the stereoscopic trick behind some anamorphoses whereby two images are superimposed into one that consequently acquires new relief—relief here bringing together the sacred and the profane, or the mythological and the scientific, the poetic and the rational—[3]the mode nevertheless remains on the side of postmodernism in so far as it will now and then bare its own enchanting artifice. In many cases indeed, the poetic vision, the myth, or the subjectivity interlaced with the reality described are brought forward not so much as a way to renounce altogether the very notion of reality—as is the case with extreme versions of postmodernism—but, rather, as a reminder that phenomena can appear in a kaleidoscopic manner, depending on the discursive and sensitive apparatuses which they are apprehended with. Rather than laying claim to a single, unified, and total truth, ecopoetic liminal realism cautiously negotiates the understandings that not only will the nature of reality always remain partly elusive, it moreover shifts depending on the means available to those trying to form a coherent picture of it. As readers of liminal realism, we must therefore always keep in the back of our minds the willful use and abuse of metaphor, myth, and anamorphosis at play, which are exploited for their charming, uplifting, or revealing powers. Nevertheless, the texts in my corpus bring to the fore a vision of *ecopoiesis* that ultimately banks on the very real, transformative effect of vision-shifting and artmaking, allowing humans to inhabit the world ecopoet(h)ically.

In the next chapter, my goal is to tease out the implications of the ecofeminist, anamorphic tendencies in my corpus whereby the earth is apprehended as a living superbeing, or, more precisely, a superwoman or goddess. In doing so, I am hoping to shed light on some of the many faces of Gaia made manifest in contemporary ecoliterature.

GAIAN ECOPOETICS: RECIPROCITY AND SYMBIOSIS ACCORDING TO STARHAWK, RICHARD POWERS, LYNN MARGULIS, AND THEODORE ROSZAK

The rehabilitation of images of an Earth Goddess by neo-pagan witch and ecopoet Starhawk is no doubt a good place to start. The opening line from the "Declaration of the Four Sacred Things" that prefaces her 1993 climate fiction novel states: "The earth is a living, conscious being."[4] What might be meant thereby is immediately tied to a malleable multicultural perspective making room for a multitude of equally valuable mythologies and epistemologies: "In company with cultures of many different times and places, we name these things as sacred: air, fire, water, and earth. / Whether we see them as breath, energy, blood, and body of the Mother, or as the blessed gifts of a Creator, or as symbols of the interconnected systems that sustain life, we know that nothing can live without them." From the start, the matter of perspective is thus foregrounded ("Whether we see them as. . . . or as symbols . . . "), together with the possibility of adopting a double vision, scientific and poetic at once. This double vision is articulated in *The Fifth Sacred Thing* via the Goddess archetype:

> Maya could feel the earth under her, alive like a beating heart. Or perhaps, she thought, I'm feeling my own throbbing feet? Still, it was good, at the place of the dead, to acknowledge that One to whom she had pledged herself long ago, the aliveness at the heart of things, the ever-turning wheel of birth, growth, death, and regeneration. It had occurred to Maya lately that calling *that* the Goddess, even though she'd fought for the term all her life, was—what? Not so much a metaphor, more in the nature of a private joke.[5]

As one of the pillars of the fictional ecofeminist community, and as a storyteller, Maya's character easily reads here as an avatar of Starhawk herself, propounding from the incipit of the novel her own nuanced and reflexive take on the notion of a Mother Earth Goddess. While befitting Bron Taylor's important concept of "dark green religion," which he defines as a set of practices and beliefs that holds nature as sacred, possessing inherent value, and thus deserving reverent care,[6] Starhawk's ecopoet(h)ics corresponds even better to his notion of "Gaian Earth Religion," which "understands the biosphere [. . .] to be alive or conscious," and which "relies on metaphors of the sacred to express its sense of the precious quality of the whole."[7] Yet, like that of most of the ecofeminist writers in my corpus, Starhawk's non-dualistic poetics complicates Taylor's division between categories opposing spiritual and naturalistic stances. It thus offers a relevant example of how various types of dark green religion, that is, "spiritual animism," "naturalistic animism," "Gaian spirituality," and "Gaian naturalism" are effectively permeable.[8]

Maya's hesitation as to whether the pulse which she attributes to the earth could in fact be her own moreover highlights the inextricable entanglements between inner and outer reality. Phenomena, we are reminded, arise from the intra-action between the world and our subjective corporeal experience of it. At the same time, Maya's proprioception here expresses a sense of rhythmic harmony between her individual body and that of the earth, both being animated by continuous, somewhat synchronized pulses. Significantly, the above passage upholds a crucial tenet of ecofeminism, that is, that, before it can be tentatively conceptualized, one's sense of connection with the earth, or with "the aliveness at the heart of things," is first and foremost felt through one's body. In their essay "Naming the Sacred," ecofeminist scholars Judith Plaskow and Carol P. Christ sum up the nondualist goals driving the rehabilitation of Goddess imagery: "Philosophically, God's transcendence is frequently understood to mean that God is different from humanity and nature because God is pure spirit uncorrupted by a physical body. The human body with its connection to nature then is said to keep us from God. The mind must rule over the passions of the body, and humans must rule over 'brute' nature."[9] Attempting to revise a whole system of thought that devalues the materiality of the body, and more specifically that of women, animals, and the rest of nonhuman nature, subsequently dispossessing the sensitive of its fine intelligence,[10] the texts in my corpus activate an ecopoet(h)ics of earthly connection that follows in the steps of ecofeminism, with its new materialistic turn toward "a more *immanental*, or earth and body-centered, direction."[11]

In her nonfiction, Starhawk further unravels the ecopoetic stance from which she projects the image of "the Goddess" onto the world. "The image of the Goddess," she explains, "strikes at the roots of estrangement. [. . .] Nature is seen as having its own inherent order, of which human beings are a part. Human nature, needs, drives, and desires are not dangerous impulses in need of repression and control, but are themselves expressions of the order inherent in being."[12] While Starhawk recognizes that "[many] people will prefer the concept of immanence without the symbol attached," she justifies calling onto the ecopoetic vision because of its affective and somatic charge: "I prefer the symbol to the abstraction because it evokes sensual and emotional, not just intellectual, responses."[13] Further disambiguating her self-conscious, secular reenchantment project, and her references to "the divine,"[14] she writes: "Let us be clear that when I say *Goddess* I am not talking about a being somewhere outside of this world, nor am I proposing a new belief system. I am talking about choosing an attitude: choosing to take this living world, the people and creatures on it, as the ultimate meaning and purpose of life, to see the world, the earth, and our lives as sacred."[15]

Embracing a vision of the absolute, inherent value of all life aligned with Deep Ecology, one of Maya's ceremonial stories in *The Fifth Sacred Thing* leads her to reassert their community's rather pragmatic conception of the sacred. This, we learn, grew out of "the Uprising"—a time of revolution against a dictatorship after which the inhabitants of San Francisco (note the implied reference to the Patron Saint of Ecologists) founded their resistant city. During that time, the characters underwent a long period of drought and hunger, having to toil the earth and grow their own food: "It was a long dry season. But we had pledged to feed one another's children first, with what food we had, and to share what we had. And so the food we shared became sacred to us, and the water and air and the earth became sacred. / When something is sacred, it can't be bought or sold. It is beyond price, and nothing that might harm it is worth doing."[16] As for the syncretic, hodgepodge, and fluid Goddess symbolism underpinning Starhawk's anticapitalistic novel, and more generally her special brand of patchwork ecofeminism, it is self-consciously described as "an eclectic mixture of traditions":

> The upper slopes of the hill were dotted with shrines to Goddesses and Gods, ancestors and spirits. [. . .] A cairn of memorial stones crowned a green mound dedicated to the Earth Goddess, who could be called Gaia, or Tonantzin, or simple *Madre Tierra*, Mother Earth. Kuan Yin had a shrine and so did Kali and Buddha and many bodhisattvas, along with devis and devas, African orishas, and Celtic Goddesses and Gods. Some formed natural clusters: The Yoruba Oshun, Love Goddess, Goddess of the River, stood near Aphrodite and Inanna/Ishtar/Astarte [. . .]. Farther down the hill, the Virgin of Guadalupe overlooked the Stations of the Cross.[17]

Making room for the cultural traditions of a cosmopolitan, multiethnic community, characters in the opening chapter alternatively invoke many different versions of the Goddess, including Yemaya, the "Yoruba Sea Goddess," "the pregnant fish-tailed mermaid, the great mother, Goddess of the Sea."[18] Foregrounding the heterogenous elements of a spiritual practice exposed to cultural relativism and deconstructionism, Maya, the ceremony oratress, serves as a mouthpiece for Starhawk when she repudiates naïve and narrowly patriarchal views on the Mother-Earth archetype. Via Starhawk's postmodernist and kaleidoscopic (re)vision, the Goddess is first and foremost a multifaceted anamorphic image conveying a profound sense of interconnectedness and ecological responsibility:

> "*Este es el tiempo de la Sagadora*, the Time of the Reaper, she who is the end and the beginning, scythe to the grain. The Crone, the Goddess of Harvest. [. . .] The Crone, the Reaper, is not an easy Goddess to love. She's not the nurturing Mother. She's not the Maiden, light and free, not pretty, not shiny like the full or crescent moon. She is the Dark Moon, what you don't see coming at you; what you don't get away with, the wind that whips the spark across the fire line. Chance, you could say, or, what's scarier still: the intersection of chance with choices and actions made before. The brush that is tinder dry from decades of drought, the warming of the earth's climate that sends the storms away north, the hole in the ozone layer. Not punishment, not even justice, but consequence. [. . .] This is the Age of the Reaper, when we inherit five thousand years of postponed results, the fruits of our callousness toward the earth and toward other human beings. But at last, we have come to understand that we are part of the earth, part of the air, the fire, and the water, as we are part of one another."[19]

Ultimately, and in a way recalling Carolyn Merchant's take on the ethical value of Mother Earth mythologies, the ritual honoring of the many faces of Mother Earth goes hand in hand with moral, ecological, and political obligations that can translate into action: "What is sacred becomes the measure by which everything is judged. And this is our measure, and our vow to the life-renewing rain: we will not be wasters but healers. / Remember this story. Remember that one act can change the world. When you turn the moist earth over, and return your wastes to the cycles of decay, and place the seed in the furrow, remember that you are planting your freedom with your own hands."[20]

Before delving further into some of the faces of Mother Earth etched throughout my corpus, I would like to consider for a moment how scientists James Lovelock and Lynn Margulis may have influenced the resurgence of Gaia as a powerful, anamorphic figure of the Earth in contemporary ecopoetics. Invoking the Greek Goddess, Lovelock and Margulis have transfigured acceptable representations of geophysiological and symbiotic processes in the scientific world. While Lovelock was studying negative feedback loops interlinking the earth and the atmosphere, he found out that the biosphere was homeostatic. As paraphrased by Lynn Margulis, "[just] as our bodies, like those of all mammals, maintain a relatively stable internal temperature despite changing conditions, the Earth system keeps its temperature and atmospheric composition stable."[21] Striving to find a correct and simple enough way to refer to a planet exhibiting, in Lovelock's terms, "the behavior of a single organism, even a living creature,"[22] the chemist eventually turned for inspiration to novelist William Golding. As Lynn Margulis recounts, it is he who first suggested Gaia, the ancient Greek word 'Mother Earth' which "provides an etymological root of many scientific terms, such as geology, geometry," or geography.[23]

Lovelock was cautious, however, to make the distinction between the personifying, mythical view of Gaia, and the planetary, systemic organicism which could be established via scientific evidence. This is visible in the quotation by Lovelock chosen by Richard Powers as an epigraph to his 2018 tree novel, *The Overstory*: "Earth may be alive, not as the ancients saw her—a sentient Goddess with a purpose and foresight—but alive like a tree. A tree that quietly exists, [. . .] endlessly conversing with the sunlight and the soil. Using sunlight and water and nutrient minerals to grow and change."[24] As reformulated by Margulis, Lovelock's Gaia hypothesis encapsulates the organicism of the planetary biosystem, which ultimately depends on myriad symbioses and exchanges between the different parts that together make up and regulate the whole: "aspects of the atmospheric gases and surface rocks and water are regulated by the growth, death, metabolism, and other activities of living organisms."[25]

Despite Lovelock's effort to justify his poetic call upon the Greek Goddess to convey a scientific reality, both himself and Margulis—who stood by Lovelock and advocated the validity of the Gaia hypothesis and concept all along—had to deal with ire and adamant rejection from part of their scientific community. By and large, the staunch criticism which the Gaia hypothesis fired up can be accounted for by two interrelated reasons. On the one hand, as Margulis stresses, Gaia disrupts the "venerable position of humans as the exact center of the universe":[26] "We *Homo sapiens sapiens* and our primate relations are not special, just recent: we are newcomers on the evolutionary stage. Human similarities to other life-forms are far more striking than the differences. Our deep connections, our vast geological periods, should inspire awe, not repulsion."[27] Exploding all notions of human exceptionalism, together with the myth of our domination and control of nature, Gaia has forced Moderns to a rather uncomfortable, dramatic shift away from humanist and mechanistic representations of the world: "We are a recent, rapidly growing part of an enormous ancient whole."[28] Taken seriously, Gaia propels a radical rethinking of our economic models and of part of humanity's misguided belief in infinite growth: "The planet will not permit our population to continue to expand. [. . .] We people are just like our planetmates. We cannot put an end to nature; we can only pose a threat to ourselves."[29] Hence a growing ecophobia made manifest in the recent boom in apocalyptic fiction, symptomizing at once Modernity's reluctance to let go of its founding myths, political and economic systems, together with mounting anguish triggered by the increasingly widespread awareness that collective denial of our belonging to nature is no longer tenable.

Meanwhile, as ecofeminists and ecopsychologists rebounded on Lovelock and Margulis's work to explore what their discoveries might entail in the domains of the human psyche, mythography, spirituality, and even politics, dread propagated throughout the scientific community that rehabilitating Gaia might lead backward to pre-scientific animistic beliefs. "Many scientists are still hostile to Gaia," Margulis explains, "both the word and the idea, perhaps because it is so resonant with anti-science and anti-intellectual folks. In popular culture, insofar as the term is at all familiar, it refers to the notion of Mother Earth as a single organism. Gaia, a living goddess beyond human knowledge, will supposedly punish or reward us for our environmental insults or blessings to her body. I regret this personification."[30] Correcting things through the scientific lens, Margulis goes on:

> Gaia is not an organism. Any organism must either eat or, by photosynthesis or chemosynthesis, produce its own food. [. . .] No organism feeds on its own waste. Gaia, the living Earth, far transcends any single organism or even any population. One organism's waste is another's food. [. . .] The Gaian system

recycles matter on the global level. Gaia, the system, emerges from ten million or more connected living species that form its incessantly active body. Far from being fragile or consciously petulant, planetary life is highly resilient. [. . .] Gaia itself is not an organism directly selected among many. It is an emergent property of interaction among organisms, the spherical planet on which they reside, and an energy source, the sun.[31]

The way I see it, these heated controversies arose in great part owing to a rampant lack of understanding on the part of many scientists as to how ecofeminists and ecopoets themselves deliberately and consciously manipulate metaphors and myths for their enchanting anamorphic power, without ever losing sight of the poetic process at play.

Moreover, while Margulis claims to "regret this personification of Gaia," she herself reverts to ecopoetic personification that possess the merit of carrying meaning across and mobilizing our affects more effectively than arid technical language may do. This is the case for instance of her humorous phrase when addressing planetary resilience and nature's indifference, in the long run, to the petty interests of humankind: "Gaia is an ancient phenomenon. Trillions of jostling, feeding, mating, exuding beings compose her planetary system. Gaia, a tough bitch, is not at all threatened by humans. Planetary life survived at least three billion years before humanity was even the dream of a lively ape with a yearning for a relatively hairless mate."[32] Although Margulis warns against the fallacy and pitfalls of lending Gaia Goddess-like intentions toward her human creatures—"Gaia is neither vicious nor nurturing in relation to humanity,"[33] she insists, or "My Gaia is no vague, quaint notion of a Mother Earth who nurtures us"—she herself slips every now and then, giving into anthropomorphic images when referring to nonhuman agency: "Life, especially bacterial life, is resilient. It has fed on disaster and destruction from the beginning. Gaia incorporates the ecological crises of her components, responds brilliantly, and in her new necessity becomes the mother of invention."[34] What it ultimately comes down to is that, as Bruno Latour underlines, "[myth] and science [. . .] speak languages that are distinct only in appearance."[35]

Richard Powers stages a similar conceptual struggle with Patricia. This scientific character weighs the pros and cons of kindling the poetic imagination of her fellow human beings to offer them an inkling of that part of reality accessible to her. How to rightly convey the notion of something akin to an ecosystemic consciousness remains a moot point: "'A forest knows things. They wire themselves up underground. There are brains down there, ones our own brains aren't shaped to see. Root plasticity, solving problems, and making decisions. Fungal synapses. [. . .] Link enough trees together, and a forest grows *aware*.'"[36] // "There are brains down there!"[37] exclaims the biologist, ostentatiously reverting to metaphor as she is striving to picture the underground network allowing for sylvan sentience and interspecies communication. Putting in a nutshell her discoveries of the symbiotic nature of nature, Patricia contends that "'[to] see green is to grasp the Earth's intentions.'"[38] Lending volition to the planet here presented as a superorganism, Patricia encapsulates the life drives, multispecies entanglements, and homeostatic tendencies underlying the resilience and mechanisms of more-than-human life at the level of a tree, a forest, and the planet.

As a matter of fact, while the personification of Gaia forms an ecopoetic trope, it may be compatible with science all the same, as long as we keep in the back of our minds that Gaia is but a powerful, anamorphic lens—a robust cosmovision, or a seeing instrument,[39] which reveals, when considered from the right perspective, the many enmeshments between all lifeforms that together animate our planetary *oikos*, or ecosystem. If the Earth may not be said to be endowed with personhood in the sense of what conventionally defines a conscious organic being, its *being-like*, or at the very least its *behaving-like* a Creation Goddess may nevertheless provide necessary food for thought. Gaian ecopoetics disrupts anthropocentric paradigms and reveals the Earth as, indeed, a sentient and responsive entity:

> Just who is this Gaia? / Lovelock replies that Gaia requires no consciousness to adjust to the planetary environment. [. . .] Gaia, as the interweaving network of all life, is alive, aware, and conscious to various degrees in all its cells, bodies and societies. Analogous to proprioception, Gaian patterns appear to be planned but occur in the absence of any central 'head' or 'brain.' Proprioception, as self-awareness, evolved long before animals evolved, and long before their brains did. Sensitivity, awareness, and responses of plants, protocists, fungi, bacteria and animals, each in its own local environment, constitute the repeating pattern that ultimately underlies global sensitivity and the response of Gaia 'herself.'[40]

Like recent discoveries on tree sentience and communication, Gaian science and ecopoetics may revolutionize the way we think in fundamental, reenchanting ways. First, it forces us to adopt a biocentric perspective. Second, it oppugns the dichotomy separating matter and mind. Third, it sheds a different light on animistic worldviews, no longer to be approached with the disenchanted Eurocentric condescension of our forebears, but to be respected as cognizant of other-than-human forms of proprioception and intelligence, and of the symbiotic nature of life which makes all beings and things in the world interconnected. Fourth, and in a way that could give totally new direction and meaning to human affairs, it replaces the Western myth of the survival of the fittest—of the business of life as competition and natural selection—with the notion of life as evolution and resilience via multispecies cooperation.

It is precisely because, in Margulis's terms, "Gaia resonates with our longing for significance in our short Earth-bound lives"[41] that her many faces tend to reappear in ecopoetic works hinging on a postmodern reenchantment. Gaia serves an ecopoetic function similar to Patricia's notion of "giving trees"—"the great ancient trunks of the uncut forest, the ones who keep the market in carbons and metabolites going."[42] As Patricia fumbles for the right name to designate those ancient trees that real-life forester Suzanne Simard on her end calls "mother trees," she reflects: "The reading public needs such a phrase to make the miracle a little more vivid, visible. It's something she learned long ago, from her father: people see better what looks like them. *Giving trees* is something any generous person can understand or love."[43] At the same time, the anamorphic vision meant to trigger affect is superimposed onto the scientific picture: "Before it dies, a Douglas-fir, half a millennium old, will send its storehouse of chemicals back down into its roots and out through its fungal partners, donating its riches to the community pool in a last will and testament. We might call these ancient benefactors *giving trees*."[44] The difficulty of wording observations of the more-than-human world in meaningful human terms is underlined ("we might call," "donating its riches to the community pool in a last will and testament"), keeping us aware of the double vision involved in picturing nonhuman worlds by means of human words, images, and concepts. Thus, we are reminded that even scientists often work alongside, if not *as*, ecopoets. Many of them indeed produce anthropomorphic anamorphoses as they inevitably summon metaphors to bring us to a stance which they calculate as offering the right perspective from which to contemplate those aspects of reality which they are striving to present via a coherent, organized picture.

In *the Voice of the Earth*, Theodore Roszak lucidly addresses the accusations of obscurantism and mysticism that Lovelock and Margulis had to ward off as many of their peers opposed vehement resistance to Gaia:[45]

> Initially, Lovelock was at great pains to explain that Gaia is *merely* a metaphor—as if metaphor were not a powerful agency of the mind, one of our most precious ways of understanding the world, and therefore perhaps related to whatever we take the 'truth' to be. Challenged or stung by the criticism of his colleagues, he sought to find some chance-based, nonintentional mechanism to explain the actions of Gaia. [. . . But] what Lovelock was up against is a profound issue in the rhetoric, and ultimately the metaphysics of science. The issue has to do with the increasingly large role that *systems* have come to play in our understanding of the universe.[46]

Claims to pure objectivity notwithstanding, Roszak scrutinizes the contradiction inherent in much scientific discourse, which, despite its mechanistic and reductionist frames and precepts, exhibits overall a strong tendency toward anthropomorphic rhetoric. "At times," writes Roszak, "it is almost comic to see scientists struggling with the problem of the purposiveness of nature, trying to pretend it's not there even while it infuses every word they say."[47]

The books in my corpus work toward an ecopoetics of reenchantment oriented by the postmodern shifts in scientific paradigms from individualism, competition, and human exceptionalism to cooperation, symbiosis, and multispecies organicism. In terms of evolution, as Margulis has shown, even *Homo sapiens* owe their late emergence in the long run to the fundamentally symbiotic nature of the most ancient bacteria. In a way that is coherent with new forestry studies, or, on a different plane, with the relational philosophy of Alan Watts, Margulis's research has established that the concept of the individual is but a human myth without any biological foundation. In nature, all organisms are "metabolically complex communities of a multitude of tightly organized beings."[48] As relayed by Powers in his novel, the discoveries on sylvan life achieved by his fictional scientist similarly invite us to the joyful contemplation of life-systems that can no longer serve to advocate ruthless competition models and the survival-of-the-fittest ethics prevailing in the nineteenth century:

> The things [Patricia] catches Doug-firs doing [. . .] fill her with joy. When the lateral roots of the two Douglas-firs run into each other underground, they fuse. Through those self-grafted knots, the two trees join their vascular systems together and become one. Networked together underground by countless thousands of miles of living fungal threads, her trees feed and heal each other, keep their young and sick alive, pool their resources and metabolites into community chests . . .[49]

As underlined by the protagonist, even science fulfills an ecopoet(h)ic quest as it revisions the pictures we form of the natural world. From those are derived the cosmopolitical conclusions and ethical principles which might no less than determine our long-term capacity to dwell with the Earth: "It will take years for the picture to emerge. [. . .] Her trees are far more social than even Patricia suspected. There are no individuals. There aren't even separate species. Everything in the forest is the forest. Competition is not separable form endless flavors of cooperation. Trees fight no more than do the leaves on a single tree. It seems most of nature *isn't* red in tooth and claw, after all."[50] As the novel underscores, should we hope to survive as such a large species, then we had better learn to abide by these natural principles of multispecies cooperation which our current ways of life have radically strayed from. To put it in a nutshell, as Margulis phrases it, "'Gaia is just symbiosis as seen from space.'"[51]

The resurfacing of Gaia in postmodern environmental sciences, arts, and humanities at once gestures to the collaborative aspects of nature which Charles Darwin had, in fact, pinpointed early on. As science writer Colin Tudge insists, while Darwin's *Origin of Species* famously bolstered the role of competition, he "stressed too that we also see collaboration in nature—he made a particular study of the long-tongued moths that alone are able to pollinate deep-flowered orchids: two entirely different creatures, absolutely dependent on each other."[52] The reclaiming of Gaia thus checks the nineteenth-century models of nature which focused exclusively on competition. Counterbalancing the Social Darwinist ethics promoted by Herbert Spencer's recuperation of science in support of his "the survival of the fittest" theory, Gaia changes the face of evolution from models of power-over to models of power-with. According to Riane Eisler, the rejuvenation of Goddess mythologies and traditions at the turn of the century "signals a way out of our alienation from one another and from nature."[53] Like many ecofeminists, Eisler cogently deconstructs patriarchal ideologies of conquest from an intersectional standpoint. She consequently pleads for the rehabilitation of Mother Earth imagery and tales which entail radically different sociological principles: "Let us teach our sons and daughters that men's conquest of nature, of women, and of other men is not a heroic virtue; that we have the knowledge and the capacity to survive; that we need not blindly follow our bloodstained path to planetary death; that we can reawaken from our 5,000-year dominator nightmare and allow our evolution to resume its interrupted course."[54] Reclaiming Gaian ecopoet(h)ics clearly offers a way to "reclaim our lost sense of wonder and reverence for the miracles of life and love," and to "fulfill our responsibility to ourselves and to our Great Mother, this wondrous planet Earth."[55]

FROM MOTHER EARTH TO ANTHROP-O(BS)CENE IMAGERY IN BARBARA KINGSOLVER'S *ANIMAL DREAMS*

Now zeroing in closer on the Gaian ecopoet(h)ics in my corpus, I would like to delve first into the complex multicultural perspective offered by Barbara Kingsolver's 1990 novel *Animal Dreams*. Throughout, the land gets conceptualized via various prisms conflating the earth with a human body and with incarnations of a Mother Earth Goddess—images that foreground notions of organicity and ecological interrelatedness. Contesting the demarcation line separating humans from their "environment," Kingsolver envisions humans as intricately enmeshed within the more-than-human world. Shifting our perspective away from a humanist worldview where the more-than-human world is perceived as a mere backdrop for human matters, Kingsolver's ecoliterature, like all the texts in my corpus for that matter, substitutes the notion of an environing nature with images of entanglements between human and other than human worlds.[56] In her dialogic novel, Cosima, the main character, is a white female subject trained in medicine and employed as a biology teacher in her little Arizona hometown. Meanwhile she gets initiated to indigenous worldviews by a boyfriend with mixed, Native ancestry. Braiding different cultures, the ecofeminist vision of the earth which her character forms is thus shaped in part by biological and ecological reading grids, and in part by various indigenous stories of places. As with most of Kinsgolver's oeuvre, the novel is much concerned with poet(h)ics of dwelling in a way that questions modern notions of belonging. By and large, Kingsolver's literature harks back to the notion that humans belong to the earth, rather than the contrary, and that our true home resides in our earthly *oikos*.[57]

During excursions to ancient dwelling places such as Kinishba, Arizona, Cosima's anamorphic vision highlights the humility underlying Pueblo architecture: "I couldn't stop running my eyes over the walls and the low, even roofline. The stones were mostly the same shape, rectangular, but all different sizes [. . .]. There was something familiar about the way they fit together. In a minute it came to me. They looked just like cells under a microscope."[58] Polishing the organic vision seen through the biologist's eye, the following dialogue stresses the naturcultural[59] embeddedness underpinning the Pueblo design of dwellings: "'It doesn't even look built,' I said. 'It's too beautiful. It looks like something alive that just *grew* here.' / 'That's the idea. [. . .] No Washington Monuments. Just build something nice that Mother Earth will want to hold in her arms.'"[60] As the couple later visit Canyon de Chelly on Navajo land and contemplate the enfolded cliff dwellings, the description superimposes autopoietic forms of nature onto the intricate designs of animal homes: "The walls were shaped to fit the curved hole in the cliff, and the building blocks were cut from the same red rock that served as their foundation. I thought of what Loyd had told me about Pueblo architecture, whose object was to build a structure the earth could embrace. This looked more than embraced. It reminded me of cliff-swallow nests, or mud-dauber nests, or crystal gardens sprung from their own matrix: the perfect constructions of nature."[61]

Later in the book, the characters stay in the village of Santa Rosalia Pueblo, a fictional place not unlike Acoma Pueblo in New Mexico. The place is described in terms underlining again the intended harmony between human constructions and their natural environment: "The village was built on a mesa and blended perfectly with the landscape, constructed of the same stones as the outcroppings that topped all the other, empty mesas. [. . .] It was a village of weathered rectangles, [. . .] the houses all blending into one another around a central plaza."[62] The materials and colors are integrated in the natural landscape, reflecting the rich shades and textures of the earth itself: "The stone walls were covered with adobe plaster, smooth and appealing as mud pies: a beautiful brown town. The color brown, I realized, is anything but nondescript. It comes in as many hues as there are colors of earth, which is commonly presumed infinite."[63] It becomes clear that such biomimesis is meant to coalesce with the Pueblo's reverence for the earth as "the mother of invention," as Margulis has it, providing the inspiration for how to dwell ecopoetically. Averring that our true home is our encompassing earthly *oikos*, Loyd apprises Cosima that in his culture of reciprocity, "'the greatest honor you can give a house is to let it fall back down into the ground,' [. . .]. That's where everything comes from in the first place.'"[64]

Shifting perspectives on the very notion of dwelling, Loyd's vision foregrounds a vital connection to the land which clashes with Cosima's learned concept of a homeland: "I remembered Loyd one time saying he'd die for the land. I thought he meant patriotism. I'd had no idea."[65] Through their dialogues, Cosima comes to ponder the rift separating the anthropocentric capitalist take on the land's value from the vision of indigenous peoples possessing ecological consciousness. On the one hand, Loyd introduces Cosima to the philosophy underlying the ceremonial drumming, chanting, and dances of his community—the value and point of Kashinas, of deer dances and corn dances for instance, halfway between prayers, thanksgiving, and a way of knowing one's place in the overall fabric of life: "'It has to do with keeping things in balance. [. . .] The spirits have been good enough to let us live here and use the utilities, and we're saying: We know how nice you're being. We appreciate the rain, we appreciate the sun, we appreciate the deer we took. Sorry if we messed up anything. You've gone to a lot of trouble, and we'll try to be good guests.'"[66] Cosima and Loyd thus mediate two widely different perceptions of how humans might fit into the larger scheme of things: "I laughed because I understood 'in balance.' I would have called it 'keeping the peace,' or maybe 'remembering your place,' but I liked it."[67]

Interweaving her characters' heteroglossic discourses with what forms the gist of Lynn White Junior's investigation into the role of Abrahamic religion in the ecological crisis,[68] Kingsolver spurs readers to question the perspective commonly adopted through Judeo-Christian views:

> "It's a good idea," I said. "Especially since we're still here sleeping on God's couch. We're permanent houseguests."
>
> "Yep, we are. Better remember how to put everything back the way we found it."

It was a new angle on religion for me. I felt a little embarrassed for my blunt interrogation. And the more I thought about it, even more embarrassed for my bluntly utilitarian culture. "The way they tell it to us Anglos, God put the earth here for us to use, westward-ho. Like a special little playground."[69]

After learning from Loyd about various Pueblo, Navajo, and Apache traditions, places, and stories—both their similarities and differences—and their more or less shared notion of the land as sacred Mother Earth, Cosima tries to envision the landscape that incorporates the mine from Loyd's point of view: "I wondered what he saw when he looked at the Black Mountain mine: the pile of dead tailings, a mountain cannibalizing its own guts soon to destroy the living trees and home lives of Grace. It was such an American story, it was hardly interesting."[70]

As in much of her writing, Kingsolver's multicultural setting and characters offer a stereoscopic vision of place that forces us to reflect on the various prisms and paradigms available to us, while it simultaneously reveals the "slow violence" and environmental injustices affecting poor and minoritized groups in the United States.[71] Essentially, it sheds light on the combined hubris and hamartia driving the ecological tragedy of the "Anthropocene," while simultaneously pointing to the fact that, despite what the term infers, not all humans on the planet can be held equally responsible for ushering the Earth into this new geological era. On reading *Animal Dreams*, it may be tempting to align with those who prefer the term "Capitalocene."[72] Indeed, seeing simultaneously through her own and through indigenous eyes, Cosima takes a new, hard look at the Black Mountain mining enterprise and at the beautiful Jemez Mountains, in New Mexico, where Loyd has taken her to secret hot springs, and which, he informs her, are being "mined savagely" for pumice utilized in the denim industry: "To people who think of themselves as God's houseguests, American enterprise must seem arrogant beyond belief. Or stupid. A nation of amnesiacs, proceeding as if there were no other day but today. Assuming the land could also forget what had been done to it."[73] As Cosima's analyses of water samples from the river will reveal, far from being inert matter deprived of agency, the land is in fact inhabited by many other lifeforms that bear the brunt of human extraction and pollution.

As Cosima lectures highschoolers on ecological, industrial, and consumerist responsibilities, she gets her meaning across via a somewhat anthropomorphic image of the material "memory" of water and land: "People can forget, and forget, and forget, but the land has memory. The lakes and the rivers are still hanging on to the DTT and every other insult we ever gave them."[74] While undercutting the Western myth of humans meant to behave "like masters and possessors of nature," as Descartes had it, Cosima's scientific view combines with Loyd's Native perspective, inviting us to shift paradigms and gesturing to the ancient, indigenous understanding of humans' role as that of nature's guardians. In this sense, Kingsolver's writing chimes in with philosophical debates taking place as she was writing her novel as to the urgent need to supersede Descartes' anthropocentric take on nature in favor of a worldview where humans have come to exist, rather, as the "masters and protectors of nature."[75] Diffracting radically divergent representations of the place and role of humans in relationship to the rest of the world, Kingsolver's ecopoetic writing serves as a magic wand correcting a vision of the natural world which has been distorted for centuries. "'I'm teaching [my students] to have a cultural memory. [. . .] I want them to be custodians of the earth,'"[76] says Cosima the biologist who, bringing Rachel Carson's *Silent Spring* into the classroom, reads as a mouthpiece for both the famous scientist and Kingsolver herself. As custodians of the earth, Kingsolver's writing intimates, we humans must take ecological responsibility.

Thanks to Cosima's scientific lens, the biotic organicism of an ecosystem can ultimately be viewed in a stereoscopic way superimposing animistic and naturalistic worldviews. Looking at the life of the river under a microscope, Cosima draws attention to the question of perspectives and to the technological and metaphorical apparatuses allowing humans to perceive matter otherwise invisible to the human eye:

> I drew huge, fantastic pictures in colored chalk of what we could expect to see in this river water: strands of Nostoc like strings of blue pearls; multi-tentacled hydras, rotifers barreling into each other like hyperactive kids. I demonstrated the correct way to put a drop of water on a glass slide, coverslip it, and focus the scope. [. . .] They couldn't see anything. At first I was irritated but bit my tongue and focused a scope myself, prepared to see the teeming microscopic world of a dirty river. I found they were right, there was nothing. It gave me a strange panic to see that stillness under powerful magnification. Our water was dead. It might as well have come from a river on the moon.[77]

Cosima's scientific take on the aliveness of the land or river thus provides ballast to the animistic stories told by Loyd, stories whereby Spider Rock in Canyon de Chelly is for example the sacred home of the Navajo Creation Goddess Spiderwoman.

Cosima moreover resorts to many comparisons—a figure of speech allowing us to see otherwise-elusive dimensions of reality and helping us focus on and comprehend matter and form via their being-like qualities ("Nostoc like strings of blue pearls"; "multi-tentacled hydras, rotifers barreling into each other like hyperactive kids"). In lieu of a *genius loci*, what is revealed here is the presence or absence of microscopic lifeforms that normally animate a healthy ecosystem: "The pH, which we tested from some areas came just a hair higher than battery acid. I couldn't believe the poisoning in the mine had gone this far. Protozoans are the early-warning system in the life of a river, like a canary in a mine. And this canary was dead."[78] While unveiling some of the invisible ecocidal effects of inconsiderate anthropic activity, Cosima's investigation invites us to ponder how the empirical notion of "biotic death" gives a different relief to animistic conceptions of elemental agency and sacredness. Besides, the teaching session betrays how even science is performed through apparatuses and eyes that are enmeshed with human perception, which itself is always to a certain extent culturally trained ("I drew huge, fantastic pictures in colored chalk of what we could expect to see in this river water: [. . .]; I demonstrated the correct way to put a drop of water on a glass slide, coverslip it, and focus the scope"). Anticipating Karen Barad's theory of agential realism, Kingsolver addresses how the scoping out of phenomena can never be entirely disentangled from situated knowledge, lenses, and material-discursive practices.[79]

In Kingsolver's short story "Homeland," which is narrated from the point of view of a young girl moving in-between the dominant white Christian culture of the South and her Cherokee heritage on the side of her maternal grandmother, the land is further presented as a living organism, with veins and flesh: "Even the earth underneath us sometimes moves to repossess its losses: the long deep shafts that men opened to rob the coal veins would close themselves up again, as quietly as flesh wounds."[80] The underlying, extended metaphor is consonant with the ecofeminist rehabilitation of images of the earth as a woman's body—an anamorphosis meant to trigger affects encouraging human restraint, reverence, and measure in our cultivation enterprise, while it often represents industrial extraction as a form of rape, or wounding, of the earth. As Carolyn Merchant argues, "[one] does not readily slay a mother, dig into her entrails for gold or mutilate her body."[81] In *Animal Dreams* too, the mining industry is represented as "[chasing] a vein of copper" in the land.[82] Cosima's anamorphic vision moreover highlights the question of perspective, making us see the land as wounded and dying from industrial despoliation: "They pretended the dam was some kind of community-improvement project, *but from where Viola and I stood* it looked like exactly what it was–a huge grave. Marigold-orange earth movers hunched guiltily on one corner of the *scarred plot of ground*."[83] The hypallage attributing guilt to the construction machinery standing over the site catalyzes the injurious role of the Black Mountain mining company—a name that proves to be rather appropriate as, beyond the color of coal, it may evoke a dead mountain while adumbrating dark designs.[84]

Seen from an ecofeminist perspective, the Anthropocene appears as the age when humans' exploitation of nature has reached obscene proportions—a notion which has inspired me to coin the portmanteau term "Anthrop-o(bs) cene."[85] Hence Cosima's vision of the mine's smokestack erected over the landscape as a sign of its anthropic domination: "And up the head of the canyon was the old Black Mountain copper mine. On the cliffs overlooking the valley, the smelter's one brick smokestack pointed obscenely at heaven."[86] Standing as a phallic monument testifying to the violence and hubris of the Anthrop-o(bs)cene, the smokestack metonymically embodies the whole extractivist mindset and economy of Modernity. In a single brushstroke, the reference to "heaven" via the sky-pointing industrial chimney additionally paints a cultural background of religious patriarchal ideologies of dominion which sustain the instigation of capitalistic industries. Whereas the term "Anthropocene" has been criticized for the way it could glorify the anthropocentric paradigm—potentially satisfying part of humanity's craving for dominion, allowing us to revel in the spectacle of our destructive might—,[87] the foregrounding of the obscenity at stake means to forestall techno-phallocentric complacency and induce a more humble overall stance.

One of the main reasons why ecofeminist thinkers sometimes advocate a redemption of Goddess traditions and mythologies is because these stories impart a powerful sense of our interdependence with earthly matters. A common critique among ecofeminists points to the deprecation of both female and earthly flesh and powers of creation which goes hand in hand with the advent of Abrahamic cultures.[88] As implied in much of Kingsolver's writing, and explicated by ecofeminist theorists, our societies have gradually lost reverence for the agencies of the earth and of women owing to the powerful male sky and war Gods that were introduced over time to obliterate the Earth Goddess imagery that was very much alive until then in all so-called pagan societies, including in Europe.[89] In turning away from the earth and from women's bodies which were no longer envisioned in a sacred light, patriarchal mythologies took to projecting notions of the divine into a celestial sphere, both separated from and above the earth. The masculinization of the divine thus led to the desacralization of the earth, of the animal and vegetal worlds, and of the feminine. In turn, it paved the way for the deterrestrialization of modern spirituality and the subsequent disenchantment of the more-than-human world. Through the phallic imagery associated with the mining industry in Kingsolver's prose ("the smelter's one brick smokestack pointed obscenely at heaven"), we may gain important clues about the underlying ecofeminist critique of an entire techno-phallogocentric ideology instating the correlated domination and abuse of women, of indigenous peoples, and of the earth in the name of cultivation and control for the expansion of the White Man's civilization.[90]

Resisting such unworldly concepts, Kingsolver's *ecopoeisis* reclaims the wonderful agency and resilience pertaining to the earth and, abiding by similar laws for that matter, to animals, plants, insects, rivers, mountains, and humans. This transpires for instance in the story "Homeland," where the workings of nature extend to human naturcultural resilience: "We lived in Morning Glory, a coal town hacked with a sharp blade out of a forest that threatened always to take it back. The hickories encroached on the town, springing up unbidden in the middle of dog pens and front yards and the cemetery. The creeping vines for which the town was named drew themselves along wire fences and up the sides of houses with the persistence of the displaced."[91] Metonymically, the forest and morning glories stand for nonhuman nature and for the Cherokee, both of whom have been violently "displaced" to make way for European conquest and capitalist exploitation ("a coal town hacked with a sharp blade"). Modern civilization is metonymically symbolized by the "dog pens," the neatly organized and built "front yards," "houses," and "cemetery," and the "wire fences" meant to keep wilderness squarely at bay. Yet, like indigenous peoples, the forest and creeping vines tenaciously resist and threaten to reverse colonial and industrial encroachment. As I have elaborated on elsewhere, the mythopoetic and magical realist dimensions of this dialogic short story eventually remap naturcultural territories and representations, demonstrating how ecopoetic literature can both deterritorialize and reterritorialize the world.[92]

Whether reverting to mythopoeic or scientific lenses, Kingsolver's overall ecofeminist vision is one that traces the material continuum between humans and the greater-than-human world. Meanwhile, it heightens our perception of the universal renewal and fertility principles that underlie the natural cycles of life and death, as well as the capacity for resilience through interconnection which rules both human and other-than-human realms.[93] In Kingsolver's first novel for instance, *The Bean Trees*, rhizobia foster the extended metaphor through which the resilience of the child protagonist, initially molested and abandoned, is to be comprehended.[94]

At the end of *Animal Dreams*, Cosima is pregnant. This provides an occasion to highlight the life-bearing miracle women are invested with: "From the outside you remain ordinary, no one can tell from looking that you have experienced an earthquake of the soul. You've been torn asunder, invested with and ancient, incomprehensible magic. It's the one thing we never quite get over: That we contain our own future."[95] To prevent any misprision of Kingsolver's valorization of women's fertility, it must be underlined here that if her plots often hinge on positive visions of the female engendering potential, motherhood is by no means idealized in any monologic way in her writing. Some of her female characters choose paths toward love or happiness that do not involve motherhood. That is the case of Hallie, in *Animal Dreams*, whose mother died from childbirth complications. Embodiments of fertility in the same novel furthermore include Cosima, who, stuck with an unwanted teenage pregnancy, has gotten rid of her own fetus by starving it to death. Loyd moreover shares the somber view of twins as "bad luck" in Pueblo culture, whereby "[when] twins are born people say there'll be a poor rainy season of grasshoppers or some darn thing. In the old days you had to let twins die."[96]

Although the notion of death ominously pervades the beginning of *Animal Dreams*, the book structure, with its twenty-eight chapters, points to the natural cycles of regeneration and fertility tying the moon to women's menstrual cycles—phenomena that provide incontrovertible evidence of the world's tendency to create and renew itself, while pointing to microcosmic and macrocosmic links. The composition of the book thus holds from the start a promise of regeneration inspired by cosmic biological phenomena. Indeed, the premature deaths of Cosima's mother, of her unborn baby, and later, of her sister, are eventually transcended via non-dualistic, ecofeminist imagery reaching beyond life and death dichotomies, while emphasizing both community and biotic continua. At the end of the novel, looking at her life from a much wider angle, Cosima considers the death of her mother no longer as "a tragedy," but, as "[just] a finished life,"[97] with its own part in the greater, dynamic fabric of an ever-becoming world.

The dysphoric bone leitmotif so prevalent at the beginning of the book turns out in the end to symbolize fertility and continuation, rather than simply death and curtailment. This is in fact hinted at in the opening chapter, via Cosima's memories of the thousands of cows dying in the fields of Northern France at the end of World War II, the casualties of German occupation: "In the sudden quiet after the evacuation the cows had died by the thousands in those pastures, slowly, lowing from pain from unmilked udders. But now the farmers who grew sugar beets in those fields were blessed, they said, by the bones. The soil was rich in calcium."[98] Following a similar logic of resilience, and focusing on onomastics, Kingsolver's choice of names for Cosima's sister, "Halimeda" turns out revealing. While the narrator explains that the name means "thinking of the sea," it also provides the biological name for an alga that is specifically rich in calcium. When Hallie gets murdered in Nicaragua and her body cannot be returned to her family—partly duplicating what happened in real life to Ben Linder—Cosima remembers that her sister had mentioned that, should anything happen to her, she would like to be buried in Nicaragua. Again, Kingsolver uses an image of ecosystemic resilience enfolding human lives: "The land could use the fertilizer."[99]

It might be pointed out that, especially in her writing from her long Arizona period, Kingsolver's pictures often resemble the anti-dualistic, multispecies celebration of life characterizing Georgia O'Keeffe's ecopoetic and colorful desert paintings, many of which bring together bones and flowers, life and death, animals, vegetal, and elemental lifeforms.[100] As a matter of fact, the cemetery scenes in *Animal Dreams* where the characters celebrate the Day of the Dead in the novel are sprinkled with references to the bright marigold of pollen. Thus, Kingsolver's *ecopoeisis* frequently relies on images evoking a multispecies life-death continuum, whereby flower seeds, dead tree trunks, fruit stones, animal and human bones all end up going back to the ground where they get decomposed and, eventually, where they take part in enriching the soil, making it fertile and thus participating in breeding new life in the long run. Kingsolver's postcolonial Congo novel, *The Poisonwood Bible*, likewise opens and closes on metonymical references to cosmic renewal, focusing on the forest, "sucking life out of death [. . .], that eats itself and lives forever."[101]

MOTHER EARTH IN *JOY OF MAN'S DESIRING*: JEAN GIONO AS AN AVANT-GARDE ECOFEMINIST?

Read in a similar ecofeminist light, the ending of Jean Giono's *Joy of Man's Desiring* redeems Bobi's death, returning him to Mother Earth in a final embrace. It may be that this French novel, published in 1935, translates a vision that could be regarded as avant-garde in many ways, at least from the point of view of ecopoetics and liminal realism. The death of prophet-like Bobi at the end of the story brings a conclusion to a long inner battle opposing wonder and joy to "un-joy," life drives to death drives.[102] Eventually, Bobi sticks to his praise of the open stance to the world proper to children and ecopoets: "'Ma is a word. The word for your mother. When you called it she used to come. Now you call it and she comes no longer.' [. . .]/ 'No,' he said, 'now I know. I have always been a child; but I am right. [. . .] Ma!' he cried. / The lightning planted a golden tree between his shoulders."[103] Bobi's last cry sounds like a call to Mother Earth, who takes him back into her bosom as the man is struck by lightning. It may be that Bobi's tragic flaw, driving him away from the Gremone Plateau community after young Aurore's suicide, initially resided in his misguided utopian attempt to repress sorrow, loss, and death altogether in his monologic euphoric vision of the dance of life. The lesson he learns then, and eventually comes to embody, is that death, failure, and sorrow, are inescapably part of it all. The lightning finally brings him an epiphany as to the complementary nature of opposites, making it absurd to strive for a life springing purely from joy and shielding one from death, grief, violence, decay, or loss. The lightning bolt then returns Bobi to a place of peaceful acceptance, where joy and sadness, life and death, are tightly interwoven as part of one and the same continuum, which the novel realistically upholds.

As a matter of fact, it is only toward the end of Giono's novel that death finally appears in this non-dualistic light, as part and parcel of Mother Earth's processes. After Aurore's violent suicide, the crude allusions to the smell of her decomposing and messy body—she has blown her brains out—finally go hand in hand with images evoking her natural return to the ground: "'Aurore is dead. They are going to bury her today. [. . .]' / Aurore is like the house, no more than the house. In the world. [. . .] See how the house has become a grove of trees. Next year it will be hidden by the lilacs. It will become a mountain of lilacs, and when it is all in bloom, the people who pass by along the hill paths [. . .] will catch in the wind the odour of lilacs.'"[104] In a manner that may remind us of Kingsolver's *Animal Dreams*, such a view of death as a going back to the earth evokes the notion of a wholesome life-death continuum interweaving human naturcultures within the fertile earth. In Giono's humble, integrative vision, as in Kingsolver's or Linda Hogan's for that matter,[105] ecophobic tendencies that lead to the denial of the more dysphoric, violent, and chaotic aspects of life's creative processes can only lead to tragic consequences. It is as if Bobi's quest, and potentially Giono's, ultimately aligned with that of later ecofeminists who, in Carol Christ's terms, hanker for "a wholeness that unites the dualisms of spirit and body, rational and irrational, nature and freedom, spiritual and social, life and death, which have plagued Western consciousness."[106]

In his full reclaiming of our bodily natures,[107] Giono thus eventually highlights not human exceptionalism, but, rather, the material continuity between humans and the greater stuff of the more-than-human world. While Giono's initial ending was cut from the published version of *Joy of Man's Desiring*, its focus on the multispecies cooperation allowing for the return of one's body to the earth can in fact appear in a more euphoric light than that apprehended by most death-repelled critics:

> The skin on Bobi's face is fissuring. Near his mouth, there is a clear liquid oozing from the crack, wetting the earth like a smashed egg. [. . .] On the ground, filing up in brooks, ants climb over Bobi's hands, over his face, his eyes, his mouth, inside his nostrils. Flies are glued to the tear, near his mouth. Bobi is opening up in other places. The insects enter him and do their work. At that moment, Bobi is in full *science*. He is spreading out into the vaster dimensions of the universe.[108]

While Giono's frank take on the messy process of physical decay may feel upsetting to some, the oxymoronic vision he offers here nevertheless bustles with life—hence the comparisons between the files of ants and brooks, or between Bobi's bodily juice and the gooey stuff inside an egg. Decomposition is overall positively described as part of a harmonious earthly process, in accordance with the laws of the wider cosmos. Slowly turning into humus thanks to multispecies cooperation, Bobi paradoxically achieves full grandeur via this humbling encounter with death. Freeing him, and readers in his wake, of his delusional belief in his own exceptionalism, Bobi's death is thus transcended, as he, like all vulnerable beings, is returned to the cyclical body of Mother-Earth whence new life will spring.

As opposed to Jacques Noiray's reading of Marthe's character as tainted with a dapple of misogyny on the part of the writer,[109] I would argue that Giono may to the contrary have been an early ecofeminist. In the passage where Bobi expresses a distrust of women combined with an arrogant faith in his own potential as a messiah,[110] it is crucial to distinguish the diegetic levels separating Bobi's partly narcissistic character from Giono the writer. It appears rather early on that the embodied experience of Giono's female characters best fulfills the project at the heart of Bobi's ecopoetic gospel, extolling the earthly, sensuous intelligence uniting humans with the greater-than-human world. While I will later expand on the characters' fine attunement to the pulse of the earth through their bodies, we might for now turn to Marthe, Zulma, and Madame Hélène as cases in point.

Half-woman, half-goat, large and voluptuous Zulma, the goat keeper, dresses in sheepskin and wears her plaited, straw-like hair braided together, topped with a crown of interwoven fescue ears. Living on the threshold between human and animal worlds, she is characterized as a kind of Cybele, or shaman, whose singing produces an enchanting effect, "as if all the lamps of the sun had illumined the harvests."[111] Sometimes treated like a simpleton, Zulma's deeply oxy-*moronic* character and poetic expression reveal the vanity of dualistic thinking. Hence Marthe's epiphany in Zulma's company—if Marthe is struck at first by the girl's "monstrous body, enlarged by the voluminous sheepskin clothing," her vision gradually adapts to the girl's liminality, allowing her to appreciate the wholesome complementarity of opposites which the girl glows with: "I am beginning to see that you are beautiful, that perhaps everything is beautiful, that the black sun has light, that you are perhaps really the fine weather."[112] Taking in the cold winter weather and wild country all around her, Marthe realizes: "Here was freedom. Suddenly, all grew out of bounds, escaped the measure of Man, and was trying to regain its natural stature. / 'Oh!' thought Marthe. 'This is paradise on earth!'"[113] If her epiphany may seem oxymoronic from a Cartesian point of view, her anti-dualistic wisdom nevertheless leads her to warn Bobi of the main flaw in his project, eluding as it does the interdependence between joy and sorrow, life and death, growth and loss, and so on. "'I think now that you can't make joy last. [. . .] I even believe that we ought not to desire it.'"[114]

Marthe's superior embodied wisdom is also discretely highlighted earlier as she discusses with Bobi and Jourdan the pleasure she finds in feeding and watching the birds: "She [understood that Bobi] was right, but she was thinking of the joy of feeding. It was not her fault. It was her woman's desire.

Reason cannot prevail against the desires of the body. [And when it strives to do so, it is misleading.]"[115] In a way that might today be regarded as aligned with an ecofeminist stance, and presumably because they cannot elude their own materiality, fertility, and finitude, which are made manifest in their biological rhythms, Giono's women seem more accepting of the life-death, cyclical aspects of nature. Both childless Marthe or widowed, menopausal Madame Hélène have learned to live with loss, sorrow, and frustration. Both seem to understand better than Bobi that those are the flip side of joy, or, in other words, that one cannot embrace only the sweet and comforting life-sustaining parts of Gaia. Hence Madame Hélène's sensuous relationship with the wind, which seems to appease her in her menopausal restlessness, providing her with a connection to the forces of the earth that soothe her aching. The wind is described eco-erotically, as if making love to her in a way bringing her peace and comfort, despite her loneliness, her unfulfilled craving for human flesh, and her hormonal confrontation with her own finitude:

> Madame Hélène had no peace save in touching her breasts with her cold hands. It made her sad to find them hard and burning. The moaning of the wind pierced her flesh and wailed against her heart. The window was loosely joined and its panes shook and rattled. [. . .] It was a warm wind. It pressed gently on Madame Hélène's body. It treated her like the foliage of a tree; it penetrated deep inside her; it brough its warm disturbance to the center of her being, where she had always thought herself strongest. The wind moaned. [. . .] Madame Hélène's sadness was mixed with peace and came from nothing human. [. . .] "All in all, I am very well, at peace." She repeated to herself: "At peace, at peace," with the blowing of the wind; and the moaning brought still more gentle sadness and warmth. And it was a very desirable sadness.[116]

Where Bobi grows restless and vainly tries to dodge the sadness, failure, and loss that are an unavoidable part of harmonious living, Marthe, Zulma, and Madame Hélène never give in to the illusory Manichean pitting against each other of earth/paradise, life/death, joy/"un-joy," desire/reason, bodily/spiritual yearnings, or men/women. In fact, Aurore's tragic death may appear as the consequence of Bobi's unrealistic promise of an inalterable joy, making frustration and sadness seem unbearable to those under the spell of his misguided hankering for only the bright side of Gaia. As opposed to Bobi, Giono's female characters can find peace early on by acknowledging that the finitude of their own individual selves is no tragedy, but, rather, finds meaning in the multispecies continuum driving all life. As the narrator contemplates toward the end: "In all the empty places left on earth by the death of an animal, there were three or four newly born."[117]

OF SPIDER WOMAN AND WEBS: BARBARA KINGSOLVER AND LESLIE MARMON SILKO

In Kingsolver's 2000 *Prodigal Summer*, the dominant imagery does not explicitly call upon Mother Earth archetypes. However, the descriptions of nature are studded with instances of personification, while they simultaneously tend to feature the image of the web, thus abiding by the master metaphor of ecological discourse.[118] While straying from Gaian figures, the nonanthropomorphic image of the web essentially symbolizes the many entanglements interconnecting all lifeforms and matter.

In *Prodigal Summer*, the inference of a cosmic order tying all lifeforms together is reinforced by the emphasis on the seasonal pull connecting individuals on a microcosmic level to the macrocosm:

> [. . .] rhododendrons huddled in the cleft of every hollow. [. . .] But for now their buds still slept. Now it was only the damp earth that blossomed in fits and throes: trout lilies, spring beauties, all the understory wildflowers that had to hurry through a whole life cycle between May's first warmth—while sunlight still reached through the bare limbs [. . .]. On this path the hopeful flower heads were so thick they got crushed underfoot. In a few more weeks the trees *would* finish leafing out here, the canopy would close, and this bloom *would* pass on. Spring *would* move higher up to awaken the bears and finally go out like a flame, absorbed into the dark spruce forest on the scalp of Zebulon Mountain. But here and now spring heaved in its randy moment. Everywhere you looked, something was fighting for time, for light, the kiss of pollen, a connection of sperm and egg and another chance.[119]

Highlighting other-than-human agency, the mountain and its vegetation are personified (the mountain possesses a "scalp," and the plants "huddle"; buds "sleep," tree branches have become "limbs," and flower heads are "hopeful"). The repeated modal "would"—both epistemic and radical in its usage here—predicates a driving life force, the will to live ingrained in DNA and activated by seasonal rhythms that move individuals, packs, and entire species toward reproduction. The whole mountain forest is driven by a unifying material agency, the propensity of which is to breed more colorful, luscious life. Set on a par with botanical reproduction, animal and human romance and sexuality take part throughout in one and the same dance of life (with references to "the kiss of pollen," or spring "heaving in its randy moment").

Kingsolver's lush descriptions throughout underscore the ecosystemic web of life formed by the mountain, disclosing an interconnected biotic community with an inherent and intelligible order. As encapsulated in the opening and closing paragraphs of the novel, the metaphor extended throughout to capture the organicity of life is packed in the closing lines, featuring every single action as "a tug of impalpable thread on the web pulling mate to mate and predator to prey."[120] Kingsolver thus visibly threads an ecological tapestry into the dialogic fabric of her book. This bespeaks the influence on Kingsolver's work of Charles Darwin and Rachel Carson, both of whom mobilize the same web metaphor. In *The Origin of Species*, Darwin writes of "plants and animals, remote in the scale of nature," and yet "bound together by a web of complex relations."[121] As for Carson—herself much inspired by Darwin's study of the toil of earthworms in the soil—, this is how she put it: "The soil community, then, consists of a web of interwoven lives, each in some way related to the others."[122] The vision of life as a web is thus engrained in ecology, a science that specializes in the study, as Carson insists, of "interrelationships," "interdependence," "the web of life—or death."[123] If the image of the web is less anthropomorphic than Gaia, the principles of symbiotic dwelling and multispecies interconnection it encapsulates nevertheless come down to the same. This is highlighted by Robin Wall Kimmerer's scientific description of how "trees in a forest are often interconnected by subterranean networks of mycorrhizae, fungal strands that inhabit tree roots. The mycorrhizal symbiosis enables the fungi to forage for mineral nutrients in the soil and deliver them to the tree in exchange for carbohydrates. The mycorrhizae may form fungal bridges between the trees, so that all the trees in a forest are connected." Wrapping up these phenomena with a meaningful image, Kimmerer writes that these networks "weave a web of reciprocity, of giving and taking. [. . .] Through unity, survival. All flourishing is mutual. Soil, fungus, tree, squirrel, boy—all are beneficiaries of reciprocity."[124]

I would like to insist at this juncture that if Kingsolver's overall sanguine *ecopoiesis* rests on Mother Earth or web tropes epitomizing life-oriented, multispecies entanglements, by no means does her writing skimp on the darker sides of what such meshworks also entail, nor does it simply wax lyrical on the exuberant creativeness at work in the world. For all the ecofeminist imagery turned toward hope in a fertile future, Kingsolver's fiction clearly never eschews human responsibility in the induction of toxic, transcorporeal phenomena that may destroy both human and other-than-human lives.[125] On the contrary, although certainly an ode to life in many ways, *Prodigal Summer* nonetheless stages the pathetic degeneration of life caused by chaotic, genetic or cellular growth, resulting in Down's syndrome on the one hand, and in a lethal case of cancer on the other. Although the novel does not directly indict the widespread use of pesticides in the rural area where these characters' abnormal developments, degeneration, and pathetic deaths take place, the potential causal relationship is nonetheless hinted at in the background activities of the characters. The web image thus carries a sense of ecological responsibility: "'Everything alive is connected to every other by fine, invisible threads. Things you don't see can help you plenty, and things you try to control will often rear back and bite you, and that's the moral of the story,'" one character in the book, Nannie Rawley, apprises her pro-pesticide neighbor.[126] Similarly, in *Animal Dreams*, the protagonist's father is afflicted with a precocious case of Alzheimer's disease—a disease the origin of which could be correlated to the environmental poisoning provoked by the mining company's industrial dumping nearby their hometown.

In his early study of apocalypse as a master trope, Lawrence Buell first scrutinizes some of the most common metaphors used to describe nature in a Western context, particularly that of the web as favored by John Bruckner, Charles Darwin, Rachel Carson, and Wendell Berry. He then moves on to consider how this might relate to the figure of Spider Woman, or Thought Woman, cast in Leslie Marmon Silko's *Ceremony*.[127] Buell underlines how Silko's hybridized and mythopoeic writing has adroitly fused together the ecological concept of life as a web with the archetype of Spider Woman.

As it is, Spider Woman—a Native American version of Gaia to some extent—has weaved its way into Kingsolver's ecofeminist vision of the world's creativity. She appears for instance in her short story "Stone Dreams"[128] as well as in *Animal Dreams*. In the latter, Loyd's cursory mention of Spider Woman stresses her polymorphism, appearing first as Spider Grandmother who raised abandoned twins in Pueblo mythology, and then again in Navajo stories, as a creation goddess living on top of Spider Rock, and who is said to have "lassoed two Navajo ladies with her web and pulled them up there and taught them how to weave rugs."[129] As Loyd emphasizes for Cosima's sake, beyond the cultural variations on the story of the creation goddess, "it's the same Spider Woman. Everyone kind of agrees on the important stuff."[130] The essential value of Spider Woman as an archetype can be deciphered from her liminality. Half-woman and half-spider, the world-spinning arachnean goddess embodies the earth's creative powers, whereby every living thing is connected one way or another to every other lifeform within the intricate web of life: "She is the Old Woman Spider who weaves us together in a fabric of interconnection," Paula Gunn Allen explains. To "those people of the Americas who kept to the eldest traditions," she is identified "as the Sacred Hoop of Be-ing." She is "celebrated in social structures, architecture, law, custom, and the oral tradition," as thought-woman, who has thought, or sung, the world into being.[131] "There is a spirit that pervades everything," Allen clarifies. It is a "power of intelligence," "that is capable of powerful song and radiant movement, and that moves in and out of the mind. [. . .] Old Spider Woman is one name for this quintessential spirit. [. . .] Her variety and multiplicity testify to her complexity."[132] Spider Woman symbolizes the creative interlacing power of sentience and spirit which infuses the world's constant birthing and is breathed into ecopoetic ways of dwelling.

As Leslie Marmon Silko stresses in the incipit of *Ceremony* (1977), Spider Woman embodies the ongoing dynamic agency inherent in ecopoetic storytelling. Potentially an avatar for the storyteller herself, Spider Woman possesses the power to breathe new life into old discourse, to revitalize disanimated matter, birth new visions, and, ultimately, to reweave word to world. Silko's dialogic novel specifically tells a multifaceted story of healing, braiding issues related to postcolonial hybridity and to surviving trauma within an ecopoetic quest on both individual and collective planes. Interweaving many different types of textuality—from myths, poems, or chants, to prose—in the fabric of her postcolonial, dialogic novel, Silko's creative activity itself seems inspired by the archetypal figure of Spider Woman, the Pueblo creation Goddess spinning the world into a web of sacred relations out of pure thinking. As words materialize onto the pages of Silko's novel, framed as it is by poems that read halfway between chants, poems, and myths, the impression is that the entire fabric of the literary work is being birthed from Spider Woman's spinning: "Ts'its'tsi'nako, Thought-Woman,/ is sitting in her room/ and whatever she thinks about /appears./ [. . .] Thought-Woman, the Spider,/ named things and/ as she named them/ they appeared./ She is sitting in her room/ thinking of a story now/ I'm telling you the story /she is thinking."[133] Here the name "Ts'its'tsi'nako" itself sounds like the song of a spider stridulating in her web.

In the highly metatextual, eponymous song-poem that follows, Silko then highlights the sustaining value and ongoing dynamism of stories. As we hold on to and regenerate them, stories in turn nurture us. They provide food to thrive on even in the midst of destruction: "[Stories] aren't just entertainment. [. . .] They are [. . .] all we have to fight off illness and death./ You don't have anything if you don't have the stories./ [. . . .] He rubbed his belly. I keep them here [. . .]/ Here, put your hand on it/ See, it is moving. There is life here for the people./ And in the belly of this story/ the rituals and the ceremony are still growing."[134] The novel ultimately suggests that literature can perform healing the way a ceremony does, reinstating relationships of interdependence and reciprocity between humans and the wider world.

As Paula Gunn Allen elsewhere explains, although Spider Woman's name translates to Thought Woman in English, in Keres cosmogony, it comes closer to "Creating-through-Thinking Woman."[135] In their influential writing from a few decades ago, both Silko and Allen have helped unravel the onto-epistemological and material ties between thinking, storytelling, and mattering—between wording and worlding. Hence the spiderweb as an anamorphic cosmovision allowing us to apprehend worldly entanglements in a coherent way, while remaining aware of the discursive intricacies of trying to refer to the world by means of concepts and signifiers that are always part and parcel of an entire, elaborate system of thinking and worldview. This would be concordant with the elder's untranslatable notion of the world as being "fragile," caught within linguistic and discursive threads much like spider webs.[136]

Preventing too simplistic a gendered take on Spider Woman and quoting from Anthony Purley's study on "Keres Pueblo Concepts of Deity," Allen insists that Spider Woman extends beyond female attributions. Much more than just a "fertility Goddess," and as "the supreme Spirit, she is both Mother and Father to all people and to all creatures."[137] The female aspect of Spider Woman as Creation Goddess, Allen stresses, should *not* mislead one into limiting female power to maternity. This, according to Allen, is a but a "demeaning" patriarchal notion that "trivializes the tribes and trivializes the power of woman" as it diminishes "the power inherent in femininity."[138] As a matter of fact, the ecofeminist reclaiming of Earth-Goddesses as a source of reempowerment goes hand in hand in many cases with a rehabilitation of the metaphor of weaving. Evoking a craft often practiced by women, the centrality of the trope appears from the titles of pioneering anthologies of ecofeminist theory such as *Weaving the Visions: New Patterns in Feminist Spirituality*, edited by Judith Plaskow and Carol P. Christ (1989), or *Reweaving the World: The Emergence of Ecofeminism*, edited by Irene Diamond and Gloria Feman Orenstein (1990). In one essay, Brian Swimme refers to Charlene Spretnak's important work on this trope. A leading ecofeminist, Spretnak has connected the scientific discoveries of cosmic expansion and phenomenal entanglements with the metaphor of weaving long honored in many traditional cultures. "In fact," Swimme contends,

> nothing is more obvious than Spretnak's assertion that weaving is a fundamental dynamic of this universe. Picture it: from a single fireball the galaxies and stars were all woven. Out of a single molten planet the hummingbirds and pterodactyls and gray whales were all woven. What could be more obvious than this all-pervasive fact of cosmic and terrestrial weaving? Out of a single group of microorganisms, the Krebs cycle was woven, the convoluted human brain was woven, the Pali Canon was woven, all parts of the radiant tapestry of being. Show us this weaving? Well it is impossible to point to anything that does not show it, for this creative interlacing energy envelops us entirely. Our lives in truth are nothing less than a further unfurling of this primordial ordering activity.[139]

As Swimme has it, one of the reasons why this enchanting metaphor of weaving makes so much sense from an ecofeminist point of view may be that "[women] are beings who know from the inside out what it is like to weave the Earth into a new human being."[140] Referring to both Spretnak's and Allen's seminal work, Swimme hopes that in bringing together indigenous, ecofeminist, and scientific worldviews, we may "teach our children at a young age the central truth of everything: that the universe has been weaving itself into a world of beauty for 15 billion years.[. . .] We will teach that their destinies and the destinies of the oak trees and all the peoples on Earth are wrapped together. That the same creativity suffusing the universe suffuses all of us, too."[141]

At this stage of my enquiry into the different faces of Gaia in contemporary ecopoetic literature, I would like to return briefly to some of my previous work on Annie Proulx's Wyoming short story cycle, which comprises three, interrelated collections, respectively published in 1999, 2004, and 2008.

DARK REENCHANTMENT: ANNIE PROULX'S WYOMING STORIES AND THE MORE SOMBER FACE OF GAIA

As I have shown in previous work dealing with Annie Proulx's mythopoeic writing, the latter takes issue with colonial and patriarchal ideologies. Her mordant resort to a postmodernist type of magical realism is not without feminist and environmental overtones.[142] Proulx's overall ecofeminist stance and poet(h)ics becomes salient as a Wyomite comments on the filthy yellowish atmosphere: "'Pollution. It's smog. Comes from that goddamn Jonah infill methane gas project. One well ever ten acres. Never seen that smog before in Wyomin. You're seeing her start to die. The whoremasters got ahold of her. They got her down on her knees and any tinhorn with five bucks in his jeans comes by they put the prod pole to her and say, 'suck his dick.'"[143] Proulx's blunt personification of the land forces us to take a hard look at the Anthrop-o(bs)cene. Through metaphoric mapping, the feminine earth is degraded into a subaltern prostitute, violently submitted to exploitation by industrial developers and capitalists in the position of abusive, humiliating clients.[144]

The obscene image strikes an ecofeminist cord, resonating with Carolyn Merchant's analysis of the immense influence of Francis Bacon's bombastic patriarchal rhetoric over Modern, techno-phallogocentric representations of nature and women, as well as politics. As Merchant details, Bacon—considered by many as a "father of modern science"[145]—instrumentalized the imagery of the earth as female the better to "[transform] the earth as a nurturing mother and womb of life into a source of secrets to be extracted for economic advance. [. . .] As woman's womb had symbolically yielded to the forceps, so nature's womb harbored secrets that through technology could be wrested from her grasp for use in the improvement of the human condition."[146] Treating "nature as a female to be tortured through mechanical inventions,"

Bacon "transformed the magus from nature's servant to its exploiter, and nature from a teacher to a slave."[147] The above image of the Wyoming land as a whore thus resonates with Bacon's frequent descriptions of matter as a "common harlot [. . .] not devoid of an appetite and inclination to dissolve the world and fall back into old Chaos," and which, Bacon concluded, must be "restrained and kept in order by the prevailing concord of things."[148] As Merchant shows, Bacon's violent symbolism of the forceful penetration of nature through science ("searching into the bowels of nature"), technology, and mining helped bolster what I propose to call Anthrop-o(bs)cene imagery and ideology: "By the art and the hand of man," nature should be "forced out of her natural state and squeezed and molded." Bringing human hubris to a paroxysm, Bacon thus celebrated a vision where "human knowledge and human power meet as one."[149] This, in turn, "[legitimated] the exploitation and 'rape' of nature for human good."[150]

Although Proulx's protagonist in the above passage takes umbrage at the crude metaphor, many of Proulx's brutal tales unapologetically flaunt coarse language and obscene images at their readers. A bleak vision of the capitalistic overexploitation of land comes through ecoterrorist Wade Walls' nasty assessment in the story "The Governors of Wyoming": "But he knew all about the place, the fiery column of the Cave Gulch flare off in its vast junkyard field, refineries, disturbed land, uranium mines, coal mines, trona mines, pump jacks and drilling rigs, tank farms, contaminated rivers, pipelines, methanol-processing plants, ruinous dams, the Amoco mess, railroads, all disguised by the deceptively empty landscape."[151] One cannot fail to notice here the disturbing musicality and internal rhymes of the long, stretched out sentence listing all the forms of violence against the land. As the polysyndeton piles up nominal clauses naming the many injurious aspects of industrial extractivism, the syntax and tonic stress pattern build up a rhythm the tempo of which evokes the pounding, thumping, pumping, and drilling— potentially the beating, or raping—of Wyoming, which throughout Proulx's stories, stands in as a metonymy for Gaia.

Come to think of it, it makes perfect sense for feisty ecofeminists to invoke Gaia in their rebellion against patriarchal systems of oppression and dominion. Indeed, one of the original many faces of the complex deity goes back to the Greek Goddess as initially depicted by Hesiode in his *Theogony*, written around 700 B.C. Hesiode's Gaia is Mother Earth in her mineral body, with its mountains, rocks, and plains. From her terrestrial womb, Ouranos is born and becomes a Sky-God that takes Gaia as his mate. Together they procreate, and Gaia gives birth to a great number of children, the famous Titans and Cyclopes. As a patriarchal despot, Ouranos however rules over Gaia and their children, imposing his power over them by means of brute force. Growing tired of having to submit to Ouranos's lust for power, cunning Gaia secretly forges a mineral unbreakable scythe which she hands over to their son Chronos, whom she connives with to disempower the tyrannical patriarch. As Ouranos lies over Gaia to impose sexual intercourse, Chronos springs out of hiding and cuts off his father's genitals—a phallic dismemberment potentially symbolizing the castration of male authority and putting an end to abusive patriarchal ruling over the Earth.

My contention here is that Proulx resorts to symbolic networks and ecopoetic justice throughout her stories in a way that delineates the darker face of Gaia, ruthless in her response to a mode of dwelling which has degenerated into human parasitism. Encapsulating the symbiotic relationship between humanity and Gaia, the parasite trope first surfaces in Proulx's gruesome story of "The Half-Skinned Steer." Recycling Alfred Llyod Tennyson's famous vision of nature as "red in tooth and claw," Proulx stages the grim death of a character whose son is conveniently named "Tick." Tick's father gets eviscerated by a "waspy emu" who has attacked him with "its big razor claws."[152] Tick and his family, the plot suggests, are cursed through some sort of ecopoetic retribution for their prolonged meddling with Gaia in the most stupid and inconsiderate ways. Among other things, they have taken part in transforming the failed family ranch into a hyperreal tourist trap, quite ironically called "Wyoming Down-Under," with emus imported all the way from Australia.

The grotesque story is moreover interlaced with a tall tale wherein a character named Tin Head goes about butchering a live, knocked-out steer:

> He ties up the back legs, hoists it up and sticks it. [. . .] When it's bled out pretty good he lets it down and starts skinning it, starts with the head, cuts back of the poll down past the eye to the nose, peels the hide back. [. . . H]e gets the hide off about halfway and starts thinking about dinner. So he leaves the steer half skinned there on the ground and he goes into the kitchen, but first he cuts out the tongue which is his favorite dish all cooked up and eat cold with Mrs. Tin Head's mustard in a forget-me-not teacup.[153]

After dinner, as Tin Head intends to finish up his botched job of killing the steer, he finds out that the creature has magically vanished:

> It's gone. Only the tongue, laying on the ground all covered with dirt and straw, and the tub of blood [. . .]. He looks around for tire marks or footprints but there's nothing except old cow tracks [. . .], but way over there in the west on the side of the mountain he sees something moving, stiff and slow, stumbling along. It looks raw and it's got something bunchy and wet hanging down over its hindquarters. [. . .] And just then it stops and it looks back. And all that distance Tin Head can see the raw meat of the head and the shoulder muscles and the empty mouth without no tongue open wide and its red eyes glaring at him, pure

teetotal hate like arrows coming at him, and he knows he is done for, and all of his kids and their kids is done for.[154]

The wild half-skinned steer provides an obscene synecdoche for Man's abuse and silencing of nature—the dirty cut-off tongue, the "empty mouth" and the "mute bawling" of the flayed animal providing metaphors of Nature's raw, speechless voice. Moreover, through complex narrative processes that involve intermingling, mise an abyme, and metaleptic contagion, the murderous emu in the frame narrative ends up transforming into a double for the eponymous half-skinned steer in the embedded story, accomplishing Biblical-sounding retaliation on the second and third generations of ranchers.[155] Through synecdoche and anamorphosis, the two mishandled animals thus merge into a humbling, frightening, and reenchanted vision of vengeful Gaia, red in tooth and claw indeed,[156] and bent on getting rid of her human ticks:

> Then the violent country showed itself, the cliff rearing at the moon, the snow smoking off the prairie like steam, [. . .], the sagebrush glittering and along the creek black tangles of willow bunched like dead hair. [. . .] As he walked, he noticed one from the herd inside the fence was keeping pace with him. He walked more slowly and the animal lagged. He stopped and turned. It sopped as well, huffing vapor, regarding him, a strip of snow on its back like a linen runner. It tossed its head and in the howling, wintry light, he saw he'd been wrong again, that the half-skinned steer's red eye had been watching him all this time.[157]

Reinforcing this reading, in Proulx's later story "What Kind of Furniture Would Jesus Pick?" the narrator claims that the wild country "[wants] to go to sand dunes and rattlesnakes, [wants] to scrape off its human ticks."[158]

Throughout her fiction, Annie Proulx depicts the relationship between humans and the earth not as a balanced symbiosis tying symbiont to host, but as a form of parasitism whereby humans are endangering both the equilibrium of their host, and, in turn, their own capacity to survive. As a matter of fact, the parasite trope lies at the heart of French philosopher Michel Serres' theory of a "natural contract" tying humankind to the Earth: "the symbiont admits the rights of its host, where the parasite—our current status—condemns to death the being that it plunders and within which he dwells, without acknowledging that in the long run, he is dooming himself. / The parasite takes everything and gives nothing: the host gives everything and takes nothing. The right to mastery and ownership comes down to parasitism."[159] Along the same lines, the vision that transpires through Proulx's stories is that of a pervaded symbiotic relationship between Gaia and humans where, because most humans have lost track of the difference between usage and abuse, between dwelling and overexploiting, they have been living as parasites, running the risk that autoregulatory, itchy Gaia might react to "scrape off its human ticks."

Ringing a bell with the notion of a "tickly" Gaia formulated by Belgian physicist and science studies philosopher Isabelle Stengers[160]—a major influence on Bruno Latour's recent work on the intrusion of Gaia—Proulx's ecopoet(h)ics of reenchantment stages the animation of Gaia in a rather foreboding way:

> You stand there, braced. Cloud shadows race over buff rock stacks as a projected film, casting a queasy, mottled ground rash. The air hisses and it is no local breeze but the great harsh sweep of wind from the turning of the earth. The wild country–indigo jags of mountain, grassy plain everlasting, tumbled stones like fallen cities, the flaring roll of sky–provokes a spiritual shudder. It is like a deep note that cannot be heard but is felt, it is like a claw in the gut."[161]

The pathetic fallacy at play (with "cloud shadows rac[ing]," or "the air "hiss[ing]") presents a dysphoric, agential nature that appears beastly and indomitable. "[The] great harsh sweep of wind from the turning of the earth" reads like's Gaia's sulfurous breath, almost as if she were getting ready to buck. Suffering from a "ground rash," the description suggests, itchy Gaia could turn against human exploitation and throw off its human ticks—a threat hovering via the "tumbled stones like fallen cities," which obliquely evoke the collapse of earlier human civilizations. And of course, the "claw in the gut" felt by the viewer—with an ambiguous, unidentified focalizing agent ("You stand there, braced")—echoes the earlier literal evisceration of Tick's father by an emu and its "razor claws." As a result, and when regarded from an overarching perspective encompassing all of Proulx's Wyoming stories across three collections, the emu serves as a metonymic agent of ecopoetic justice enacting Gaia's rebellion against human greed and stupidity.

Elsewhere, in "mima mound county," the characters recoil as they see "to the west a fanged landscape that [seems] to be coming at them."[162] In Proulx's short stories, the weather and terrain are always looming large over humans' fate. They too appear as ominous and malevolent agents of Gaia:

> December was wretchedly cold, made worse by violent winds. [. . .] The few clouds drew out as fine and long as needle threads and the wind-damaged sky showed the same chill blue as a gas flame. *The [dry and bitterly cold] wind set its teeth* into the heavy log house and shook it with terrific gusts. In the early mornings it ceased for a few hours, then as the sun climbed over the aspen, *it returned, brutal and avid*, sweeping into the air what little loose snow remained. It never really stopped.[163]

Such animated descriptions of "a fanged landscape" or of a wind presented as "brutal and avid," "bitterly cold," and [setting] its teeth" into the log house set the decor à la Tennyson, thus symbolizing the savage, ruthless quality inherent in Gaia. Ultimately, Wyoming/Gaia proves indifferent, Proulx's stories insist, to the fate of her human offspring. To echo Lynn Margulis, one might say she turns out to be a tougher bitch than some might like to think.

Some may feel wary of the risk that such *ecopoeisis*, in its staging of Mother Earth turning malevolent toward humans, or calling the biosphere "a tough bitch" as Margulis does and Proulx implies, may exacerbate our ecophobic streak.[164] It may nevertheless be argued that Proulx's caustic, ecofeminist, and grotesque writing mobilizes Gaian ecopoet(h)ics the better to confront her readers with the futility and perilousness of myths conveying a sense of humans as masters of an inert, mute, and submissive other-than-human nature. Proulx thus performs a dark reenchantment, one that reminds us again and again that nature will always retain its own agency, and that humankind may indeed end up paying the price for the obscene ways in which we have been overexploiting her. As Timothy Morton recapitulates, the Anthropocene has brought about "the gradual realization by humans that they are not running the show, at the very moment of their most powerful technical mastery on a planetary scale. Humans are not the conductors of meaning, not the pianists of the real."[165]

Warning against the foolish ways in which misguided tales of the disenchantment of nature may backlash against humankind, Proulx's humbling epiphanies do overall provoke a "spiritual shudder," spurring us to respond to the "deep note" which, as in the above quotation, may be part of the ominous, wild music composed by the earth. Proulx's fiction thereby invites a starkly disillusioned dark reenchantment of the world. Often tinged with a sense of cosmic, retributive justice, her grotesque, reiterative pattern of brutal story endings indeed works as shock therapy against the state of anthrop-o(bs)cenic delusion and anesthesia characterizing the still dominant Western attitude toward the ecological crisis. Like many ecofeminists and scientists, Proulx, it might be argued, strives to shake that "naïve faith which many still have in an allegedly disanimated 'material world.'"[166] As with Starhawk or Paula Gunn Allen, we can see that Gaian ecopoetics in its more somber aspects call attention to ecological responsibility, making the notions of reciprocity and consequence deeply felt—indeed "like a claw in the gut."

ECOFEMINIST REENCHANTMENT AS RESISTANCE AGAINST THE AGONY OF GAIA IN ANN PANCAKE'S *STRANGE AS THIS WEATHER HAS BEEN*

To unpack one last ecofeminist, mythopoeic take on Gaia in my corpus, I would finally like to consider the powerful transmedia *ecopoiesis* at play in Ann Pancake's *Strange as this Weather Has Been* (2007). Foregrounding the intersemiotic intertextuality undergirding her ecopoet(h)ic approach to Mountain Top Removal (MTR), Pancake's book cover stages "The Agony of Gaia" via a photograph of the eponymous multimedia sculpture by Appalachian artist Jeff Chapman-Crane. Cueing viewers to adopt an ecofeminist perspective, the human-size sculpture anthropomorphically embodies Gaia in the form of a dying mountain mercilessly submitted to MTR. Achieved over a period of sixteen years, Crane's powerful work of art involves clay, dust, Styrofoam, paint, moss, twigs, sand, resin, and miniature plastic toys—a variety of materials representing the naturally grown and human-made components that matter in today's world. The artwork has travelled worldwide as an indictment of the ravages brought about by strip-mining.[167] As I hope to show, both works of art negotiate a form a reenchantment stemming from experiences of pain and revolt against the disenchantment of the earth in the Anthrop-o(bs)cene.

With an aposiopesis for a title that calls the reader in to fill the silence trailing in the wake of the left-off sentence, Ann Pancake's *Strange as This Weather Has Been* has aptly been called "a novel that voices the silence."[168] With this oxymoronic phrase in mind, I have demonstrated elsewhere how this ecofiction substantiates Michel Serres's claim that the land experiences both motion and emotion, that it speaks to us and moves us humans to respond to it.[169] Steeped in the poetry of West Virginian vernacular and interview-inspired life stories of families and communities living in the disastrous wake of Mountain Top Removal coal mining in the Appalachian range, Pancake's neoregional, ecopoetic prose translates the suffering and crying out of the muted earth—with its strangled ecosystems and uprooted fauna and flora—of its soil and rivers contaminated by toxic waste. The tragic schema underpinning the novel follows an only too true-to-life apocalyptic scenario, which is no climate fiction, but, rather, comes close to reportage fiction dealing with the catastrophic consequences of MTR strip mining practices. In fact, prior to writing this novel, Ann Pancake, originally from Romney, West Virginia, conducted interviews on MTR in the Appalachian range. She then codirected with her sister, Catherine Pancake, a documentary film titled *Black Diamonds*, issued in 2006. It can be inferred that much of the novel's language, ideas, and background draw from this firsthand nonfictional material.[170]

Pancake interweaves in her dialogic novel myths commonly invoked to support the ideology behind the industrial pillage of her rural homeland together with myths that, on the contrary, can rebut it. To this day, managers, shareholders, or miners working in the MTR industry will endorse a literal reading of the Bible to defend their exploitive activity. Such presumed determination to fulfill God's command is relayed in the novel through the local "preachers preaching, 'God gave man dominion over the earth' and 'The good Lord put this coal under the ground for us to use.'"[171] Counterbalancing such facile anthropocentric interpretation of Genesis, one of the women activists fighting the industrial companies with the help of an organized environmental group points to an alternative ethics of care which could also be derived from scriptures and are more aligned with Baird Callicot's interpretations: "'Bullcrap and foolishness [. . . .] Anybody with a grain of sense can see we're destroying what God made. 'The Earth is the Lord's, and everything in it,' Psalm 24. He wants us to fight for it, and I pray every day for God's help in this fight.'"[172] Diffracting such tales of power-over, of a male creation God, together with a *poiesis* rehabilitating the opposite figure of Gaia—and portraying its Anthrop-o(bs)cene agony—Pancake's novel forcefully tinkers with different mythological views of the earth, which are then reframed and reweaved from a resistant, ecofeminist stance.

Hinting at a parable of Modernity, Pancake characterizes ten-year old Corey—Lace's son and Bant's brother—as an embodiment of the fascination with control and machines. The boy moreover fits constructed ideas of masculinity. On discovering the gigantic dragline—actually the most humongous machine ever built in terms of size, weighing between 8,000 and 13,000 tons, with a dragline boom between 45 to 100 meters long—excavating the coal from the beheaded mountain behind their hollow, the child revels in fantasies of dominion and control that speak volumes in terms of the culture he metonymically symbolizes: "And finally, he'd scale Big John. That vast mountain-handling piece of gorgeous machinery. And as Corey climbs it, the smell of its fluids, the good grease he'd get on his clothes. [. . .] He'd crawl in, settle in the seat, take a look at how it ran, push his legs to the pedals, grip sticks and handles. [. . .] This, his body in that gigantic body, his body running that body, and the size, the power of that machine: inside Big John, Corey can change the shape of the world. Corey Can."[173] This passage is not without erotic imagery expressing lust and suggesting penetration and violent handling. The rather blunt series of monosyllabic words in the last two sentences produce a pounding effect engaging the prose in imitative harmony: coming forth with alliterations in [k], [tʃ] and [ʃ], and with visual anaphoric rhymes in <c>, <ch> and <sh>, the rhythm sounds like a mighty, anthropic drilling power: "Corey can change the shape of the world. Corey Can." As little Corey has incorporated the Modern hubris and thirst for power, the chapters focalized through his point of view relay a voice that is engaged in a deadly anthropophonic song. He thus basically substantiates Theodore Roszak's understanding that "the way the world shapes the minds of its male children lies somewhere close to the root of our environmental problem."[174]

The name Corey is derived from Gaelic, "*coire*," meaning a cauldron, a seething pool, or a hollow. It signifies "dweller in or near a hollow," which applies to the child's entire family. By the end of the novel however, onomastics acquires a more somber, ironic meaning. Indeed, Corey's death by drowning takes places as he is recklessly driving a four-wheeler, landing him at the bottom of a catchment pound, dragged under by the weight of the machine. Not only does this innocent death of one of the main characters make the reader feel the pathetic, direct human cost of MTR over the locals' lives, silencing in its wake one of the novel's main voices that goes extinct together with many a nonhuman lifeform, it moreover contains a tragic morality as to the fate of Modernity. Challenging some of the lethal attitudes that are the byproducts of highly industrialized cultures, this pre-teen character is indeed doomed by his thirst for power and his "talent for machines," quite significantly envied by his more effeminate brother as "man-talent."[175] Corey is described through his brother's eyes as "metal-made," as if "welded" to the four-wheeler, "that hard little body moving with the machines the way other people move with horses."[176] The boy in that sense embodies the "ideal of manly prowess" that Roszak criticizes. Deconstructing the "virulent power images" such as those supported in blockbuster movies like *Robocop*, Roszak argues that "these tempt but do not assuage the most violent appetites of the death instinct."[177] Read through the lens of ecopoetic justice, Corey's technophallogocentric death-drive thus ominously gestures toward what future generations growing up under the spell of a disenchanted world might be headed for.

As it is, both Chapman-Crane's sculpture and Pancake's literary text attack mythologies celebrating the advent of "the machine in the garden," as Leo Marx first put it.[178] In Chapman-Crane's sculpture, the humanized land morphs into a woman identifiable as agonizing Mother-Earth. Its round, voluptuous contours evoke the rolling hilltops of the Appalachia, a mountain range that has undergone great erosion since the time of its formation about 450 million years ago. The lying mountain-*woe*-man is submitted to relentless extraction, literally dominated by the parasitical tractors and draglines strip-mining into her nakedness as she curls up under the pain in a quasi-fetal position. The topsoil-skin of the mountain is being sheered in the process of deforestation, while her earth-flesh is being blasted, her insides spilling out.

From between her exposed, hacked ribcage flow blood-rivers the color of which intimates the now-evidenced oxidization of creek beds contaminated with heavy metals from the slurry impoundments. Those are actually incriminated in a kind of deadly, ecopoetic incantation that is bitterly slammed in Pancake's novel by one of her informed, activist characters: "Mercury. [. . .] Lead, arsenic, cadmium, copper, selenium, chromium, nickel."[179] In Crane's intense, affect-triggering work, tears stream down Gaia's face, while her bosom is crumbling under the machine blows, the bulldozers hard at work, gutting her live. Her face is tucked in her hands, contorted in pain, and presumably to avoid the spectacle of her own raw devastation which we viewers can hardly turn away from.

Serving as an ecopoetic threshold before Pancake's novel, this invitation to contemplate Chapman-Crane's "Agony of Gaia" sets from the start an ecofeminist picture, taking up the typical analogy between the despoiling of the earth and the desecrating of a woman's body. As Bant and her father Jimmy first trespass onto the mining site to gauge how badly the company has encroached upon their mountains, and to thereby assess how far it could be responsible for the latest flood, they end up staring at one of the poorly kept sediment ponds: "The sides of the hollow, as we got further in, more naked and scalped, more trees coming down, and up above, mostly just scraggly weeds, the ground deep-ribbed with erosion, and I told myself, yes, this is where the floods come from."[180] Through Bant's anthropomorphic gaze ("the sides [. . .] naked and scalped," "the ground deep-ribbed,"), the vision of the land proves highly reminiscent of Chapman-Crane's sculpture on the book cover. Throughout, the adolescent's testimony taps into images extending the metaphor of the earth as a pillaged body:

> [Nothing] had got me ready for how the top of Yellowroot was just plain gone. Where ridgetop used to be, nothing but sky. [. . .] The closest I'd ever seen to it was the Summersville Dam, but this was bigger, darker, and looser. [. . . And] then it dawned on me exactly what I was standing under–*Yellowroot Mountain, dead*. I knew from Lace and Uncle Mogey that after they blasted the top off the mountain to get the coal, they had no place to put *the mountain's body* except dump it *in the head of the hollow*. So there it loomed. *Pure mountain guts*. Hundreds of feet high, hundreds of feet wide. Yellowroot Mountain blasted into bits, turned inside out, then dumped into Yellowroot Creek.[181]

At this point, Bant's syntax falters as her heart skips a beat, forming a language the rhythm of which takes shape in part through her many entanglements with the earth: "I followed him. Hit ground [. . .]—sudden silence, clawed-up earth, sky shifting towards rain [. . .], sudden dead spots in what should be green."[182] Again evoking Chapman-Crane's "Agony of Gaia," the "straight gray line" carved into Cherryboy by the mine is said to start a the mountain's "east flank."[183] Already at this juncture, Bant's animal response ("[a] prickle moved under my hair"[184]) to the preying on the mountain betrays the bodymind continuum knotting her sense of self with the existence and fate of her mountains.

In addition to the anthropomorphizing at play in such descriptions, the figure of an embracing Mother-Earth is perceptible in the relationship tying Bant's mother, Lace, to the land. As a freshwoman in college, Lace pines for the comforting feeling her mountains yield: "[that] feel backhome hills wrap you in."[185] The enfolding she describes recalls motherly attachment, protection, and solace: "Grow up shouldered in them, them forever around your ribs, your hips, how they hold you, sit astraddle, giving you always, for good or for bad, the sense of being held. It had something to do with that hold."[186] Filtered through her daughter's perceptions, the mountain furthermore acquires gestational force, forming a kind of concentric, second womb bearing both Bant's pregnant mother and her *in-utero* self. Bant has thereby retained a symbiotic attachment to the mountains that sheltered and sustained her mother's pregnancy: "I'd been running this path since before I was born. I'd started running this mountain since I was still inside Lace—and they carried me back up just weeks after I came out. If I said it out loud, Lace would say I couldn't remember, but I could, the ground moving below me, dead-leaf-colored, how many colors of brown. The smell of November rain on beginning to rot leaves."[187] As occurs frequently in the texts I have been exploring, the character's deep attachment to the land is felt through her nose. Bant's periphrastic "smell of November rain on beginning to rot leaves" points to our lack of vocabulary for smells—an important factor partly accounting for our impoverished sense of smell. For lack of a better word to name the smell of forest humus in the fall, Bant must rely on circumlocution.

If, by and large, our olfactory attachment to people and places often remains subliminal, the close attention paid to earthly smells in my corpus is all the more significant in the light of studies proving the function of smell in the production of certain hormones that are involved in our feelings of attachment. It is now well-known that the bond between infant and mother—or any early caregiver, most likely—is extensively dependent on early olfactory imprinting. According to Robin Wall Kimmerer, smell likewise plays a great part in the feelings of attachment one may feel for land: "Recent research has shown that the smell of humus exerts a physiological effect on humans. Breathing the scent of Mother Earth stimulates the release of the hormone oxytocin, the same chemical that promotes bonding between mother and child, between lovers. Held in loving arms, no wonder we sing in response."[188]

Highlighting the fundamental biosemiotic connections tying Bant and the other characters to their multispecies mountains, Pancake's Gaian *ecopoiesis* acts as "a pointer to an ongoing "becoming-with" between individuals and their *oikos*. In Haraway's terms, "[the] partners do not precede their relating: all that is, is the fruit of becoming with."[189] From the time of her *in-utero* development, Bant's other-than-human companion species intra-act with her sense of time, of food, of her naturcultural identity, and her resilience in times of need: "I helped my grandma from the time I could walk. *Good little helper, Bant. Such a good helper*, creasies, Shawnee, poke, ramps, molly moochers in spring, blackeberries in summer, mayapple and cohosh, then ginseng and nuts, hickory, black walnut, butternut, chinquapin, beech—in the fall. Yellowroot after the sap went down. Sumac and sassafras in November, come Christmas, holly and greenery. I knew these things before I could read."[190] Anticipating my last part dealing with the song of the earth, we may already appreciate the rhapsodic quality of Bant's litany of the local plants she and her people could forage. As a result, her lyrical narrative rings with poetic echoes of an endangered mountain. Invoking the biodiverse community currently imperiled by MTR mining throughout the Appalachian range, the chanting quality of the prose resonates with an *ecopoiesis* of resistance. The endemic plants are both the alphabet of the language Bant speaks, and the lyrics of the song of the earth that she has learned as other children learn lullabies. Spelling out her local vegetal song enables her and her grandmother to both recite/re-site their land's co-becoming and, on a concrete level, to forage Mother Earth for food: "*You can live off these mountains*, Grandma'd say. *And in bad times*, she'd say, meaning layoffs, strikes, but also I knew, the year I was born, *we did*."[191] As the novel emphasizes, to destroy Gaia is eventually to destroy ourselves, both physically and spiritually.

Bant's relationship with her *oikos* is decidedly fusional and reminiscent of the mother-infant dyad: "I still spent a lot of time up here. [. . .] I mostly just sat in my places. Those places where if you sat quiet, the space dropped away between you and the land. [. . .] Like the heart of a rhododendron thicket, the limbs bendy and matty and strong, it was like being inside some kind of body there. It felt animal live."[192] The enduring, comforting, and womblike quality of the earth is accentuated here by the poetic breach of conventional linguistics. Creatively transcending the agrammatical tendencies of Bant's rural sociolect, the polysyndeton ("bendy and matty and strong"), personification (the rhododendron "limbs"), and weird adjectival construction ("animal live") allow for the emergence of an Appalachia-grounded *ecopoiesis*. The highly oxymoronic, partly asyntactical, and vernacular language Bant uses indeed ecopoetically challenges habitual partitions between animal, vegetal, and human realms, while translating the haptic sensation of symbiosis between the girl and the mountain. Presenting the mountain as a kind of ecosystemic womb, mother, and companion for these humans, Pancake thus broadens Donna Haraway's notion of "companion species" to include the immense, organic lifeform of an entire mountain, with all the other-than-human "messmates" engaged together in the Gaian "dance linking kin and kind."[193]

Bant's manyfold entanglements with the mountain make her psychically and physically porous to its disintegration: "I did feel the hurt for myself. I understood. It was like they were knocking down whatever it is inside you that holds you up. Kicking down the blocks that hold up your insides; kicking, until what the blocks kept up falls and leaves you empty inside."[194] With a sense of self that is inextricable from her mountain matrix, Bant empathetically responds to its harming in her innermost self as well as through various sensations experienced in her own flesh: "What I saw punched my chest. Knocked me back on my heels. At first I saw it only as shades of dead and gray, but I pushed my eyes harder, I let come in the hurt, and then it focused into a cratered-out plain. Whole top of Yellowroot amputated by blast, and the dragline hacking into the flat part left. Monster shovel clawed the dirt and you felt it in your arm, your leg, your belly [. . .]. / My tongue moved in my mouth. It had lost all water, tasting what I saw."[195] Through Bant's anamorphic vision, the machines are animalized ("monster shovel clawed the dirt") in a grotesque, beastly way. Meanwhile, the mountain is anthropomorphized in pathetic terms evoking violent amputation ("Yellowroot amputated by blast") and obscene treatment of the landscape, in ways activating our mirror neurons and, consequently, an empathetic response: "And past where Yellowroot had been, miles of mountain stumps, limping all the way over to what used to be horizon, and what would you call it now? The ass-end of the world."[196] Referring to the "weird brittle grass," the toxicity of which is indicated by the fact that the deer will not even eat it, Bant scopes out "the killed ground"—a manmade landscape looking like "the moon upside down."[197]

Grieving her pulverized mountain, Bant tries to wrap her head around what has been destroyed. She ponders how Yellowroot was much more than just a mountain top, wholesomely enmeshed as it was with the inhabitants' culture, imagination, and health:

> how lucky Grandma died, I thought. [. . .] I was remembering what Yellowroot had been. Yellowroot, shaped like a rabbit with its ears laid back, Grandma showed me that. It took its name from goldenseal, she said, that was the real name of yellowroot. *Yellowroot's the country name*, Grandma said. *Now, yellowroot's what you use for a sore throat, gargle that, nothing better for it. Turn your mouth yellow, your throat yellow, too. Everything in these woods was put here for a reason.*[198]

As becomes apparent from Bant's tale of agony and subsequent call for reenchantment, Pancake sows the seeds for an earthly rerooting of human's naturcultures and languages. Meanwhile—and this will appear to be true of all the books in my corpus—what Pancake's novel reveals is a sense of place that is attentive to the "invisible landscape" which Kent Ryden has drawn attention to. Indeed, Pancake's Appalachian narrative of inhabitation is layered with "experiential, narrative reality," constituted by "a wide variety of mythic, legendary, historical, and personal meanings overlaying what may have seemed [to cartographers] to be largely a neutral chunk of geography."[199] All the books in my corpus in fact give evidence of the "invisible landscape" they inhabit, relaying a sense of place with "an unseen layer of usage, memory, significance."[200] In other words, taking up E. V. Walter, these books refer to "the history of what we learn through our feet. [They grasp] the world that meets the eye, the [place] we know through our legs, [. . .] in our hearts, in our guts, in our memories, in our imaginations. [They] include the world we feel in our bones."[201]

Bant's intense reaction to the devastation of the earth brings to light a view formulated in Western cultures by ecopsychologists, i.e., that "each of us now experiences in some way—physically, psychologically, economically, or politically—the pain of the Earth."[202] Bant's sympathetic response to the earth's destruction notably waters down the Western view of "the 'self' as a bounded, masterful agent who is separate from and prior to the outside world, including the natural world." As Sarah A. Conn bemoans, "[in] psychotherapy, we have pathologized and individualized personal pain, viewing any 'pain for the world' as a probably pathological experience that has been projected outward."[203] Pancake's writing embraces the view that is central to ecopsychology, which, as Roszak put it, "proceeds from the assumption that at its deepest level the psyche remains sympathetically bonded to the Earth that mothered us into existence."[204] Ultimately, Bant's character embodies the notion that to hurt the earth is really to hurt ourselves. The relationships between characters and land in Pancake's novel eventually coalesce into a diffracted worldview evoking the renewal of the Sacred Hoop envisioned by Cherokee spiritual leader Dhyani Ywahoo: "Your consciousness is not just in your body. It is in everything. Everything is related. The mountain, too, is your body, so all the better to treat it with respect. / As long as you are walking upon the Earth you are like a child in the womb, being fed by this Earth."[205]

As the decapitating of her home mountains gains ground, Bant starts displaying symptoms of the dissociation, or "psychic numbing,"[206] which individuals can experience as a defense mechanism against the trauma of being brutally severed from their earthly attachments:

> The memory picture of Yellowroot faded fast. And the feeling it left behind scared me worse than the mine site did. Because what I was feeling again was nothing. The distance between me and the land had set in, complete, but this time, I didn't even have any want in me to cross it. Nothing. Just like you couldn't measure the site cause it was nothing, you couldn't feel for it either, because there was nothing to feel for. Nothing stirs nothing. And it came to me for the first time: was it worse to lose the mountain or to lose the feeling that you had for it?[207]

Bant later exhibits an exponential incapacity to feel, signaling the psychic amputation that comes from being forcefully uprooted from one's homeland: "Now Yellowroot was completely gone. I'd figured that's how it would be. But for some reason, I hadn't let myself think how if Yellowroot was finished, they'd been taking Cherryboy next. And they were. They'd sheared part of it away to leave a tall flat naked face. I looked at it. I looked at the dead man-made cliff, and at first I felt nothing for that cut on Cherryboy. Then I felt nothing except I had to get in the woods."[208] Throughout, Bant fights off the feeling of dislocation that comes from being violently unmoored from one's land. She and other characters thus embody the "solastalgia" concept proposed by Glenn Albrecht, that is, "the emplaced and existential melancholia produced by the lived experience of negatively perceived transformation of a loved home environment."[209] In his more recent book, *Earth Emotions*, Albrecht further defines

> "solastalgia" as the pain or distress caused by ongoing loss of solace and the sense of desolation connected to the present state of one's home and territory. It is the existential and lived experience of negative environmental change, manifest as an attack on one's sense of place. It is characteristically a chronic condition, tied to the gradual erosion of identity created by the sense of belonging to a particular loved place and a feeling of distress, or psychological desolation, about its unwanted transformation. In direct contrast to the dislocated spatial dimensions of traditionally defined nostalgia, solastalgia is the homesickness you have when you are still located within your home environment.[210]

On the one hand, Pancake provides insight into the mechanisms that, according to ecopsychologist Joanna Macy, may lead one experiencing anxiety, a feeling of helplessness, and ecological despair to repress one's overwhelming sense of loss and fear by practicing psychic numbing. Thus, after a second slurry impoundment breaks in nearby Kentucky, Bant strives to protect herself from solastalgia: "Yeah, the impoundment bust scared me, scared me bad, but worse, it made me even more helpless than before. And from helplessness, I had learned, what a short step it was to I don't care. How else could you grow up, how could you walk around in your body every day, unless you learned not to care."[211]

On the other hand, inspired by her mother's example, Bant's development also illumines the way grief can otherwise be acknowledged, the better to reclaim a sense of earthly interconnection leading to healing, compassion, and active care.[212] Through her rough initiation, Bant indeed learns the hard way how differently people might respond to despair—either hitting the rock bottom of disenchantment, or, potentially, dedicating oneself to the reenchantment and restoration of one's world. Bant witnesses how some will act anesthetized and paralyzed, choosing to give up and withdraw their energies altogether—her father opts for quitting and moves to the city with her remaining siblings—while others, like her mother and herself, will both acknowledge the vastness of their losses and work on rebuilding a sense of connection through their human and biotic communities. Hence Bant's impulse to go shake off her numbing by running in the woods to reconnect with her ill-treated mountain: "A feeling closer to the trees all around. I took off again, really running this time, the curve and dip of the ground echo-shaping the curve and shape of my body."[213] Described in terms evoking echolocation, Bant's running is bat-like. The movement of her body entangles itself with the forms of nature, coproducing a kind of resilient, multispecies choreography. Pancake here corroborates Merleau-Ponty's revolutionary contribution to our understanding of the ongoing dance between humans and their environment which David Abram sums up as follows: "The sensing body is not a programmed machine but an active and open form, continually improvising its relation to things and to the world."[214]

Like Chapman-Crane's "Agony of Gaia," Pancake's novel creatively helps us channel the negative feelings triggered by vast environmental destruction so that, rather than repressing them altogether, we may in fact welcome them as manifestations of our interconnectedness with all forms of life. Thus, one of Lace's fellow activists confides his profound sense of loss: "The hardest thing of all about living through this, hasn't been the blasting or the dust or the flooding or the fires or how they broke the community. It's looking up there each morning, at a landscape you had around you every day of your life. And seeing your horizon gone."[215] Meanwhile, grounding herself, Lace reaches down for the power concentrated in the mountain, reclaiming it as a source of reempowerment against despair which fuels her determination to keep up the fight:

> Maybe it was something about the mountains' layers. Something about everything layered in them dead. All that once-lived stuff, strange animals and plants, giant ferns and ancient trees, trapped down there for 250 million years, captured, crushed, and hard-squeezed into—power. That secret power underground, that

> sleepy force lying all around, contagious somehow, catching, setting off the power pulls on top, the trickery and thievery, that violence and the loss, the way power will fight for power. The power under here, I told myself, if it can cause all that, it must also put a hold on us. Not greed for coal, not that kind of hold, we'd never got the profits from that. No. But just the pull, the draw, of so much power in the ground, and the kind of hold that makes.[216]

Although less esoteric, Lace's tapping of inner strength from the land evokes the "grounding" of Starhawk's shamanic characters, whereby they channel the energies of the earth. Besides, the overall attitude adopted by Pancake's heroines promotes a similar, active recognition of one's negative emotions, as if adhering to Starhawk's mantra, which forms a leitmotif in both her fiction and nonfiction: "Where there is fear, there is power." As Pancake's novel suggests, the flip side of working through, rather than against, fear and grief is one's regained capacity for interconnection, for wonder, and for hopeful, creative resistance. Against disenchantment and solastalgia, Pancake devises an *ecopoiesis* of reenchantment that follows in the tracks of many ecofeminists before her engaged in "dreaming the dark," as Starhawk once put it, which consists in facing and transforming the dark: "We can know the dark, and dream it into a new image."[217]

Following in her mother's tracks, Bant joins the local group organizing resistance against MTR. In the end, the adolescent's story reads as an initiation into today's complex world, the injustices of which her family must endure. Despite the formidable amount of loss and sorrow, Bant's story nevertheless gestures toward resilience. In many ways, her development parallels what happens to the land. A biological symptom of her puberty, the acne that craters out her facial skin provides an analogy with the disfigurement of the land, mirroring the inverted "moonscape" of the blasted, hacked-out mountain behind her home.[218] Moreover, Bant loses her virginity to an abusive miner from Ohio. An outsider to the state that he plunders for a living, the man embodies the reckless industrial pillage that is often hypocritically rationalized as long as it does not take place in one's backyard. Drawing a parallel between land abuse and women abuse, their first sexual encounter—one that the older, manipulative miner pushes onto Bant—is described in terms evoking rape, rather than mutual desire and consent:

> His finger first, a hurt both dull and sharp. Raggedy nail edge. [. . .] We were undressed only where it mattered, his jeans peeled off his hips, no underwear, my shorts dangling off one ankle. We wore our shirts and we wore our shoes. Me forced up the passenger door, the window handle in my back, the armrest, too, and he felt like a muscle between my legs. The sweat helped, the slickness there. The slickness, the glare of the cab, and a pounding that had nothing to do with any way I'd ever moved or'd want to.[219]

As it is, the man's rough handling of the young girl is not without evoking the Anthrop-o(bs)cene pounding, digging, and drilling of the mountain by the dragline that he operates: "That dull pounding after the hurt, my back against the door, my spine crunched up, jammed again and again."[220]

And yet, despite the child's brutal loss of innocence and the many ways she and her family are victimized, Bant and her mother turn to activism all the same. They refuse to be moved, silenced, or crushed by intimidation or any other demonstration of (mostly white male industrial) power. Rather than letting despair disempower them, the two women stick together and to their land. They constructively turn their fear and rage into organized political action to defend themselves and their beloved mountains. Hence Bant's epiphany as she gains awareness, late in the story, of "what Lace already knew. What Grandma knew. What a lot of people around here had to have known for a very long time"—that "[in] times like these, you have to grow big enough inside to hold both the loss and the hope."[221] Tenaciously struggling and speaking up against the injustices imposed upon them, Bant and her strong-headed mother still stand for hope in the end, as beyond grief, they go on cultivating their resilient naturecultures and their ways of dwelling within an enchanting *oikos*.

Looking toward potential healing in the future, Bant's mother tries to contemplate past the present season in her daughter's life, envisioning a time to come when she has grown out of adolescence, when her skin has healed and scarred. As Lace ponders that it "is a luxury to heal,"[222]—it is inferred she might be thinking of those relatives and friends of hers dying of cancers caused by environmental pollution, and, obviously, of her own child loss, which shattered her family—one might consider, by comparison, the potential future restoration of the mountains, living on a different geological timescale. Scientists estimate that if creek beds might purify faster, restoration of a healthy biodiversity in those 450-million-year-old mountains may take between 300 and 500 years. Moreover, the adamant voices and the irredeemable losses in the novel make it clear that certain extinctions are forever irreversible, whether that of a child or a species, and that everybody's backyard is in fact enmeshed within the "planetary garden,"[223] making it even more urgent to take full responsibility and action in the face of the ongoing, multifaceted ecocide. By enjoining us to take part in the song of the earth—an aspect of her prose that I will get back to in Part Three—, Pancake's narrative proposes ecofeminist reenchantment as an antidote to Anthropo-o(bs)cene disenchantment.

While redeeming the rural culture, language, and worldview of minoritized Appalachians, Pancake's ecopoetic novel demonstrates how grounded in place the evolution of humans and their languages is. With its polyphonic narrative strategy, its ecofeminist poetry and stance, and with its incursion into liminal realism which I will shortly tackle, Pancake's novel induces a radical paradigm shift that impresses the protagonists' fictional experiences of the living world onto the reader's embodied consciousness. Confronting us with the agony of Gaia in the Anthrop-o(bs)cene, the novel succeeds in bringing us to "think like a mountain," as Aldo Leopold famously encouraged, that is in a biocentric way, as well as, crucially, to *feel for* the mountains. Moreover, the rhapsodic quality of Bant's narrative turns Pancake's ecopoetic prose into a song of resistance relaying the voices of the earth. At the end of this chapter dealing with the different ways ecopoets may restore the many faces of Gaia, we might call upon Robin Wall Kimmerer's concept of "metaphoray": "When botanists go walking the forests and fields looking for plants," Kimmerer explains, we say we are going on a *foray*. When writers do the same, we should call it a *metaphoray*, and the land is rich in both."[224] Giving her pun a slight tweak, it might be said that just as humans have always foraged the earth for healthy sources of food, so have we always *metaphoraged* for a wholesome, ecopoetic language that might truthfully express and encourage our dwelling in symbiosis with the earth.

NOTES

1. Faris, *Ordinary*, 65. In my second part, I go over various definitions of magical realism, the better to situate my own concept of "liminal realism."

2. Ibid, 64.

3. While my second part is entirely devoted to the definition of "liminal realism" and the demonstration of how the mode operates in my corpus, my third part starts with a chapter on the art of anamorphosis. Readers can turn to this chapter for a precise definition of the concept, clarification of how it is tied to metaphoric thinking, and central to an ecopoetics of reenchantment.

4. For a reading of this novel as climate fiction, see Claire Perrin's PhD thesis on drought and climate change fiction. As co-advisor for her PhD, I insisted that Claire include this ecofeminist novel in her cli-fi corpus, which was otherwise overwhelmingly dystopic. For a plot summary and comprehensive study of *The Fifth Sacred Thing* as ecofeminist utopia, readers can also turn to Neşe Şenel's master's thesis.

5. Starhawk, *Fifth*, 13–14.

6. Taylor, *Dark*, 10.

7. Ibid, 16.

8. Ibid, 14–16.

9. Plaskow and Christ, "Naming," 95.

10. See Susan Griffin's seminal contribution, *Woman and Nature: The Roaring Inside Her,* 1978.

11. Plaskow and Christ, "Naming," 95.

12. Starhawk, *Dreaming*, 10.

13. Ibid, 11.
14. "The evidence of our senses and our experience is evidence of the divine—the moving energy that unites all being." Ibid, 10.
15. Ibid, 11.
16. Starhawk, *Fifth*, 18.
17. Ibid, 11.
18. Ibid, 14–15.
19. Ibid, 16–17.
20. Ibid, 18.
21. Margulis, *Symbiotic*, 117–18.
22. Lovelock, *Gaia*, quoted in Roszak, "Psyche," 13.
23. Margulis, *Symbiotic*, 118.
24. Powers, *Overstory*.
25. Margulis, *Symbiotic*, 2.
26. Ibid, 3.
27. Ibid, 4.
28. Ibid, 120.
29. Ibid, 128.
30. Ibid, 118.
31. Ibid, 119.
32. Ibid, 119.
33. Ibid, 120.
34. Ibid, 120.
35. Latour, *Gaïa*, 115.
36. Powers, *Overstory*, 453.
37. Ibid, 142.
38. Ibid, 455.
39. I have situated those concepts in my introduction. See Adamson, *Middle*, "Avatar"; and Adamson and Monani, "Cosmovisions."
40. Margulis, *Symbiotic*, 126.
41. Ibid, 120.
42. Powers, *Overstory*, 220.
43. Ibid, 221.
44. Ibid, 221.
45. "Yet, in Gaia, we have a hypothesis that, for all its mathematical precision, seems bent on smuggling a barely disguised version of the *anima mundi* back into polite scientific society. [. . .] In a typical response, one critic characterized Gaia as "pseudo-scientific myth-making,' an 'almost medieval' idea that rings of 'obscurantism, wishful thinking and mysticism.' The charge is largely valid. The hypothesis does echo elements of mysticism and medieval thought." Roszak, *Voice*, 146.
46. Ibid, 146–47.
47. Ibid, 148. Roszak unpacks the anthropomorphic metaphors that invariably crop up even in some of the most systems-oriented approaches to science such as biology and evolutionary genetics. He also emphasizes that the mere study of nature as a

"mechanism" hinges on metaphoric thinking and, in fact, implies teleology and the existence of a maker, 148–52.

48. Margulis and Sagan, 117.

49. Powers, *Overstory*, 142.

50. Ibid, 142.

51. Margulis attributes credit for this phrase to Greg Hinkle, a biology professor at the University of Massachusetts, *Symbiotic*, 2. As I will rebound on later, some ecopoets such as Proulx bank on Lord Alfred Tennyson's vision of a "nature red in tooth and claw," turning Gaia into a reckless, itchy symbiont.

52. Tudge, *Secret*, 67.

53. Eisler, "Gaia," 34.

54. Eisler, "Gaia," 34. It is interesting to note in that respect that where ecofeminists insist that humans are *of* Gaia, and cannot be apprehended disjointly from her, white male philosophers who are perfectly cognizant of the symbiosis entailed by Gaia will still tend to promote images of humanity *at war* with Gaia. Precision must be made that such militarist imagery can only make sense from the point of view of a humanist culture whose interests may indeed compete with those of Gaia. In a similar vein, Latour's title "Face a Gaia" (in English, "Faced with Gaia") problematically antagonizes humans and Gaia, pitting them face to face as planetary opponents—unless Latour possibly means to confront us with the representation of Gaia which, when apprehended from the right perspective, forces us to face the fact that we are but a contingent part of her.

55. Ibid, 34.

56. For more on entanglement tropes in Kingsolver's *Prodigal Summer*, see Meillon, "Measured Chaos,"

57. See Meillon, "Homeland."

58. Kingsolver, *Animal*, 129.

59. Recuperating Donna Haraway's "natureculture," I prefer eliding the e at the end of "nature," merging the two original words into a more likely English neologism.

60. Ibid, 129.

61. Ibid, 211.

62. Ibid, 227.

63. Ibid, 227.

64. Ibid, 235.

65. Ibid, 240.

66. Ibid, 239.

67. Ibid, 240.

68. White Jr. "Historical."

69. Kingsolver, *Animal*, 240.

70. Kingsolver, *Animal*, 240. For interesting discussions on Kingsolver's deconstruction of the Western myth and recycling of indigenous topoi in *Animal Dreams*, see Naomi Jacobs's paper, "Anti-Western," Thesa Wrede's "Ecofeminist," and Robin Cohen's "Wild."

71. I am relying on Rob Nixon's *Slow Violence and the Environmentalism of the Poor*, where he studies how areas inhabited by economically and culturally

marginalized groups are often more exposed to environmental degradation and catastrophes. Kingsolver's story draws attention to the way minoritized communities made of rural or laboring people as well as Natives from various tribes have been getting displaced as part of colonial and industrial settlement, how dumping sites are more likely to be situated on Native land, how mine workers get exploited and risk early mortality caused by long-term exposure to contaminants, how people of modest means are perniciously encouraged to trade their water rights in return for money, and how the media and political institutions, including the EPA, can simply overlook or downplay the many factors of environmental injustices plaguing communities on the lower ends of the social and economic spectrum—a situation sardonically summed up by one of the local women: "Jimmy called the newspapers half a dozen times. [. . .] Nobody's interested in a dipshit little town like Grace. They could drop an atom bomb down on us here and it wouldn't make no news in the city. Unless it stirred up the weather over there and rained out a ball game or something," 162. See also pages 110–11, 161–62. The historiographic background of the novel includes a denunciation of the economic and political exploitation of Nicaragua by the U.S. government. As Cosima's sister, Hallie—a fictional avatar of Ben Linder, whom the novel is dedicated to—goes to Nicaragua to help poor rural populations oppressed by the Contras and help them learn efficient farming techniques, her epistolary exchanges with her sister reveal how, after DDT spraying got banned in the United States because of its recognized toxicity for the environment and human health, it was nevertheless still being produced and sold to Latin American countries: "Central America was becoming the toilet bowl of agricultural chemicals, she said, because of war-strained farming economies and dumping from the First World. In the seventies, when Nicaragua was run by U.S. Marines and Somoza, it was the world's number one consumer of DDT," 223.

72. See work by Andreas Malm or Donna Haraway, who have inspired Armel Campagne, Christophe Bonneuil and Jean-Baptiste Fressoz. All insist that it is not so much humankind that should be blamed for becoming the first geological force influencing the biosphere, but capitalism, which has fashioned the excessively destructive nature of anthropic activities.

73. Kingsolver, *Animal*, 239.

74. Ibid, 255.

75. This was the title of a paper written by French philosopher François Guéry, "Maîtres et protecteurs de la nature," as well as of a collection of essays coedited by Alain Roger and François Guéry.

76. Kingsolver, *Animal*, 332.

77. Ibid, 109–10.

78. Ibid, 110.

79. Barad, *Meeting*, 132–85.

80. Kingsolver, "Homeland," 2.

81. Merchant, *Death*, 3.

82. Kingsolver, *Animal*, 161.

83. Ibid, 161, emphasis added.

84. As the local population has organized to prove and contest the contamination of their river and soil by industrial dumping, rather than curtailing pollution, Black

Mountain has negotiated a deal to finance a dam diverting the poisoned river out of the inhabited Canyon, which could deprive the community of Grace from the water needed for their orchards.

85. I first used the word at a 2017 Quimper conference on magical realism and environmental arts, as I was discussing Annie Proulx's Wyoming short stories. Since then, the term has appeared in my writing in "Chant," and in a piece of creative writing ("Toutes"). It has been noted before that the term "Anthropocene" indiscriminately holds all humans responsible for the overexploitation and pollution of the earth, overlooking the very disparate economies and lifestyles proper to different societies and countries worldwide. Some have suggested to me that the right term might be "andro-o(bs)cene," to stress the patriarchal underpinning at stake. I shall stick to my initial coining because, even though women have obviously had less agency in the shaping of the patriarchal political, economic and philosophical systems of the West, and although women in such cultures have been treated in some of the same foul ways as the earth, it would seem rather unfair to exonerate women altogether from sharing responsibility in our crisis—especially today, when so many of us women are complicit in sticking to lifestyles and values that ought to be abandoned—and to lay the blame instead entirely on the masculine half of humanity.

86. Kingsolver, *Animal*, 3.

87. Fearing that the term "Anthropocene" might reinforce ecophobia, Simon Estok writes: "The much-vaunted term 'Anthropocene' starts to seem like yet another affirmation of the heroic (or antiheroic) human subject and of our obsession with ourselves. Indeed, we have to wonder about the hubris perhaps implied in the vey term 'Anthropocene': as Neimanis, Åsberg, and Hedrén suggest, 'calling an epoch after ourselves does not necessarily demonstrate the humility we may need to espouse,'" *Ecophobia*, 57.

88. See Merchant, *Death*, or Griffin, *Woman*.

89. It is believed that during the Neolithic, earth goddesses were replaced by the patriarchal gods of Indo-Europeans who invaded Greece during the second millennium BC. See Spretnak, "Ecofeminism," or Eisler "Ecofeminist."

90. I am aware that the postcolonial critique is not salient in this quotation. However, ecofeminist, postcolonial, and environmental justice issues are by and large intertwined throughout Kingsolver's writing. See Theda Wrede's paper "Ecofeminist." Readers interested in Kingsolver's postcolonial approach may also look at my paper dealing with her short story "Jump-Up Day," set in Saint Lucia, "Revisiting," or that on her Congo novel, *The Poisonwood Bible*, "Aimé."

91. Kingsolver, "Homeland," 2.

92. See Meillon, "Barbara," and "Literary."

93. Initially, the concept of resilience comes from physics, where it refers to the capacity of metal to resist outside pressure and return to its original state after having suffered distortion. The concept has since been applied in ecology to account for the long-term capacity of an ecosystem to restore itself after having undergone dire stress via a natural or industrial catastrophe. This capacity inherent in natural systems and nonhuman matter to resist damage has inspired psychologists in their studies of human resilience, that is, humans' capacity to recover from trauma, and, more specifically,

some children's capacity to grow up unhindered in their development despite trauma or growing up in extreme adverse conditions. The theory of human resilience was first imported from natural ecosystems by child psychologist Emmy Werner in the 1980s, as she was working with Hawaiian children. In France, it was popularized by psychiatrist and neurologist Boris Cyrulnik. See Cyrulnik, *Merveilleux*.

94. See my master's thesis, "Feminine."
95. Kingsolver, *Animal*, 341.
96. Ibid, 207.
97. Ibid, 342.
98. Ibid, 7.
99. Ibid, 303.
100. I have touched upon this intertextuality in my paper "Literary" and in my PhD. This is an inspiration that Barbara Kingsolver herself has confirmed in the final report she wrote as a member of my PhD Committee. Quite astonishingly, the "Ladder to the Moon" painting by Georgia O'Keeffe which, because it is coherent with the oxymoronic quality of Kingsolver's poetic short stories, I had chosen to illustrate the front cover of my PhD, turned out to be the inspirational painting Kingsolver has in her home study.
101. Kingsolver, *Poisonwood*, 5, 537.
102. Giono, *Joy*, 454.
103. Ibid, 458–59.
104. Ibid, 443–44.
105. In a similar spirit, the return of one's body to the earth is a common topos in Linda Hogan's fiction, where, after they die, some of her indigenous characters are offered to vultures and wolves entrusted with bringing them back into the life-death continuum. This is the case of Ruth's father in *People of the Whale*, and Angel's grandmother in *Solar Storms*.
106. Christ, *Diving*, 8.
107. French ecocritic Alain Romestaing has written at great length about the centrality of embodied experience and knowledge in Giono's writing. His thorough study *Jean Giono: Le corps à l'œuvre* includes many cogent pages approaching Giono in the light of Merleau-Ponty's phenomenology. Romestaing's analyses on the porosity of the body and how the latter allows for a kind of communion with the world beyond the self are particularly consonant with my readings exposed throughout this book (227–30). If my interpretation of Giono's *Joy* ties in with Romestaing's analysis of the "sensuous intuition" at work in Giono's writing, my reading nevertheless diverges from his as regards the meaning of Bobi's death, which I am reading in an ecofeminist light. Romestaing's insight into the horror of decomposing bodies in Giono's oeuvre as linked with Julia Kristeva's notion of the "abject" may be worth pursuing through an ecofeminist reading of the ecophobia that is potentially at stake here.
108. See the notes to Giono's unpublished last chapter, quoted in Romestaing, 261–62 (emphasis in the original, translation mine).
109. Noiray, "Utopie," 73.

110. As Bobi's desire keeps swiveling between Joséphine and Aurore, the character thinks to himself: "No [. . .], that will not come from women. Nothing can come from women but the desire of women. All will come from me." Giono, *Joy*, 173.

111. Ibid, 326.

112. Ibid, 327.

113. Ibid, 325.

114. Ibid, 331.

115. Ibid, 285. I have tweaked the translation slightly here to keep it closer to the original text.

116. Ibid, 347–48.

117. Ibid, 405.

118. See Buell, *Environmental*, 280–88. Buell observes that "in western culture, the order of nature has been variously imagined as, for example, an economy (from the Greek *oikos*, household), a chain or scale of being, a balance, a web, an organism, a mind, a flux, a machine," 280. He then zooms in on the web metaphor, which calls for a conversation that I will get back to shortly.

119. Kingsolver, *Prodigal*, 8–9, emphasis added.

120. Ibid, 444.

121. Darwin, *Origin*, 73. Quoted in Buell, *Environmental* (note 8, page 516).

122. Carson, *Silent*, 56.

123. Ibid, 189.

124. Kimmerer, *Braiding*, 20.

125. I am implicitly relying on Stacy Alaimo's definition of "trans-corporeality," which captures the way bodies are porous contact zones allowing for material exchanges between different bodies: "Imagining human corporeality as trans-corporeality, in which the human is always intermeshed with the more-than-human world, underlines the extent to which the substance of the human is ultimately inseparable from 'the environment.' [. . .] By emphasizing the movement across bodies, trans-coporeality reveals the interchanges and interconnections between various bodily natures" (*Bodily*, 2).

126. Kingsolver, *Prodigal*, 216.

127. Buell, *Environmental*, 280–84.

128. See Meillon, "L'implicite."

129. Kingsolver, *Animal*, 210.

130. Ibid, 210.

131. Allen, Sacred, 11.

132. Ibid, 14.

133. Silko, *Ceremony*, 1.

134. Ibid, 2.

135. Allen, *Sacred*, 98.

136. Silko, *Ceremony*, 35–36.

137. Allen, *Sacred*, 14–15.

138. Ibid, 14–15. In a way that is coherent with Starhawk's insistence on the many faces of the Goddess, Paula Gunn Allen argues: "Woman bears, that is true. She also destroys. That is true. She also wars and hexes and mends and breaks. She creates

the power of the seeds, and she plants them." Evoking how Spider Woman presides over all cycles of life and death, Allen quotes Laguna writer Anthony Purely: "'She is mother of us all, after Her, mother earth follows, in fertility, in holding, and taking us back to her breast," 14.

139. Swimme, "Lobotomy," 20.
140. Ibid, 21.
141. Ibid, 22.
142. Meillon, "Silent" and "Digesting." Readers interested in Proulx's debunking of the Frontier myth can turn to Mark Asquith's *The Lost Frontier*.
143. Ibid, 191.
144. The shocking picture of power abuse is reinforced by the rampant violence, both sexual and physical, suffered by women and gay men throughout Proulx's Wyoming stories. I have dealt at length with her feminist indictment of patriarchal norms and violence in my paper "Deconstructing."
145. Merchant, *Death*, 164.
146. Ibid, 165, 169.
147. Ibid, 169.
148. Bacon, quoted in Merchant, *Death*, 171.
149. Ibid, 171.
150. Merchant, *Death*, 171.
151. Ibid, 213. On "the myth of emptiness," see Wendy Harding's illuminating study by the same name. Although her corpus does not include Annie Proulx's fiction, her take applies to the debunking of the Frontier myth and ideology at work in these stories.
152. Proulx, *Close*, 22–23.
153. Ibid, 32.
154. Ibid, 35–37.
155. For more detail on this analysis, see Meillon, "Silent."
156. As I have shown in my paper "Silent," the teeth and claws of un unforgiving nature form a recurrent leitmotif throughout Proulx's stories. Via several postmodernist self-reflexive tricks, Proulx gestures to the implied demiurge writer retributing ecopoetic justice onto her arrogant, stupid, or violent characters.
157. Proulx, *Close*, 39–40.
158. Proulx, *Bad*, 111.
159. Serres, *Contrat*, 67, translation mine. See also 61, and 64–65, where Serres spells out his trope.
160. Stengers, "Gaïa," and *Résister*.
161. Proulx, *Close*, 99.
162. Ibid, 214.
163. Proulx, *Bad*,184, emphasis added.
164. See Estok, *Ecophobia*.
165. Morton, *Hyperobjects*, 164.
166. I am here quoting Bruno Latour again.

167. In 2016, it was estimated that the extent of MTR plundering in the Appalachian range amounted to over 2,700 ridge tops impacted in 2016. See https://www.skytruth.org/2016/06/skytruth-mountain-top-removal-analysis/.
168. Jones, "Pancake's."
169. Meillon, "Agony," and "Chant." Parts of these two papers have been reworked here.
170. See the interview by Robert Gipe, "Straddling," as well as Pancake's essay, "Creative."
171. Pancake, *Strange*, 347.
172. Ibid, 347.
173. Ibid, 164.
174. Roszak, *Voice*, 242.
175. Pancake, *Strange*, 341. While the perceived effeminacy of the one contrasts with the perceived masculinity of the other, Pancake deconstructs gender stereotypes. Uncle Mogey and Dane both prove on the high end of gentleness and sensitivity, making them more receptive to and aware of their surroundings.
176. Ibid, 341–42.
177. Roszak, *Voice*, 242.
178. See Marx, *Machine*.
179. Pancake, *Strange*, 266.
180. Ibid, 16.
181. Ibid, 19–20, emphasis added.
182. Ibid, 19.
183. Ibid, 19.
184. Ibid, 19.
185. Ibid, 4.
186. Ibid, 99.
187. Ibid, 34.
188. Kimmerer, *Braiding*, 236.
189. Haraway, *When*, 17.
190. Pancake, *Strange*, 34–35.
191. Ibid, 35.
192. Ibid, 35.
193. Haraway, *When*, 16–17. For a definition of Haraway's "companion species," see also pp 64–65.
194. Pancake, *Strange*, 103.
195. Ibid, 165.
196. Ibid, 165.
197. Ibid, 165–66.
198. Ibid, 165–66.
199. Ryden, *Mapping*, 42–43.
200. Ibid, 40.
201. Quoted in Ryden, 40.
202. Conn, "Who Responds?" 161.

203. Ibid, 161. Reversing the understanding of what may or may not be a pathological relationship with other-than-human nature, Sara Conn sums up the process of dissociation from the natural world which, according to ecopsychologists—in the wake of seminal work carried out by Roszak and Joanna Macy—may account for the high rate of depression in our Western societies: "The same process of numbing and constriction occurs with our loss of connection to a sense of place in a viable, thriving ecosystem. Many of us have learned to walk, breathe, look, and listen less, to numb our senses to both the pain and the beauty of the natural world, living so-called personal lives, suffering in what we feel are 'merely personal' ways, keeping our grief even from ourselves. Feeling empty, we then project our feelings onto others, or engage in compulsive, unsatisfactory activities that neither nourish us nor contribute to the healing of the larger context. Perhaps the high incidence of depression is in part a signal of our bleeding at the roots, being cut off from the natural world, no longer as able to cry at its pain or to thrill at its beauty" (171).

204. Roszak, "Psyche," 5.

205. Ywahoo, "Renewing," 274–75.

206. First coined by Robert Jay Lifton, the term was expanded upon from the standpoint of ecocriticism by Scott Slovic (*Going* 143–163). Laura Sewall defines "psychic numbing" as "a psychological defense against witnessing the world's pain. It is a form of denial that shields us from fully experiencing the latest reports on ozone depletion, increasing pollution, toxicity, poverty, illness, and the death of species. Full awareness hurts. In response we build defenses, [. . .] we turn down the volume. We become numb to our feelings, to what we might hear and see" ("Ecological" 202). See also Scott Slovic and Paul Slovic's *Numbers and Nerves*.

207. Pancake, *Strange*, 167.

208. Ibid, 354.

209. Albrecht, "Psychoterratic," 249.

210. Albrecht, *Earth*, 38–39.

211. Pancake, *Strange*, 346.

212. See Joanna Macy's paper, "Working Through Environmental Despair," where she elaborates on the crucial ways in which environmental despair manifests our fundamental interconnectedness. Macy moreover diffracts the tropes of the Earth as matrix, as a mother, and as a web. In one and the same sentence, she writes: "The web of life both cradles us and calls us to weave it further," 255.

213. Pancake, *Strange*, 37.

214. Abram, *Spell*, 49.

215. Pancake, *Strange*, 309.

216. Ibid, 312.

217. Starhawk, *Dreaming*, xvi.

218. "*Moonscape*, that's what many said after they'd seen it, but I saw right away this was something different. [. . . This] was the moon upside down. A flake of the moon's surface fallen to the earth, and in that fall, it had kept its color, nickel, and beige, kept its craters, its cracks. But then it landed not up, but moonside down." Pancake, *Strange*, 165.

219. Ibid, 324.

220. Ibid, 324.
221. Ibid, 357.
222. Ibid, 333.
223. I am taking up Gilles Clément's concept of the earth as planetary garden. See Clément, *Jardin*.
224. Kimmerer, *Braiding*, 46.

Chapter Four

Sowing the Seeds of an Ecopoet(h)ics of Wonder and Enchantment

Reincorporating Language and the Human into the Flesh and Song of the World

Rachel Carson's 1962 book *Silent Spring* has become one of the most often quoted books in the arena of nature writing, both as a precursor of postapocalyptic fiction,[1] and for its powerful poetic appeal mingled with scientific data. Its ecopoet(h)ics is crystallized in the title of the book and in its allegorical opening, "A Fable for Tomorrow." Imagining "a town in the heart of America" suddenly hushed with the dying off of chickens, cattle, sheep, fish, streams, and humans, the incipit of Carson's study on the effects of pesticides over an entire ecosystem stages the sickening of birds, the disappearance of pollinator bees, and consequently, unfertilized apple blossoms unable to bear fruit. Carson's poetic vision of a bleak future allows us to unpack the complexity of her metonymical "silent spring": "It was a spring without voices. On the mornings that had once throbbed with the dawn chorus of robins, catbirds, doves, jays, wrens, and scores of other bird voices there was no sound: only silence lay over the fields and woods and marshes."[2]

Much has been written about the affective and political power of Carson's book. Ringing an alarm as to the impact of DDT spraying, it has influenced legislation and triggered the creation of the Environmental Protection Agency in the United States of America. In the wake of Scott Slovic's illuminating comments on Carson's use of a rhetoric relying alternately on rhapsody and jeremiad,[3] what interests me most here is Carson's underlying ecopoetics of reenchantment, and how the latter is tied to her finetuning of her own writing to the song of the world. Her "silent spring" relies on a hypallage—spring being neither a person nor a thing, it is not spring itself that has gone silent, but the moribund birds and vanished insects that are metonymically associated with the season of natural rebirth and growth. In this unprecedented spring—quite the reverse of the Covid spring that I opened this book with—birds and insects no longer sing and drone, making for a greatly impoverished soundscape. And yet, from the start, this bleak, disenchanting picture is nevertheless discretely imbued with a sense of ecopoetic resistance. The voice of the nearly silenced spring may indeed be heard in the alliteration in [s], in addition to the paronomasia between "spring" and "sing" that is enfolded in the poetic title, "Silent Spring." Coupling a sense of hissing with a sense of singing, Carson's layered poetic title thus resonates with both a sense of urgency—of alarm—and with a sense of resistance via poetic echoes of the endangered song of the earth. Moreover, with its heady rhapsodic texture, Carson's title also diffracts the polysemy of the word "spring"—a season, but also a source of water and life, and, potentially, the capacity to rebound, tied to the notion of resilience.

From the threshold of Carson's book then, the title "Silent Spring" does not so much *speak* as *chant* volumes as to the political power of literature—itself a source of nurture and life that may call us into awareness and thus help sustain the resilience of ecosystems. Resisting the muting of other-than-human voices and interests, this powerful formula relays the voices of the earth. It enacts and draws attention to what Scott Knickerbocker calls "sensuous poesis"—a way of "weaving world to word" that "rematerializes language, returns it from 'speech' back to 'sound.'"[4] Like the poets Knickerbocker has studied and the prose ecopoets I will be looking at here, Carson employs sensuous *ecopoiesis*. With her "silent spring," she "[rematerializes] language specifically as a response to nonhuman nature" in a way that "unapologetically embrace[s] artifice—not for its own sake, but as a way to relate meaningfully to the natural world."[5]

In her book *The Sense of Wonder*—an ode to nature that is decidedly on the side of the rhapsodic and of en-*chant*-ment—Carson writes: "A child's world is fresh and new and beautiful, full of wonder and excitement. It is our misfortune that for most of us that clear-eyed vision, that true instinct for what is beautiful and awe-inspiring, is dimmed and even lost before we reach adulthood." The best gift she can think of for a child is "a sense of wonder so indestructible that it would last throughout life, as an unfailing antidote against the boredom and disenchantments of later years, the sterile preoccupation with things that are artificial, the alienation from the sources of our strength."[6] In her 2001 book *The Enchantment of Modern Life*, Jane Bennett makes the case for a form of enchantment that "entails a state of wonder." These are experienced as moments of "pure presence" and "moments of joy" that, Bennett argues, may move humans to ethical generosity.[7] "To be enchanted," writes Bennett, "is to be struck and shaken by the extraordinary that lives amid the familiar and the everyday." "Starting from the assumption that the world has become neither inert nor devoid of surprise but continues to inspire deep and powerful attachments," Bennett explores how "life provokes moments of joy," and how "that joy can propel ethics."[8]

This perspective may shed light on the fact that many ecopoets, whether writing in prose or in poetry, weave into their stories moments of epiphany. These are moments when the encounter, or merging, with the more-than-human world is permeated with a heightened sense of wonder, grace, humility, and bliss. This touches upon a quintessential aspect of an ecopoet(h)ics of reenchantment. Let us for now turn to Scott Knickerbocker's monograph on ecopoetics, where he deems that "the greatest capacity for wonder and gratitude [. . .] are the first requisites of any enduring environmental ethic."[9] For Knickerbocker, wonder is simply "the most important quality of ecocentrism."[10]

If the feeling of enchantment incorporates that of wonder, my take on ecopoetic reenchantment is however tied further to the notion of incantation. The etymology of the word enchantment, from *incantare* in Latin, potentially refers to the practice of magic and ritual, to the action of casting a spell with sounds, of making spellbound through song, of arousing wonder through chanting. Or, possibly, it may mean to sing along with the earth, to sing our human selves back into the earth. As a matter of fact, as Berendt notes, the Huichol Indians in Mexico refer to the magician, or shaman, as the "cantor." Sustaining my reading of *ecopoiesis* as a form of consciousness-changing magic that musically reincorporates us into the sensitive earth, it turns out that "the words for 'poet,' 'singer,' and 'magician' go back to the same linguistic root not only in Latin but also in many other languages."[11]

In her phenomenological approach to enchantment, Bennett underscores how enchantment "requires active engagement with objects of sensuous experience." It is "a state of interactive fascination" involving the bodymind:[12]

> Thoughts, but also limbs [. . .] are brought to rest, even as the senses continue to operate, indeed, in high gear. You notice new colors, discern details previously ignored, hear extraordinary sounds, as familiar landscapes of sense sharpen and intensify. The world comes alive as a collection of singularities. Enchantment includes, then, a condition of exhilaration or acute sensory activity. To be simultaneously transfixed in wonder and transported by sense, to be both caught up and carried away—enchantment is marked by this odd combination of somatic effects.[13]

Activating sensuous *poiesis*, the prose *ecopoiesis* I am looking at both relays and triggers enchantment. Grounding my work in Merleau-Ponty's grasp of the "flesh of the world" wherein humans and nonhumans are inextricably co-related, interwoven, I argue that *ecopoiesis* possesses the power to reenchant the world as it calls attention to what it best performs, that is an aural and sensuous embodiment and continuation of the world it *articulates with*. To take up ecopoet Charles Bernstein's wording: "Sound is language's flesh [. . .]. We sing the body of language, relishing the vowels and consonants in every possible sequence."[14]

Merleau-Ponty's concept of a "flesh of the world" is grounded in the interlacing of perception and world, of the perceiver and the perceived. Much of his work rests on sight and touch: "the flesh I am speaking of is not matter. It is the coiling of the visible around the seeing body, of the tangible around the touching body, which is attested in particular when the body sees itself, touches itself as it is seeing and touching things. [. . .] Flesh—whether that of the world or mine—is not contingence, chaos, but texture—a texture that folds back into itself and suits itself.[15] In *La Nature*, Merleau-Ponty writes that "[perceived objects are] the correlations of an embodied subject, replicas of its movement and its sensing, embedded in its internal circuitry; perceived objects are of the same stuff as the perceiving body: the sensitive is the flesh of the world, i.e., sense within the outside. The flesh of the body makes us comprehend the flesh of the world."[16] Let us note the resonance with Karen Barad's take on phenomena as entanglements between perceiving subject and perceived object: "We don't obtain knowledge by standing outside the world; we know because we are of the world. [. . .] The separation of epistemology from ontology is a reverberation of a metaphysics that assumes an inherent difference between human and nonhuman, subject and object, mind and body, matter and discourse."[17]

To Merleau-Ponty's notions of "co-perception" and of a "movement-perception interlacing"[18] may correspond Barad's "ontological entanglements," the "ongoing intra-activity" always enfolding us into the world, and I would add, always enfolding words within worlds.[19] Part of this intra-activity, I argue, involves our capacity to respond to the soundscapes of the world—the biophony, geophony and anthropophony that take part in its "flesh." For the flesh of the world is also the coiling of the audible around and through the hearing, sensing, and rhythmic body. And the flesh of the body makes us comprehend the flesh and singing of the world, itself intermeshed with the flesh and singing of language. By making us appreciate more fully the sensuous quality of the "flesh of language," to go back to Bernstein, ecopoetics reincorporates language into the bodily locus of sensuous cobecoming involving world, word, and self. Sound and rhythm thus allow us to "sing the body of language," and further, to sing ourselves into the flesh of the world. Ecopoets, therefore, play a precious role as they are the mediators of the en-*chanting* textures of both world and words.

Ecopoiesis moreover enriches the quality and kinds of experiences a reader is capable of by providing extraordinary ways of thinking and perceiving, knowing and feeling, including animal, vegetal, mineral, and elemental ones. One of my central hypotheses is that by reviving all of our senses, *ecopoiesis* might help us correct both the logocentrism, or "frontal lobotomy" as formulated by Swimme, and the "ocular hypertrophy" diagnosed by Joachim Berendt.[20] In his defense of the emergence of a New Consciousness, Berendt invites us to heed the world as sound. He contemplates that there may arise a "consciousness of hearing people"—one where "our ears will have priority over our eyes," where "the audible, the sound, will be more important than the visible." To a great extent, Berendt attributes the disenchantment of the modern world to the hegemony of sight in modernity and to the neglect of what manifests itself through the vibrations of living matter, which our ears can pick up as sound, noise, or music. "Human beings," Berendt laments, "with their disproportionate emphasis on seeing have brought on the excess of rationality, of analysis and abstraction, whose breakdown we are now witnessing."[21] While "[the] field of the visible is the surface," Berendt points out, our ears "immerse themselves deeply into the spheres they investigate by hearing."[22]

In a similar vein, bioacoustician Bernie Krause remarks on how the acoustic world eludes most of us, appearing as an invisible, formless, unbounded, and intangible entity. Hearing has become a "ghost sense," eclipsed as it is by sight.[23] This, as Murray Schafer signals, is true of most Westerners, but not of all cultures. Schafer gives the example of rural African peoples for whom hearing is the most reliable sense.[24] This no doubt applies to many indigenous cultures still predominantly dependent on orality. As David Abram has observed, "[the] sense of inhabiting an articulate landscape—of dwelling within a community of expressive presences that are also attentive, and listening, to the meanings that move between them—is common to indigenous, oral peoples on every continent."[25] Moreover, scientific studies and histories of the senses have revealed that the human ear does possess much more potential than most of us Westerners are cognizant of. In our modern cultures we have indeed come to rely principally on sight to the detriment of hearing and smell—senses we have come to shun as they have been attributed to the 'lower' instincts long deprecated as animal senses. And yet, it turns out, human subjects are nevertheless equipped with the potential to detect a much wider range of acoustic and olfactory signals than most people are even conscious of.[26]

According to neuroscientists and historians, our sensory organs, olfactory and auditory neurons and pathways can—and often do—pick up much more data than we think. Our ignorance can be explained by our lack of training, our education not having attuned us enough to the rich soundscapes and smellscapes we bathe in. This paucity is both reflected in and accentuated by the dearth of words in our vocabularies to even distinguish, identify, and name the great palette of sounds and smells that the world creatively puts out. And yet, going back to the aural texture of the world that too often eludes us, the fact remains that acoustic signals reach us from all directions at once, both horizontally and vertically. Bernie Krause describes a soundscape as a kind of acoustic, 3-D dome constituted by all the sounds emanating from geological forces and elements ("geophony"), those produced by the living organisms from a specific biome ("biophony"), and those produced by humans ("anthropophony").[27] One of the hypotheses this book seeks to test is whether some sort of ecopoetic reenchantment may potentially take place through a translation of those soundscapes, a translation that directs our attention and sensitivity to the very material and concrete, polyphonic song of the earth. Should that be the case, a subsidiary question is then concerned with assessing the role of literature from the point of view of attention restoration theory (ART).[28]

Since humankind, together with its languages and cultures, has evolved through direct contact with and adaptation to various environments, it makes sense that our very languages be shaped in part by attentive practices of echovocalizations and echolocation.[29] "What if the very language we now speak arose first in response to an animate, expressive world—as a stuttering reply not just to others of our species but to an enigmatic cosmos that already *spoke* to us in a myriad of tongues?" asks ecophenomenologist David Abram.[30] Arguing that we humans would benefit from "[tuning] our animal senses to the sensible terrain," Abram calls for a rehabilitation of our sensitive intelligence and vitality, whether smell, hearing, touch, and so on.[31]

Abram's view of language is at heart ecopoetic: "To our indigenous ancestors, and to the many aboriginal peoples that still hold fast to their oral traditions, language is less a human possession than it is a property of the animate earth itself, an expressive, telluric power in which, we, along with the coyotes and the crickets, all participate."[32] Abram then tells the story of a man in the Pacific Northwest whose ears were so attuned to "the different *dialects* of the trees" that he could identify different species while standing blindfolded inside the forest he was intimate with.[33] Following this line of thought, this book strives to verify whether contemporary ecopoets may be working to attune us to the voices and songs of the earth that many of us have become deaf to. In that sense, my work follows in the steps of Abram's claim that "the power of language remains, first and foremost, a way of singing oneself into contact with others and with the cosmos. [. . .] Whether sounded on the tongue, printed on the page, or shimmering on the screen, language's primary gift is not to *re*-present the world around us, but to call ourselves into the vital presence of that world—and into deep and attentive presence with one another."[34]

Delving into the *ecopoiesis* in my corpus, I wish to assess to what extent literature might reempower language with "the primordial power of utterance to make our bodies resonate with one another and with the other rhythms that surround us."[35] This is something that Abram himself performs in his highly sensuous, ecopoetic and self-reflexive prose, that in many ways resembles Jean Giono's in *The Song of the World* or in *Joy of Man's Desiring*:

> This tonal layer of meaning—the stratum of spontaneous, bodily expression [. . . is] the very dimension of language that we two-legged share in common with other animals. We share it, as well, with the mutter and moan of the wind through the winter branches outside my studio. In the spring the buds on those branches will unfurl new leaves, and by summer the wind will speak with a thousand green tongues as it rushes through those same trees, releasing a chorus of rustles and whispers and loudly swelling rattles very different from the low, plaintive sighs of winter. And all those chattering leaves will feed my thoughts as I sit by the open door, next summer, scribbling and pondering.[36]

As I will argue more in depth throughout this book, taking up Julia Kristeva's concepts of "the symbolic" and "the semiotic," ecopoetics may provide insight into how the *chora* of the Earth resonates within the semiotic, kinetic dimension of human languages.[37] Manifesting a pre-symbolic, deep connection with the earth, the poetic languages with which we humans consecrate our dwelling within the *oikos* might indeed offer ways of staying connected to the rhythms, the voices, and flesh of Mother Earth.[38]

In her psychoanalytical approach to linguistics, Kristeva exposes the existence of a "thetic phase"—the crucial moment when the subject breaks away from the mother's body and self, and leaves the semiotic order to enter the patriarchal symbolic order. Simultaneously, the child assimilating the symbolic order acquires the notion of a language where the signifier has been severed from the signified. In other words, it is the moment when a child is initiated into a patriarchal system of dwelling *on* Earth. As we have seen, this system is supported by an ideology that is disenchanted, that is, dualistic, mechanistic, reductionist, individualistic, anthropocentric, and logocentric— one might say here, in the wake of Jacques Derrida, "phallogocentric." It follows that by reincorporating us into the flesh and song of the world, *ecopoiesis* might provide a way to resist and reverse the phallogocentric disenchantment operated via the symbolic. As a way of holding on to and honoring our primordial psychic and bodily connections *within* the earth, *ecopoiesis* could then potentially be a fleshing out of our entanglements within our *oikos*—a sort of magical incantation that keeps us semiotically attuned to the *chora* of Mother Earth, responsive to her movements, her rhythms, her music, and even, if we follow behind Michel Serres, her emotions.[39]

Carson the scientist pleaded for the rehabilitation of the senses and affects in our exploration of the world: "it is not half as important to *know* as to feel."[40] Clearly, her own writing conforms to her awareness of the need to reconnect with the world through our senses, not simply via the facts produced by science. Inaugurating many of the debates among ecofeminists, ecopsychologists, and ecocritics as to what antidote might help fight today's generalized state of psychic numbing, Carson wrote:

> If facts are the seeds that later produce knowledge and wisdom, then the emotion and the impressions of the senses are the fertile soil in which the seeds must grow. [. . .] Once the emotions have been aroused—a sense of the beautiful, the excitement of the new and the unknown, a feeling of sympathy, pity, admiration or love—then we wish for knowledge about the object of our emotional response. Once found, it has lasting meaning. It is more important to pave the way for the child to want to know than to put him on a diet of facts he is not ready to assimilate.[41]

From my own experience of having turned away from science as it was drily taught in school, and trying to return to it later as I was made to feel and care by direct embodied contact with what we call "nature" and via ecopoetic literature, I can back up the hypothesis that esthetic and affective responses to the natural world—including those mediated by ecopoets—can be conducive to hunger for knowledge. This is something that the forefather of biocentric environmental writing Aldo Leopold established early on: "We seek contacts with nature because we derive pleasure from them."[42] In his famous plea for a land ethic based on contact, love and respect, Leopold stresses that "the evolution of a land ethic is an intellectual as well as an emotional process." As he remarked, "[we] can be ethical only in relation to something we can see, feel, understand, love or otherwise have faith in."[43]

Significantly, in his mid-century ecological meditations on how to best build a land ethic grounded in the understanding that "the raw stuff" of the world is "something to be loved and cherished, because it gives definition and meaning to [our lives],"[44] Leopold identified the ties between biodiversity in the wild and cultural biodiversity,[45] while pinpointing our sensitivity to the song of the earth: "The song of a river ordinarily means the tune that waters play on rock, root, and rapid. [. . .] This song of the waters is audible to every ear, but there is other music in these hills, by no means audible to all. To hear even a few notes of it you must first live here for a long time, and you must know the speech of hills and rivers."[46]

Sowing the Seeds of an Ecopoet(h)ics of Wonder and Enchantment 117

In my intuition that today's prose ecopoets conscientiously work as bioregional mediators of the flesh of the world in different places, as translators of its polyphonic song, and, in turn, as co-composers of the symphonies of the Earth that may flesh out the scientific worldviews available elsewhere, I am no doubt influenced both by my own, direct experiences of nature and by writings such as Leopold's. Leopold describes hearing "a vast pulsing harmony—its score inscribed on a thousand hills, its notes the lives and deaths of plants and animals, its rhythms spanning the seconds and the centuries."[47]

The present book seeks to encourage research rebraiding science with *ecopoiesis*. The hypotheses and analyses here ventured aim to abolish the disenchanting conception of academia and science, which Leopold summed up:

> There are men charged with the duty of examining the construction of the plants, animals, and soils which are the instruments of the great orchestra. These men are called professors. Each selects one instrument and spends his life taking it apart and describing its strings and sounding boards. This process of dismemberment is called research. The place for dismemberment is called a university.
>
> A professor may pluck the strings of his own instruments, but never that of another, and if he listens for music he must never admit it to his fellows or to his students. For all are restrained by an ironbound taboo which decrees that the construction of instruments is the domain of science, while the detection of harmony is the domain of the poets.[48]

Against this glum vision, writing seventy years after Leopold—and also as an ecofeminist researcher who is not afraid of reclaiming embodied and transdisciplinary academic practices through dance, photography, poetry, and film—the task I am undertaking is to move toward a scientific culture where men *and women* who are called "professors" no longer feel obligated to uphold a warped claim to full objectivity, or to write in arid analytical languages that dismember and disenchant our naturcultures. Following the lead of many ecopoets, thinkers, and researchers before me, my work is instead devoted to making room in academia for researchers who may feel honored to take part in an ecopoetic enterprise of reenchantment of the Earth—one that, to a certain extent, makes our own work a form of chanting with the Earth. For both my work as an ecopoetician and as an amateur artist, like that of the ecopoets I have embarked with on this enchanting journey, stem from the understanding that what a river mostly needs is not "more people, [. . .] more inventions, and hence more science," but, something radically different. For, as Leopold phrased it, "the good life on any river may likewise depend on the perception of its music, and the preservation of some music to perceive."[49]

Almost as if writing in defense of ecopoetics, Carson emphasized the enchanting power of corporeal experiences with nature through senses other than sight, most particularly the "delight" offered by the world's scents, or the "source of even more exquisite pleasure" that comes from listening to "the voices of the earth and what they mean—the majestic voice of thunder, the winds, the sound of surf or flowing streams."[50] It is precisely because, as Carson underlines, hearing "requires conscious cultivation," that ecopoetics might potentially play a role in attuning us to the song, the humming, and rhythms of the natural world. When Jonathan Bate formulates the very term "ecopoetics" in his seminal book *The Song of the Earth*, he argues that it is a field where one listens to "the voice of art," which itself echoes the voices and song of the earth:

> Ecopoetics asks in what respects a poem may be a making (Greek poiesis) of the dwelling-place—the prefix eco- is derived from Greek oikos, 'the home or place of dwelling.' According to this definition, poetry will not necessarily be synonymous with verse: the poeming of the dwelling is not inherently dependent on metrical form. However, the rhythmic, syntactic and linguistic intensifications that are characteristic of verse-writing frequently give force to the *poiesis*: it could be that *poiesis* in the sense of verse-making is language's most direct path of return to the oikos, the place of dwelling, because metre itself—a quiet but persistent music, a recurring cycle, a heartbeat—is an answering to nature's own rhythms, an echoing of the song of the earth itself.[51]

Whether looking at prose or verse, ecopoetics studies a special kind of writing that, to take up Bate again, "has the peculiar power to speak 'earth.'"[52] As I like to put it, it may be that, nourished via our unbreakable bonds with other-than-human nature, and influenced by the chaotic *chora* of Mother Earth, *ecopoiesis* offers amplified 'poetic echoes' of the living world.[53] As I will show, this poetic reverberation can concern the aural dimension of the world as much as the other textures that we apprehend through our various senses. Hence the crossover in my approach of ecopoetics between the study of *textuality*—from the Latin *textum*, itself derived from the verb *texere*, meaning to weave—and that of the world's different textures, which are at once interwoven between themselves, interlaced with human perception and language, and therefore intricately intertwined with ecopoetic language.

Getting back to Bennett's study, enchantment is conceptualized as "a surprising encounter" generating "(1) a pleasurable feeling of being charmed by the novel and yet unprocessed encounter and (2) a more *unheimlich* (uncanny) feeling of being disrupted or torn out of one's default sensory-psychic-intellectual disposition."[54] In line with Berendt's diagnosis of our ocular hypertrophy, or Swimme's notion of a frontal lobotomy, my interest in ecopoetics comes with the desire to delve into how language might relay the multiple, sensuous textures of the world, potentially restoring some of the cognitive and sentient powers that have been impaired in most modern body-minds. As for my take on an ecopoetics of reenchantment and the practice of wonder, it is no doubt affiliated to French scholar Yves-Charles Grandjeat's work on wonder in ecopoetic writing. As Grandjeat demonstrates, the experience of wonder opens "a breach in the cognitive process" by dint of which we may "let go of the need to control the world through logocentric mapping" and explore "uncharted cognitive [paths.]"[55] Wonder thus triggers a kind of *"dislocation*, a cognitive rift hurling humans beyond the limits of their known world."[56] Relying on Giorgio Agamben's notion of "openness" in *Laperto. L'huomo e l'animale*, Grandjeat goes on to explain how wonder "describes the feeling aroused by the unlikely encounter, in a common space, of two different beings living in distinct perceptive spaces. Access to that common space is only granted to humans who momentarily relinquish their own perceptual schemes—the cognitive frames through which they map their own worlds, and which hold them captive to these worlds."[57]

The need for ecopoetic practices of reenchantment as a crucial world-opening and resensitizing process can convincingly be defended by relying on the findings of ecopsychologists who work on attention restoration theory and practices (ART). For instance, Laura Sewall shows how in a Western culture of consumerism, hypermaterialism, hyperindividualism, scientific mechanism and reductionism, "our environmental crisis is a crisis of perception. [. . .] Our attention, entrained on objects and focused on flat screens, is far removed from the dynamic of the animated nonhuman world. We are as good as blind to the wonder at our feet or the daily spectacle of an ever-changing sky."[58] Sewall defines attention as "a cognitive capacity that selects and enhances sensory signals that are relevant or surprising." While sensation is "the initial interface between inner and outer world," ecopsychology and neuroscience have demonstrated that "a personalized perceptual reality is infused with previous experience in the form of already established, activated, and facilitated neural networks."[59] In other words, what we have learned to pay attention to in large part determines what we can perceive. Therefore, perception is formed as a filtering of what we might sensitively pick up, depending on acquired habits of attention. And, problematically, in a modern world where we are increasingly cut off from more-than-human nature and where our perception of life is more and more mediated via screens, we can hardly perceive the many ways in which we are embedded in the earth: "We do not readily see the patterns that would reveal our dependance on the natural world, nor are we commonly aware of the systems within which we are deeply embedded."[60] My contention is that *ecopoiesis* may have a part to play in remedying the rampant "nature deficit disorder" that has become characteristic of our age.[61]

Despite the great loss of human sensitivity to the natural world that scientists have revealed, reasons for hope might still lie in the discovery of neural plasticity. It has certainly been proven that our growing, collective move to urban habitats, coupled with the demise of natural history from school curricula and with the modern exposure to hypertechnological cultures have triggered "a loss of collective, embodied knowledge," and "a forgetting of the perceptive and sensual beings we naturally are."[62] This alarming "extinction of experience" Robert Pyle has brought to light may however very likely be reversed. Pyle studies the imbrication between lack of exposure, ignorance, and forgetfulness that in turn leads to an extinction of experience and collective anomie:

> Ecological ignorance breeds indifference, throttling up the cycle I call the extinction of experience: as common elements of diversity disappear from our own nearby environs, we grow increasingly alienated, less caring and more apathetic. Such collective anomie follows further extinctions and deeper impoverishment of experience, round and round. What we know, we may choose to care for. What we fail to recognize, we certainly won't.[63]

If many intellectuals are prompt to criticize the didacticism that characterizes much ecofiction, my claim is that sensuous *poiesis* might be even more powerful, as an attention-restoration practice, when it effectively retrains our noticing of the natural world—of its entangled textures, forms, names, colors, smells, sounds, and the complex ways in which these intra-act. As I intend to demonstrate, a certain dose of didacticism, when sensibly blended with artful, sensuous *ecopoiesis*, may have long-lasting effects over our dealings with the world. My claim is that *ecopoiesis* in the form of fiction might possess great power to restore and enhance attention, curiosity, empathy, and perceptive skills at once. Revitalizing both sense and sensibility, it may indeed perform a precious reinvestment of the senses in ecological knowledge.

At this juncture, I want to stress how the experience of wonder via ecoliterary texts is central to the concept of "liminal realism" that I am about to expound, and which will help me analyze some of the fundamental mechanisms of the ecopoetic reenchantment at play in my corpus, mechanisms that function on onto-epistemo-phenomeno-logical levels. Taking up Grandjeat's notion of the "cognitive rift hurling humans beyond the limits of their known world" which is opened by encounter with the more-than-human world, it may be said that the "dislocation" from the human world which Grandjeat attributes to wonder corresponds to the uplifting experience of enchantment that, in the case of liminal realism, ushers us into a world somewhere in-between human and other-than-human worlds.[64] As Grandjeat puts it, contact with nonhuman nature "wrenches the human subject from familiar cognitive categories to offer a perceptual glimpse into the many other worlds within the one he inhabits. This space of poetic, cognitive decentering can be hinted at [. . . by] the term 'wondering,'" which "brings out this odd mix of puzzlement, astonishment and marveling gained from the release of the rational intellect, in which the 'esthetic principle' (Leopold 1949, 280) is free to roam." Grandjeat thus conflates wonder with enchantment "since, like a magic wand, it projects us instantly into a world other than the one we are familiar with and call 'normal,' and ushers us into other worlds, other realities."[65]

NOTES

1. See Buell, *Environmental*, where he presents *Silent Spring* as "the book that inaugurated 'the literature of ecological apocalypse'" (285); Slovic, "Epistemology"; and Killingsworth and Palmer, "Millennial Ecology."
2. Carson, *Silent*, 2.
3. Slovic, "Epistemology."
4. Knickerbocker, *Ecopoetics*, 1–2.
5. Ibid, 2. Knickerbocker is not discussing Rachel Carson's book here, but *ecopoiesis* in general.
6. Carson, *Wonder*, 54.
7. Bennett, *Enchantment*, 3–6.
8. Ibid, 4.
9. Knickerbocker, *Ecopoetics*, 122.
10. Ibid, 10.
11. Berendt, *Nada*, 53.
12. Ibid, 5.
13. Ibid, 5.
14. Bernstein, *Close*, 21. I am indebted to Scott Knickerbocker for the discovery of Bernstein's work.
15. Merleau-Ponty, *Visible*, 189–90, translation mine.
16. Merleau-Ponty, *Nature*, 280, translation mine.
17. Barad, *Meeting*, 185.
18. Merleau-Ponty, *Nature*, 281, 283, translation mine.
19. Barad, *Meeting*, 333, 184.
20. Berendt, *Nada*, 7.
21. Ibid, 5.
22. Ibid, 6–7.
23. Krause, *Chansons*, 12. Krause borrows the phrase from sound designer Walter Murch.
24. Schafer, *Paysage*, 34.
25. Abram, *Becoming*, 173.
26. Berendt, *Nada*, and Bushdid et al. "Olfactory Stimuli."
27. Krause, *Chansons*, 17, 14–15.

28. This is an area of ecopsychology focusing on the ways in which culture and education affect our capacities to perceive the world. Following studies by Robert Pyle and Peter Kahn Jr, among others, ecopsychologists and neuroscientists have been leading experiments to better understand the phenomena relating to human perception of the natural world, the ways in which humans are attached to and affected by other-than-human nature, and the processes that are capable of restoring attention to the natural world. (Kahn et al., "Introduction"; Albretch, "Psychoterratic"; Sewall, "Beauty" and "Skill"). These areas of research help speculate as to the potential effects of *ecopoiesis* over the reading subject's relationship with the more-than-human world.

29. See Krause, *Orchestre*.
30. Abram, *Becoming*, 4.
31. Ibid, 4.
32. Ibid, 170–71.
33. Ibid, 171.
34. Ibid, 11.
35. Ibid, 11.
36. Ibid, 11–12.
37. Kristeva, *Révolution*, 22–30.
38. I am here using this time-old metaphor of Mother Earth to extend Kristeva's theory from the body of the biological mother to that of the earth itself.
39. Serres, *Contrat*, 136–37, 186–87.
40. Carson, *Wonder*, 56.
41. Ibid, 56.
42. Leopold, *Almanac*, 283.
43. Ibid, 251.
44. Ibid, 265.
45. Ibid, 264–65.
46. Ibid, 158.
47. Ibid, 158.
48. Ibid, 163.
49. Ibid, 163.
50. Ibid, 83–84.
51. Bate, *Song*, 75–76.
52. Ibid, 251.
53. In French, the words *écopoétique*, for "ecopoetics," and *échos poétiques*, for "poetic echoes," are homophones. Hence the pun that comes more spontaneously in French and that, with Bate in mind, I have made mine.
54. Bennett, *Enchantment*, 5.
55. Grandjeat, "Wondering," 39–40.
56. Ibid, 34.
57. Ibid, 39.
58. Sewall, "Beauty," 265.
59. Ibid, 266.
60. Ibid, 265.
61. See Louv, *Last*.

62. Sewall, "Beauty," 274.
63. Pyle, quoted in Sewall, "Beauty," 274–75.
64. Grandjeat, "Wondering," 34.
65. Ibid, 34.

PART II

Ecopoetic Reenchantment via Liminal Realism

Chapter Five

Why Liminal, Rather than "Magical," "Spiritual," "Mystical," "Ontological," or "Epistemological" Realism?

Over the past twenty-five years, I have explored the ways in which the immersive experience of reading can retrain our sensitive and affective "response-abilities"—sensitive abilities which, in turn, channel our intellectual inroads into the world together with our sensible, ethical sense of "responsibility."[1] In the wake of Jane Bennett's, Scott Slovic and Paul Slovic's, Alexa Weik von Mossner's, and Pierre-Louis Patoine's cogent research on the affective, empathetic, intellectual, and ethical powers of narrative art,[2] I argue that ecopoetic literature can sharpen our nerves to experience emotions and sensations again which otherwise tend to get numbed as our lives grow increasingly estranged from the nonhuman world. Enjoining us to empathize with small-scale fictional individuals—whether human or other than human—ecoliterature effectively redirects how we then respond to non-fictional realities, whether individual or collective, small or large scale. From that perspective, what can we make of literary thought-experiments in contemporary literature presenting us with worlds where a woman can enter a bee-hive and become part of it (in Starhawk's *The Fifth Sacred Thing*), where women have the ability to hear the voices and messages emitted by trees (in Richard Powers' *The Overstory*), where a man is endowed with sonar and can converse with whales while another can shapeshift into an octopus (in Linda Hogan's *People of the Whale*), or again where a little girl can jump up from under a kapok tree in the night jungle to float above ground and receive an epiphany as to her place within the world (as in Barbara Kingsolver's short story "Jump-Up Day")?

Within the current context of climate change, of accelerated biodiversity erosion, and pervasive talk of collapse,[3] ecoliterature has been taking issues with our modern representations of the world. As Lawrence Buell has put it: "If, as environmental philosophers contend, western metaphysics and ethics need revision before we can address today's environmental problems, then environmental crisis involves a crisis of the imagination the amelioration of which depends on finding better ways of imagining nature and humanity's relation to it."[4] Free as it is from the standard realist and rational constraints imposed on most thinkers and scientists by tightly defined disciplinary methodologies, categories, and terminologies, isn't art potentially one of the best tools to question and reinvent our cultural representations and perceptions of the world?

This part of my study follows in Buell's steps as he identifies "the need simultaneously to refine and reevaluate some of the basic analytical premises used by 'trained' readers of literature." As Buell demonstrates in his paramount study of environmental imagination, "environmental interpretation requires us to rethink our assumptions about the nature of representation, reference, metaphor, characterization, personae, and canonicity."[5] Accordingly, I will venture into ecopoetic fiction that entices us to shift perspectives onto the more-than-human world while attempting to renovate the possibilities of language itself. I will explore how this might therefore contribute to adapt our current epistemologies to the eco-onto-logical crisis we find ourselves in. Throughout this part, I will investigate books that prompt us to temporarily relinquish what we think we know about the ways in which the world is divided and ordered, and consequently, what we deem realistic as opposed to what commonly gets relegated to the world of fantasy. I will therefore be paying much attention to episodes from my corpus staging "magical," or possibly "supernatural," phenomena. This being a book about the co-composing of the polyphonic, multispecies song and story of the earth, as I reconceptualize what has been referred to in our critical jargon as "magical realism" to propose the concept of "liminal realism"—which basically corresponds to a specific, ecopoetic type of magical realism—my central focus is on cross-species communication between human and other-than-human lifeforms. Moving away from the term "magical realism," I nevertheless tap into the seminal work carried out by Wendy Faris and other great thinkers of this mode.[6] Before going into closer readings of the texts under scrutiny here, I shall therefore recall Faris's substantial contribution to the field.

Scholars using the term "magical realism" often refer to the merging of "magical" events within an otherwise realistic narrative, goading readers to suspend disbelief and simply go along with the magic worked into the diegesis. In *Ordinary Enchantments: Magical Realism and the Remystification of Narrative*, Wendy Faris identifies five characteristics of literary magical realism: "First, the text contains an irreducible element of magic; second, the descriptions in magical realism detail a strong presence of the phenomenal world; third, the reader may experience some unsettling doubts in the effort to reconcile two contradictory understandings of events; fourth, the narrative merges different realms; and, finally, magical realism disturbs received ideas about time, space and identity."[7] Beyond these five criteria, Faris has moreover listed several secondary aspects of magical realism that are "helpful in building magical realist rooms in the postmodern house of fiction."[8] This "postmodern house" can potentially accommodate all the texts in my corpus, specifically those exhibiting awareness of how discourse shapes history, mentalities, and, to a certain extent, reality. Moreover, most of the novels under scrutiny display a certain amount of metatextuality and metafiction foregrounding the artificiality of their own artistic, meaning-making processes. Yet, as opposed to some strands of postmodernism where tricky reality gets effaced entirely behind its representations, and where what matters most are purely discursive and intellectual issues, the texts here under study solidly ground language and storytelling in the physical, material world that *ecopoiesis* emerges from and takes part in.

The first question that must be answered before I can delve further into my corpus is: Why get rid of "magical realism" in favor of "*liminal* realism"? When magical realism first sparked my interest, I was trying to cast light onto the oxymoronic power of Barbara Kingsolver's complex, syncretic and mythopoeic writing. My take initially chimes in with Jean-Pierre Durix's hopeful hypothesis, that "perhaps the merit of the phrase 'magical realism' is to suggest a field of possibilities in which the term will no longer be an oxymoron."[9] However, over the years, I have come to realize that dualistic thinking keeps such a strong hold over our minds that even in academic contexts, I have invariably run into the same objections: how could anything be 'magical' *and* realistic at the same time? No matter how hard a case I have made for an ecopoetics that may allow for multiple interpretations of phenomena that may seem, at *first* glance, "*super*natural" and which yet might in *fact* rest upon entirely cultural *and natural* explanations, I have systematically encountered the same rebuttal from colleagues struggling to relinquish the dominant either/or mentality.

This recalcitrance to the subtle working of magical realism I have come to attribute to multiple factors. First, our either/or dualistic thinking is so deeply entrenched that to consider anything as being possibly tied to "magic" *and* to "reality," to a spiritual or psychic reality *and* to concrete worldly matters may seem heretic in standard academic thinking. Hence the need to systematically explicit the meaning of the word "magic" on using the phrase—a reflex that I have developed in the wake of Starhawk and David Abram.[10] In this respect, my use of the word "magic" rests on Starhawk's definition of it as "*the art of changing consciousness at will*."[11] While the art of magic can in some contexts refer to the power of using tricks to produce illusions, as for instance when the illusionist gives the impression that he can make a rabbit appear out of nowhere, the problem may nevertheless be reversed: what the illusionist truly does is simply to reveal that the rabbit has in fact been there all along, either inside her hat, or up her sleeve, or again hidden under the table, even though it had until then remained unnoticed by the audience. More appropriate than equating it with supernatural powers, one definition of magic may consist in circumscribing it to the art of revealing parts or aspects of the tangible world that may not be visible when seen through the usual modern lens of human eyesight or through the grids of the dominant established frame of mind.

Following that logic, in Richard Powers' fiction, *The Overstory*, the human characters endowed with the capacity to decipher the language of trees ecopoetically reveal the *fact* that trees *can* indeed communicate in their own nonhuman ways—a given about the life of forests that scientists have recently been probing into and obtaining greater evidence of. If the thought-experiment Powers has cooked up for his powerful book cannot be said to accurately reflect human-tree communication as most humans experience it today—and while much could and will be said about the ways in which most indigenous cultures always have singled out individuals, namely shamans, who have developed the ability to communicate with trees and other nonhuman beings[12]—*The Overstory* might however encapsulate within an ecopoetic proposition the various ways in which trees do emit messages that we humans *can* be attentive to, *can* read, and *could* act upon. As I will show, there are many ways for the rational mind to coopt the marvelous in this book; but doing so might be missing the point of Power's enchanting novel—to reawaken human consciousness to the wonders of trees, whose existences are inextricably entangled with ours.

Abiding by a similar logic, in Linda Hogan's ecopoetic fiction *People of the Whale*, some humans truck with whales and with a shape-shifting octopus. These "supernatural" phenomena turn out to rely on a subtle braiding of animistic and totemic worldviews with ethological knowledge.[13] Indeed, if it remains to be seen that an octopus can metamorphose into a human being, a great part of the magic in Hogan's story is compatible with scientific evidence that whale songs do form a kind of animal language, and that octopuses really can shapeshift—adapting their color and appearance the better to blend into their surroundings and thus escape predators.

Related to my first point, a second problem with the reception of magical realism is that many people overlook the poetic device of the oxymoron, with its deliberate bringing *together* of two concepts considered as antagonistic, the better to shed light on their potential coherence. J. A. Cuddon translates the Greek etymology of the word "oxymoron" as "pointedly foolish." He goes on to explain that the oxymoron "combines incongruous and apparently contradictory words and meanings for a special effect."[14] Among my favorite oxymora is Paul Eluard's "The Earth is blue as an orange"—a surrealistic proposition that could seem absurd, and which yet speaks volumes about the roundness, the fruitfulness and vitality of our mostly blue, multicolored planet—a whole entity subsuming different, yet interdependent quarters. The oxymoron meanwhile brings together two colors that, based on their physical properties, have been established as complementary opposites in the chromatic circle of primary and secondary colors. Picturing the Earth as blue as an orange can thus prismatically enfold all the colors and elements in the world's rich palette and chromatic circle.

The etymology of the word "oxymoron" suggests how the figure of speech functions and how this might extend to describe magical realist fiction. The word was coined in the sixteenth century, from the Greek *oxus*, meaning "sharp, acidic, keen" and *môrus*, meaning "foolish." The latter root is refracted in the word "moron," referring to somebody showing signs of feeble-mindedness, or, when used colloquially, to a person deemed stupid. The oxymoron rubs against the mainstay of Cartesian thinking in a world of either/ors. How could one be sharp and thick at once? Both Cuddon and Henry Suhamy relate the oxymoron to antithesis and paradox. As Suhamy argues, the oxymoron brings together two words which are *commonly held* to be semantically incompatible. As Suhamy insists, if the oxymoron relies on an unusual combination going against the grain of common sense, it nevertheless turns out to be fecund in meaning, and, thus, revealing.[15] Like the oxymoron, magical/liminal realism defies Western, common sense, and yet, as I will try to show, it might bring about a revelation as to aspects of the world which have been hidden from modern intelligence. Liminal realism is *para-doxical*: it challenges monologic mainstream perceptions and discourses commonly admitted as right, as rationally valid, when following the mechanistic and reductionist *doxa* established at the zenith of modernity. Elaborating on the conceptual paradox at the heart of the oxymoron, Suhamy identifies "an association that common sense admits with difficulty, *but the experience of which* may sometimes *prove the validity of*, or an association that the *visionary intuition of poets can impose* on the public."[16]

My investigations mean to find out whether liminal realism may be a linchpin in articulating a newly arranged, rational, ecopoetic *and* scientific, ecofeminist *and* postmodern paradigm. If liminal realism can corrode and dissolve the boundaries held up by common sense, this literary mode might then overthrow some of the barriers erected by Moderns, making room for a more intuitive, sensitive rapport with the world. Derived from the Latin root *limen*, liminality refers to a threshold situation. As I will show, liminal realism cues us to dwell for a moment in between worlds, within the possibility of conciliating takes on the world that have culturally been structured as irreconcilable—as *either* rational *or* magical, *either* scientific *or* poetic, *either* animistic *or* naturalistic, *either* discursive *or* materialistic. The greatest power of liminal realism lies precisely in its capacity to make two allegedly different takes on the world converge, by standing somewhere *in-between* poetic, mythological, ethological, biological, ecological, anthropological, geological, and/or psychological perspectives. In this respect, liminal realism supports Lévi-Strauss's effort to get ahold again of "that primitive alliance between the poetic and the rational." Our "great problem—and I'd say a source of trouble in our civilization—" observed Lévi-Strauss, "is that the realm of the rational and the realm of the poetic have become totally separated, whereas in all the so-called 'primitive' civilizations studied by ethnologists, the two realms are tightly united." While "the rational finds expression in poetic form," the anthropologist went on, "the poetic encapsulates the rational in a latent state."[17] As I hope to demonstrate, liminal realism turns ecopoetic literature into a crucible wherein to re-alloy the poetic and the rational.

Besides, the blending of different paradigms proper to liminal realism situates us at an intricate crossroads between human and other-than-human lifeforms and perspectives. Liminal realism thereby creates an immersive place for the reading bodymind to dwell in-between separate ontologies and phenomenological experiences of the pluriverse. Liminal realism, as we shall see, welcomes the bodymind in an open place that might integrate animistic, totemic, and naturalistic perspectives, reconciling indigenous and non-indigenous worldviews, sensory worlds, and epistemologies, while straddling human and other-than-human worlds.

Let me make the precision that as I use the terms "animistic" or "totemic," I am not adhering strictly to Philippe Descola's structuralist distinction between the two ontologies,[18] and this for three reasons. First, as the French anthropologist has conceded, the practices and belief systems of many North American indigenous groups defeat his attempt at establishing rigidly distinctive ontologies that might offer a reliable classification system within which any given culture might be circumscribed. In fact, the practices and worldviews regenerated by the indigenous ecopoetic literature I am interested in mostly slip through the structuralist classification that Descola has attempted to schematize.[19] Second of all, I do not propose to examine and analyze the workings of any specific social group living in the real world, but subversive ecopoetic and fictional craft by individual artists drawing from various cultures and from their own idiosyncratic ways of dwelling. It should therefore come as no surprise that an ecopoetic exploration of dwelling via the creative reweaving of various epistemologies and ontologies should defy the strict boundaries erected by anthropologists in their typically Modern attempts to break down thinking. And third, because overall, the texts in my corpus display a vision of the world where all matter—whether human, animal, vegetal, or elemental—is inherently agential, expressive, and engaged in transactions involving both trans-corporeality and intra-actions, my new materialist approach ultimately makes it hard for the distinction between interiorities and exteriorities forming the basis of Descola's classification to hold sway.

In consequence, I will refer here and there to some observations made by anthropologists, without however trying to wedge the ecopoetic stances I examine into fixed definitions of totemism and animism.[20] In applying the term "animistic" to aspects of the fiction under study, I will mainly be referring to those fictional situations where both humans and nonhumans are endowed with a form of sentience, if not personhood, allowing for communication between the two via the liminal experiences of shaman-like characters. When referring to "totemic" relationships in my corpus, I will be tackling those privileged bonds of interspecies guardianship and tight reciprocity established either between an individual and an animal or a plant, or more collectively, between a group, a family, or a community and an animal or a plant.[21]

Getting back to my discussion of the hindrances attached to magical realism, the fundamental in-betweenness of the mode is often missed out upon because of a profound lack of education for many of us who have been trained in the arts and humanities when it comes to the biological and physical processes running the complex systems and organisms outside of literature. As I have shown elsewhere, in Barbara Kingsolver's short story "Jump-Up Day" for instance, as the pre-adolescent protagonist is being led through an initiation rite by an Obeah Man, or shaman, her experience of flying over clouds is typically a magical realistic one. It is at once poetically *and* rationally valid, as it can be explained in various ways across anthropological, ecological, mythological, and psychoanalytical reading grids.[22] On a basic level, however, without any biocultural knowledge of the Saint Lucian ecosystem where the story takes place—with its jungle forest, its dramatic volcanic landscape, its mountains, its cities, its various snakes, insects, and the gigantic Kapok Tree bearing fruit that, when ripe, open and shed fluffy white silk cotton onto the jungle floor—without interweaving the material aspects of the place the story is set in with its colonial history and multiculturalism, its Patois, the pregnancy there of complex Obeah worldviews, myths, and practices, then what happens in the story might indeed be dismissed as merely a child's dream, or approached as "supernatural." However, as I have demonstrated, with enough research into, or first-hand knowledge of the ecosystem where the story is set—and Barbara Kingsolver did spend one year living on the island of Saint Lucia as a child herself—it turns out that the allegorical, mythopoetic, and psychoanalytical readings one might first call upon may be conciliated with a very *rational* reading of the *natural* phenomena at play in this complex, dialogic narrative casting an initiation rite under a Kapok tree.

Much of the precise observation of the world that is threaded in their art crafts by many artists from minoritized cultures is often overlooked by Western readers of postcolonial or ecofeminist literature. As a matter of fact, much liminal realism clusters in fiction written by postcolonial and ecofeminist writers interweaving indigenous and Western worldviews. As Lois Zamora and Wendy Faris have noted:

> Texts labelled magical realist draw upon cultural systems that are no less 'real' than those upon which traditional literary realism draws—often non-Western cultural systems that privilege mystery over empiricism, empathy over technology, tradition over innovation. Their primary narrative investment may be in myths, legends, rituals—that is in collective (sometimes oral and performative, as well as written) practices that bind communities together. [. . . Magical] realism is a mode suited to exploring—and transgressing—boundaries, whether the boundaries are ontological, political, geographical, or generic. [. . .] Magical realist texts are subversive: their in-betweenness, their all-at-onceness encourages resistance to monologic political and cultural structures, a feature that has made the mode particularly useful to writers in postcolonial cultures, and increasingly, to women.[23]

In various parts of their work, Maggie Ann Bowers and Wendy Faris have both zeroed in on the anti-dualistic take characterizing much postcolonial and feminist literature (as penned by Angela Carter or Louise Erdrich, to name just two). To quote Faris, "[the] dialogical, polyphonic, decentered forms that characterize postmodernism as it grows out of modernism correspond to what are often imagined to be female ways of being and knowing."[24] Let me insist here on the *imagined* femaleness of a certain rapport with the world—another cultural construct that this study will attempt to pulverize as we examine the writings of Jean Giono or Richard Powers. In this respect, my work seeks to prove wrong any essentialist claims that their might exist *in essence* a different and specific "female spirit," or gaze, that which Faris actually locates in female magical realism as well as in much postcolonial other-than-female works. According to Faris, "it may be possible to locate a female spirit characterized by structures of diffusion, polyvocality, and attention to issues of embodiment, to an earth-centered spirit world, and to collectivity, among other things [. . .]."[25] If Faris then intelligently includes male writing in her study, the chapter she devotes to "Women and Women and Women" may be misleading, and somewhat counterproductive, as it seeks to identify, at least according to the chapter title and subheading, "A Feminine Element in Magical Realism." In my opinion, the very formulation runs the risk of perpetuating binary thinking and gender stereotypes rather than exploding them.

It might be hypothesized rather, that liminal realism stems from an *ecofeminist perspective*—one that non-lobotomized or liberated males can adopt, and which is inherently antidualistic, nonhierarchical, incorporated, and biocentric. It is high time we critics got rid of such delusory, noxious notions that there might exist an *essential* difference between a female and a male "spirit" or a female and a male type of sensitivity (the assumption being conventionally that women are "naturally" more sensitive than men) or writing—one that would have to do with women being allegedly more connected to their bodies, more easily ruled by their bodily problems than men.[26] Where Faris only touches upon the importance of Julia Kristeva's contribution to our understanding of poetic language,[27] I take up the latter's exploration of the semiotic order to get to the heart of the liminal ecopoetics at work with both female and male, indigenous and non-indigenous writers. As broached in part 1, my aim is to demonstrate that, because we all come to our senses as we are first borne inside a mother's body, then born to be embraced by Mother Earth, then quite naturally, all ecopoets, no matter their gender or dominant ethnic affiliation, write through their own mother-and-earth-connected bodyminds.

The profound onto-epistemo-logical revolution pushed forward by an ecopoetic liminal realism cannot be apprehended properly without paying due attention to its political dimension. In Maggie Bowers's terms:

> [What] unites [magical realist writers across socio-cultural, geographical, or gender differences] is the political nature of the magical realism that is written in these locations, whether from an overtly anti-imperial, feminist or Marxist approach, or a mixture of all these, or whether the form reveals its political aspect more covertly through the cultural politics of postcolonialism, cross-culturalism, or the friction between the writing of pragmatic European Western culture and oral, mythic based cultures. What locates these writers politically is their narrative position outside the dominant power structures and cultural centres.[28]

This will prove to be the case for the texts I have chosen in my corpus, including those crafted by writers who do not belong to indigenous cultures (Ann Pancake, Richard Powers, Annie Proulx, Starhawk) or the female sex (Jean Giono, Richard Powers). And yet, as I will try to evidence as I unpack their writing, part of it is mostly steeped in worldviews that may not be endemic to the dominant culture of the social group or gender they were born into.

The texts in my corpus pave the way for a stance that is epistemologically liminal: while they propose poetic visions of the world, the authors show evidence of an indisputable, solid grasp of history, sociology, philosophy, psychology, and anthropology, as well as ecology (Hogan, Kimmerer, Kingsolver, Powers, Silko, Starhawk), botany and plant behavior (Kimmerer,

Kingsolver, Pancake, Powers), zoology, entomology, and ethology (Hogan, Kimmerer, Kingsolver, Pancake, Silko, Starhawk), musicology and bioacoustics (Hogan, Kingsolver, Powers), climatology and geology (Hogan, Kingsolver, Proulx, Silko). The natural phenomena these literary texts weave into their fictional realms may sometimes get overlooked by Western critics who tend to attribute fictional events that are surprising—that is, difficult to explain within the rational framework that they do possess—to marvelous causes, or else to the female or indigenous identity of the writer. In this respect, whereas Patrick Murphy has put forward the term "spiritual realism" to replace magical realism,[29] I remain convinced that such an approach problematically continues to "run alongside" modernism and postmodernism. Such theory continues to cleave and separate rather than stitch back together various forms of discourses and ontologies, which may prove counterproductive as regards the postmodernist and ecofeminist composting and reweaving of differentiated epistemologies at work within liminal realism.[30]

Defined by Patrick Murphy and taken up by François Gavillon after him, the notion of a "spiritual realism" poses a number of problems. To me, the main problem resides in the way both critics fail to define exactly what they mean when they refer to the "spiritual," or to "magic." This lacuna is combined with the premise that "spiritual realism" corresponds to an alternative worldview predominantly located within women writers and writers either indigenous or of color, one that, according to Gavillon, is eventually based on "faith."[31] Faith in what? And how exactly are we to measure to what extent the diegetic world crafted by these writers might accurately reflect what they hold to be true in the real world?

As Gavillon relies on Luis Leal and Alejo Carpetier to say that "magic is not in the writing but in phenomenal reality," still what is meant precisely here by "magic" or "faith" remains vague. While Murphy aptly concludes that "[the] spirituality these women authors represent, like their sense of ecological responsibility and their inhabitational orientation, forms part of an alternative reality, which in the long haul will prove to be far less illusory than what passes for realistic in the current ecosuicidal milieu of transrational consumptionist culture," and while Gavillon puts forward the precious notion of "spiritual realism" as "translation," I would like to take it one step further. Murphy formulates the hypothesis that "spiritual realism [. . .] seeks to represent a sense of ecological responsibility to a referentially recognized material more-than-human world on the part of its characters and the need for the adoption of such a sensibility on the part of its readers."[32] Following on Gavillon's hunch that the "magical," or "spiritual," realism at work in Linda Hogan's writing proposes "a translation" of certain worldviews, my insistence in defense of "liminal realism" partly rests on my conviction that what is at stake is more than simply tied to faith. Rather, liminal realism performs a poetic translation of the material textures, songs, and sensuous entanglements of the living world. In my defense of liminal realism, my contention is that ecopoets equipped with Traditional Ecological Knowledges and/or current scientific knowledges are simultaneously sensitive enough to pick up and interpret a wide array of sonorous, proprioceptive, haptic, taste, visual, nociceptive, and olfactory phenomena via *ecopoiesis*. I argue that they are doing so indeed out of a greater sense of "ecological responsibility," but primarily out of a greater *response-ability* to both the visible *and* invisible textures of the world, the fabric of which can be translated in scientific *and* poetic *and* mythical languages.

Wendy Faris and I share an interest in magical realism for its oxymoronic powers and liminality.[33] As David Abram observes, the shortcomings of dualistic and reductionistic thinking account for many anthropological misconceptions of indigenous practices and rituals:

> The primacy for the magician of nonhuman nature—the centrality of his relation to other species and to the earth—is not always evident to Western researchers. Countless anthropologists have managed to overlook the ecological dimension of the shaman's craft, while writing at great length of the shaman's rapport with 'supernatural' entities. We can attribute much of this oversight to the modern, civilized assumption that the natural world is largely determinate and mechanical, and that that which is regarded as mysterious, powerful, and beyond human ken must therefore be of some other, nonphysical realm above nature, 'supernatural.'[34]

This kind of Eurocentric, binary and hierarchical thinking also underpins Toni Morrison's rejection of the term "magical realism." In her endeavor to "capture the vast imagination of black people," and how they can "do something practical and have visions at the same time," Morrison's densely poetic prose relays a world where "all the parts of living are on an equal footing." It is a world where "[birds] talk and butterflies cry," and those are perceived as ordinary phenomena that simply "make the world larger." Yet, the problem with acknowledging this way of relating to the phenomenal world, Morrison explains, is that it will invariably be denigrated as "superstitions" by outsiders.[35] When asked about her dislike of the label "magical realism," Morrison explains: "My own use of enchantment simply comes because that's the way the world was for me and for the black people I knew . . . There was this other knowledge or perception, always discredited but nevertheless there, which informed their sensibilities and clarified their activities. It formed a kind of cosmology that was perceptive as well as enchanting."[36] Trying to look for a better oxymoron, I have elsewhere proposed the term "mystical realism."[37] Yet, the advantage of 'liminal realism' lies in doing away entirely with the potentially ethnocentric ambiguity of words such as 'mystic' or 'magic'—which some will equate with irrationality and, consequently, irrelevance.

'Liminal realism' stays true in many ways to the original definition of magical realism as first coined by Franz Roh. The term initially emerged as Roh was writing about German painting in the 1920s. The art critic was then attempting to capture the new trend in German New Realism, or Post-Expressionism—one without fantastical elements. Roh was looking at art where "the mystery does not descend to the represented world, but rather hides and palpitates behind it."[38] Situated at the frontier between the visible and the invisible, between what can be said or represented objectively and what can only be felt about the phenomenal world, Roh's phrase limns the mystery of things and beings as they can be sensed by the subject getting entangled with the penetrating tactility, colors, spatial forms, and the "smell and taste" of things.[39] Focusing on how the phenomenal world "emerges and vibrates with energetic intensity" in a work of art, Roh's definition of magical realism anticipates material ecocriticism. From "a more important viewpoint than the 'objectivity' everyone keeps evoking," Roh encouraged critics to "acknowledge that radiation of magic, that spirituality, that lugubrious throbbing in the best works of the new mode" he was scrutinizing.[40] Coalescing with Starhawk and David Abram's redefinitions of magic as a voluntary modification of our perception of reality, Roh viewed magical realism as the artistic process by which "an imitated 'reality' [. . .] makes sense if it starts from and then (consciously or unconsciously) transcends the representation of a window, that is, if it constitutes a magical gaze opening onto a piece of mildly transfigured 'reality' (produced artificially)."[41] Liminal realism is a matter of transfiguration, rather than transcendence.

In her extensive history of the concept, Faris refers to Jean Weisgerber's "distinction between two types of magical realism: the 'scholarly' type, which 'loses itself in art and conjecture to illuminate or construct a speculative universe' and which is mainly the province of European writers, and the mythic or folkloric type, mainly found in Latin America." As Faris observes, the "two strains coincide to some extent with the two types of magical realism that Robert González Echevarría distinguishes: the epistemological, in which the marvels stem from an observer's vision, and the ontological, in which America is considered to be itself marvelous (Carpentier's *lo real maravilloso*)."[42] But as Faris carefully notes, it is often delicate—if not impossible—to neatly separate one strand from the other.

One advantage of the term "liminal realism" is that it does justice to the many crossroads from which ecopoetic magical realism operates—somewhere in-between human and nonhuman worlds, in-between scientific and poetic stances, in-between objective and subjective realities, in-between different epistemological and ontological frameworks. Liminal realism operates via a fundamental slipping through of all our strictly defined categories. While no one has thought to propose the term "liminal realism" before, it must be conceded that part of what I am trying to encapsulate with this new coinage has been identified by Zamora and Faris:

> Magical realism often facilitates the fusion, or coexistence, of possible worlds, spaces, systems, that would be irreconcilable in other modes of fiction. The propensity of magical realist texts to admit a plurality of worlds means that they often situate themselves on liminal territory between or among those worlds—in phenomenal and spiritual regions where transformation, metamorphosis, dissolution are common, where magic is a branch of naturalism in its concern with the nature of reality and its representation, at the same time that it resists the basic assumptions of post-enlightenment rationalism and literary realism. Mind and body, spirit and matter, life and death, real and imaginary, self and other, male and female: these are boundaries to be erased, transgressed, blurred, brought together, or otherwise fundamentally refashioned in magical realist texts. [. . . Magical realism negotiates] between the normative oppositions and alternative structures with which they propose to destabilize and/or displace them.[43]

While getting rid of some of the tensions related to the ambiguous meaning of the word "magic," my coining of "liminal realism" moreover departs from the more generic term "magical realism," in that it refers to a specifically ecopoetic and initiatory mode situating us somewhere in-between human and other-than-human worlds. Indeed, liminal realism presents a material *and* spiritual—or physical *and* psychic—continuity between humans, animals, plants, microorganisms, and even non-biotic elements.

Banking on its ecopoetic license, liminal realism best erodes the boundaries that have been entrenched in most modern epistemological frameworks between human and other-than-human realms, while acknowledging those elusive aspects of reality that cannot be entirely circumscribed by our limited senses and purely rational minds. Doing away with the slippery concept of magic—tied as it is for most people to the just-as-slippery concept of the supernatural—liminal realism pointedly makes way for a better perception of multispecies entanglements and for the possibility toward cross-species communication. David Abram provides insight into the indigenous grasp of interspecies communication and of a multispecies form of consciousness:

> . . . in tribal cultures that which we call "magic" takes its meaning from the fact that humans, in an indigenous and oral context, experience their own consciousness as simply one form of awareness among many others. The traditional magician cultivates an ability to shift out of his or her common state of consciousness precisely in order to make contact with the other organic forms of sensitivity and awareness with which human existence is entwined.[44]

In this respect, Abram is no doubt influenced by the counterculture that was heralded in the United States by pioneers of ecopsychology and ecofeminism such as Theodore Roszak and Starhawk.

In the next chapter, to highlight the reenchanting potential of liminal realism, I hone in first on postcolonial ecopoetic novels by Linda Hogan and Leslie Marmon Silko. In the following chapter, delving further into this fundamental, onto-epistemo-logical in-betweenness that characterizes ecopoetic liminal realism across the board of indigenous and non-indigenous literatures, I examine post-pastoral novels written by Jean Giono and Ann Pancake. Finally, in the third and closing section of this part on liminal realism, I focus on a climate fiction novel by Starhawk and an intricate, postmodernist novel by Richard Powers.

NOTES

1. I am taking up the pun frequently used by Donna Haraway in *When Species Meet*.

2. Bennett, *Enchantment*; Slovic and Slovic, *Numbers*; von Mossner, *Affective*; Patoine, *Corps/Textes*.

3. See Diamond, *Collapse*; Servigne and Stevens, *S'effondrer*; and Servigne, Stevens, and Chapelle, *Autre*.

4. Buell, *Environmental*, 2.

5. Ibid, 2.

6. Much of my work has been greatly influenced by Wendy Faris's invaluable contribution to the collective volume she has co-edited with Lois Parkinson Zamora, *Magical Realism: Theory, History, Community*, and her later monograph, *Ordinary Enchantments*.

7. Faris, *Ordinary*, 7.

8. Faris, "Scheherazade," 175.

9. Durix, *Deconstructing*, 190.

10. Starhawk, "Power," 76. See also David Abram: "the most sophisticated definition of 'magic' that now circulates through the American counterculture is 'the ability or power to alter one's consciousness at will'" (*Spell, 9*).

11. Starhawk, *Dreaming*, 13.

12. For more on this issue, see Abram, *Spell*; Descola, *Lances*; and Kohn, *Forests*.

13. See Adamson, "Whale" and Meillon, "Writing."

14. Cuddon, *Dictionary,* 669.

15. Suhamy, *Stylistique*, 34, 215–16.

16. Ibid, translation and emphasis mine, 215.

17. Lévi-Strauss, "Bon."

18. To put it in a nutshell, Descola's classification distinguishes animism, as an ontology resting on the differentiation of physicalities between humans and non-humans conjugated with a similarity of interiorities, from totemism, as an ontology

whereby human and nonhuman physicalities *and* interiorities are considered identical in nature (*Par-delà*).

19. Indeed, many specific cultural situations exhibit an overlapping of the various characteristics Descola categorizes as belonging to one of the four ontologies he has established. This is a point Descola acknowledges mostly in relationship to ancient tribes Native to North America such as the Potawatomi, Chickasaw, Ojibway, Algonquin, Haudenosaunee (Iroquois), Cree, Penobscot, and Micmac nations. Descola, *Par-delà*, 291–301.

20. For Tim Ingold, the distinction between totemism and animism is the following: "With a totemic ontology, the forms life takes are already given, congealed in perpetuity in the features, textures and contours of the land. And it is the land that harbours the vital forces which animate the plants, animals and people it engenders. With an animic ontology, to the contrary, life itself is generative of form. Vital force, far from being petrified in a solid medium, is free-flowing like the wind, and it is on its uninterrupted circulation that the continuity of the living world depends." Ingold, *Perception*, 112.

21. See Durand-Rous's work on totemism.

22. See Meillon, "Revisiting.'"

23. Zamora and Faris, "Introduction," 3–6.

24. Faris, *Ordinary*, 170.

25. Ibid, 170.

26. As if men did not in great part talk, act, desire, and move as driven by their own genes, rhythms, and hormones, or as if men did not constantly have to worry about having an erection, having bowel movements, or going hungry.

27. Faris, *Ordinary*, 4, 171.

28. Bowers, *Magic(al)*, 47–48.

29. Murphy defends the idea of "a paramodernism"—"an aesthetic movement in which beliefs and cultural practices that run alongside of modernity or postmodernity are represented by means of literary strategies and techniques not based in either modernist or postmodernist aesthetics." Murphy, "Spiritual," 6.

30. While composting and weaving images may bring to mind Donna Haraway's *Staying with the Trouble,* the composting function of magical realism has been cogently theorized by Jessica Maufort who has ventured her own conception of an ecopoetic magical realism. "The compost image," she argues, "brings forth several issues: the nonhuman world as an active force, blurred ontological boundaries, and the emergence of liminal places." "Magic," 239. Looking at Alexis Wright's *Carpentaria*, Thomas King's *The Back of the* Turtle, and Thomas Wharton's *Icefields*, Maufort has opened up new paths to advance an ecopoet(h)ics of reenchantment. As she examines magical realism in the light of material ecocriticism, Maufort convincingly mobilizes the notion of "a compost poetics/aesthetics," which defines "nature as a volatile space, one of decay and renewal," (Adam Beardsworth, quoted by Maufort, 240). Moreover, according to Maufort, "this mode assembles antagonistic elements without subsuming them into a homogenous whole, and [. . .] by fostering such ambivalence, magic realism devises places of liminality," 239. Throughout this book, I lean toward images of braiding, reweaving, and entanglements as my guiding

metaphors for the onto-epistemological enterprise at the heart of the environmental literature in my corpus. However, Maufort convincingly makes the case for "the heterogeneous nature of the compost," which "precisely accounts for the numerous interpretations and meanings it offers us when reflecting upon artistic, literary, and cultural processes," 240.

31. Gavillon, "Magical," 52–54.
32. Murphy, "Women," 6.
33. These are aspects that Faris broaches in *Ordinary*, 29, 69, 151, 198. In her paper "'We, the Samans,'" Faris also connects some instances of magical realism with shamanism and with the merging of human and nonhuman worlds. Yet, her interpretations overall stick to postmodernist, postcolonial, and mythopoeic readings, without going further into the ecopoetic liminal *realism* that such interspecies and interwordly fiction can produce.
34. Abram, *Spell*, 8.
35. McKay, Interview, 153.
36. Davis, "Interview," 224–25.
37. The point is to reinstate the mysteries of existence and of causal formation that science, poetry, and myths keep trying to apprehend (Meillon, "Postcolonialism").
38. Roh, "Magic," 16.
39. Ibid, 19.
40. Ibid, 20.
41. Ibid, 20.
42. Faris, "Scheherazade," 165.
43. Zamora and Faris, "Introduction," 5–6.
44. Abram, *Spell*, 9.

Chapter Six

Postcolonial Liminality and (Re)initiation into a Multispecies World

Moving betwixt and between Human and Other-than-Human Realms in Linda Hogan's Power *and Leslie Marmon Silko's* Ceremony

Like most postcolonial artists, Chickasaw poet, essayist, and novelist Linda Hogan grew up in-between two cultures with different stories and ontologies. The sociocultural hybridity of her characters therefore often reflects her own. Consequently, as we are immersed in her characters' initiatory experiences—many of which are relayed through detailed descriptions of the protagonists' sensory perception of the phenomenal world around them—we must reconsider the epistemologies and ontologies we have learned to rely upon as we *make sense* of the world. In Hogan's writing, what may seem magical to Westerners may simply correspond, in Faris's terms, to "something we cannot explain according to the laws of the universe as we [moderns] know them."[1]

In *People of the Whale* (2008), Witka, the eldest traditionalist whale-hunter, is endowed with psychic and physical powers that jeopardize the alleged discontinuity between animal and human realms. Moving in-between the animal world of the ocean and the human world of land, Witka is gifted with the ability to converse with whales. He is presented as "a medical oddity, a human curiosity, a visionary [. . .], and a medicine man who could cure rheumatism and dizzy spells."[2] From the start, Witka is singled out as a shaman living on the threshold between the worlds of sea and land: "He was born with a job set out for him and his life was already known to them. We wore white cloth that set him apart [. . .]. He learned the songs and prayers. By the age of five he had dreamed the map of underwater mountains and valleys, the landscape of rock and kelp forests and the language of currents."[3] Presented as the "man who spoke with whales," Witka as an adult is so sea-travelled and savvy that scientists and anthropologists come to study him and learn about whales and marine life.

Read in the light of *Sightings*—a nonfiction book about gray whales which Hogan wrote with naturalist Brenda Peterson—Witka's characterization appears as a liminal realistic thought-experiment. Indeed, Hogan and Peterson explored the world of whales together, going on many whale-watching excursions and learning from scientists and ethologists about the gray whale's perceptions, language, and culture. Their extensive study has led them to try and see the world through the eyes of the cetaceans. In their chapter on empathy, the writers quote the inspiring words of naturalist Will Anderson, commenting on the long time he has spent in Baja, California, carefully observing whales: "'Baja gave me a perspective that was more like that of the whale's than of people. It's so essential that we human beings develop empathy. Once a person crosses the threshold, you start looking at all life differently. You have to wonder, What are the animals feeling? What is *their* experience at the hand of humans?'"[4]

From Hogan's "double vision of sighting the whale both as a Native person and a naturalist,"[5] Witka is a threshold character whose half-whale, half-human attributes abide by both a traditional logic and whale ethology. Indeed, his *Umwelt* extends beyond that of most humans, as his reading of the world is enlarged by perception skills specific to whales.[6] "In the minds of gray whales," writes Hogan in *Sightings*, "there is an inner geography, cartography unknown to us. It urges them along the electromagnetic directions of the Earth."[7] Whales' ways of reading the stars for direction, the weather in the air above the water surface, the underwater and coastal topography, current waves, tides, and water temperatures as cartography form what Hogan and Peterson refer to as "mental maps"—"[whales'] own vision of the world, a waterway." Hogan's liminal realism is thus a way of integrating a whale's "elemental knowledge of its world," inviting us to imaginatively enlarge our human *Umwelt*: "Theirs is a world we do not know or understand. [. . .] We follow them because, maybe as the early Europeans thought, a whale is a map."[8] Liminal characters such as Witka are cues for us to shift perspectives. They invite us to dwell in a pluriverse made of entangled human and other-than-human worlds. Exploring new ways of being in the world, the immersive experience of reading thus charts a greater world for us to inhabit, informed by ways of perceiving and knowing other than our limited modern ones.

Such extraordinary perceptions and skills abound in Hogan's writing. Many of the qualities and behaviors displayed by her characters testify to ontologies tallying with a mixture of what Philippe Descola attempts to categorize as either "naturalism," "animism"—which, in Descola's classification, encompasses shamanic practices—or "totemism."[9] Rather than being presented as paranormal, the sentient powers embodied by Hogan's characters form part of their everyday reality. Their sensory-motor capacities capture a form of knowledge that eludes Cartesian thinking—a thinking relying predominantly on sight and on the knowledge that can be acquired via the conscious, rational mind only. By contrast, like many indigenous peoples, Hogan's characters rely on corporeal and intuitive perceptions reaching far beyond sight. The characters with unusual perception have moreover been initiated, by their elders and via liminal experiences, into an organic, multispecies communitas.

The biocentric vision of communitas thereby produced involves various types of porosity and continua in both spirit and matter between human and other-than-human realms. In Hogan's novel *Power* (1998), while the sixteen-year-old homodiegetic narrator first asserts that, as a good school student, she does not "think there's such a thing as magic," she nevertheless lays claim to some kind of super sensorial access to reality. Sensing as an animal would the presence of the Florida panther watching her, she feels "a cold chill [that] passes over [her] back." Thus, she "can tell" she is being watched: "I feel it in my body, something not right; eyes watching from the trees, something stirring about. I feel it in my stomach, an animal feeling, something—or someone dangerous."[10] Her sensitive access to the world is definitely one that is not so much magical as it simply situates her in-between animal, elemental, and human worlds. As a human animal spending most of her time outdoors in the Florida swamps and mangrove forests, she has retained the capacity to trust her animal responses, her gut instinct of being watched and potentially preyed upon. It is suggested here that she is being stalked by a Florida panther: "You can feel it more than see it, feel it more than smell it. It feels like space has eyes and ears, and it watches with all its might, listens with ears that can pick up the slightest hint of sound, and it moves slowly, silently. [. . .] It makes the hair on your back and neck rise [. . .]."[11] If the protagonist's sensory gift sets her apart from most of her family and clan, who have largely adopted the modern ways and worldviews of the colonizers, her own take on it simply points to a heightened awareness of the material presences around her: "I know what I feel and there are things I know and feel and see that other people don't. [. . .] That's why my father called me Omishto. It means the One Who Watches [. . .]. But it's true, I watch everything and see deep into what's around me. [. . .] I feel lives and spirits in the woods, and I see growing things."[12]

Like Ama's—the older woman initiating the girl to the traditional ways and stories of the Panther Clan they belong to—Omishto's way of "watching" is not solely based on visual information. Indeed, her narrative teems with synesthetic references, marking her acute sensitivity to the auditory, haptic, and olfactory textures of the world. Unrestricted by the scopic hypertrophy most humans apprehend the world with, Omishto and Ama's *Umwelten* seem augmented. Let us look for example at the wild storm scene at the beginning of the novel. Described via internal focalization, the scene relays Omishto's point of view. It is striking here how much attention is paid to sounds, scents, and touch. The first hunches of the menacing hurricane come via "the birds flying in," birds described as "noisy and agitated about the coming storm." Observing their behavior, Omishto can read their reactions to the change of atmosphere. She moreover notices all at once how "the sky is growing darker" and how she "can feel the temperature drop."[13] Later, as she listens to Ama's premonitory dream about the panther and as the storm gathers, Omishto's attention constantly entwines the various sensory signals she picks up: "the strong-smelling oilcloth" in Ama's house, the "temperature [dropping] to the point where there's a chill in the air" and announcing "the rain coming," the curtains "[dancing] with the sweep of wind," "the noisy excitement in the air."[14] As "the wind starts knocking at the door like it has knuckles and fists," Omishto confides that "[her] skin loves the wet air." She enjoys "the overwhelming smell of freshness." Even though she is "afraid of storms," she "closes [her] eyes and [breathes]," connecting "[in] the distance" to "the soft outlines of the land breathing in the water."[15] Despite her fear, Omishto welcomes the rain, being aware that both the parched land and its multispecies dwellers have been suffering from drought and ensuing wildfires.

Omishto's deep, intuitive, and sensuous attunement to the natural world reveals her elemental connection with land and wind: "The wind is a living force. We Taiga call the wind Oni. It enters us all at birth and stays with us all through life. It connects us to every other creature."[16] Listening first to the "storm [strengthening], the rain beginning to billow down," to "the sky shaking with the sound of thunder," then feeling "the strong wind filling the rain-smelling room," and watching the "flapping" curtains, Omishto then listens to "Ama's dying house" that "sounds like a seashell with ocean wind blowing through it or maybe [. . .] like wind in a bottle." As Ama's old dog "sets to howling and looking for a safe place," as her own "hair on the back of [her] neck rises, and a cold chill climbs up [her] back," Omishto translates all her animal feelings into a certitude: "I stand up. It's more than just a storm. 'It's a hurricane,' I say to Ama. I'm sure of it.'"[17] Omishto's way of being in the world shows specific attention to the soundscapes she is immersed in. Her ears are constantly pricked, picking up the slightest movement of wind and animal presence around her which might signal what the next best step ought to be for her own survival. Serving the same function, her sense of smell is remarkably sharp, indicating significant changes in her immediate environment as she navigates the place.

Much of what Omishto has learned in her transactions with the more-than-human world has been passed down to her by Ama. In this respect, the novel constantly reminds us that what we can make of the world is in great part determined by the stories we have heard. Ama has been teaching Omishto tales uniting their Taiga people with the panther: "Back in the days of the first people, the beginning of wind was the first breathing of one of the turbulent Gods, they say. This God's name was Oni. It is said that this word was the owner of wind, and the panther was the one who first spoke it. [. . .] Oni, first and foremost, is the word for wind and air. It is a power every bit as strong as gravity, as strong as a sun you can't look at but know it's there. It tells a story. Through air, words are carried."[18] This origin story focusing on the creative, material quality of air partly chimes in with Saint John's creation story in the Christian tradition, which famously begins with "In the beginning was the Word, and the Word was with God, and the Word was God." A blatant discrepancy lies nonetheless in the secondary roles assigned animal, plant, and elemental agents in the dominant Christian stories, whereas those relayed in Hogan's novels involve humans and animals who, as in many indigenous traditions, are said to once have spoken the same language before splitting up into more differentiated species, each now with its own language and acoustic niche.

Omishto thus learns to heed the slightest movement of vibrant matter as both Ama and she spend much time outdoors in the wild. As a result, she develops a keen sense for the most subtle modifications in the air's texture. She pays meticulous attention to the many ways air and wind manifest themselves to her human senses:

> Usually, it is invisible. Only today I can see it. It is moving shadows. Its hands are laid down on every living thing. The plants that create it are held inside it and moved by it. In the presence of air, every living thing is moved. [. . .] It is the breath of life translated from trees. Because of this, there is no such thing as emptiness in our world, only the fullness of the unseen. [. . .] Sometimes air, full of plant and animal breathings, shows itself only in the ripples it sets moving water or the grasses it bends.[19]

Insisting on the hardly visible quality of air, the narrative paradoxically makes it apparent as a body (with "hands") that shapes water and grass—a rippling, sculpting force that readers are likely to have contemplated before. Additionally, Omishto's ecopoetic vision here transcribes the actual gas exchanges between lifeforms constantly at work through the air. In that sense, wind is presented here as much more than pure, mythical spirit. Air is foregrounded as one of the tangible elements interlacing humans within a multispecies world: "Field, forest, swamp. I knew how they breathed at night, and that they were linked to us in that breath. It was the oldest bond of survival.""[20] In this passage from *Solar Storms,* Angel the plant dreamer forms a poetically meaningful and ecologically accurate image of the interconnections between herself and plants which are transcribed in her dreams: "Maybe the roots of dreaming are in the soil of dailiness, or in the heart, or in another place without words, but when they come together and grow, they are like the seeds of hydrogen and the seeds of oxygen that together create ocean, lake, and ice. In this way the plants and I joined each other. They entangled me in their stems and vines and it was a beautiful entanglement."[21] Hogan's liminal realist narrative here echoes Native American Jack Forbes' vision of the very material continuum between all lifeforms:

> That which the tree exhales, I inhale. That which I exhale, the trees inhale. Together we form a circle. When I breathe I am breathing the breath of billions of now-departed trees and plants. When trees and plants breathe they are breathing the breath of billions of now-departed humans, animals and other peoples. As Lame Dear said, 'A human being too, is many things. Whatever makes up the air, the earth, the herbs, the stories, is also part of our bodies.[22]

Whether taking place through dreaming or in waking life, Hogan's liminal moments of revelation serve to dispel what Alan Watts has called "the illusion of oneself as a separate ego," reminding us that "differentiation is not separation."[23] With their transcorporeal stances, Hogan and Forbes provide an antidote to the modern alienation that stems, in Paula Gunn Allen's words, from "a condition of division and separation from the harmony of the world."[24]

It should appear clearly by now that Hogan's use of liminal realism is not so much magical—in the sense that it includes phenomena that hard to explain rationally, even though it does also include such elements,—as it is fundamentally liminal, situating us in-between different *Umwelten,* different ontologies, and different epistemologies. Hogan's literary mode may be said to be magical mostly in the sense that it affects our perception of dwelling. It is magical in the way David Abram or Starhawk refer to magic—that is as a willful act that transforms our awareness of what may or may not be.

Hogan's ecopoetic prose fiction is systematically and inherently liminal in the following seven ways: (1) The characters evolve in-between human and more-than-human worlds. For instance, in P*ower*, both Ama and Omishto act throughout as mediators, or shamans. Trained as they are in tracking, they both possess the ability to read the world like a cat, and to move through the Everglades as if in the animal's skin. (2) The postcolonial cultural settings of Hogan's fiction situate her characters in-between precolonial and modern ways of life. Establishing early on the liminal potential of *Power*, Omistho reflects on the elders of her tribe living above the swamp: "I think of them, always, as living at the threshold to our world."²⁵ (3) The geographical location of her stories are more often than not ecotones, that is, threshold ecosystems in between water and land, either coastal areas or marshlands. (4) Hogan's stories waver between history and fiction. Her postcolonial dialogism intertwines the characters' discourses and fates with collective ones. Her narratives reclaim the experiences of the colonized, shedding light on historical processes from the standpoint of minoritized—when not endangered—indigenous groups. Moreover, much of her fiction is partly historiographic as it dramatizes true environmental conflicts, or centers on endangered species such as the gray whale or the Florida panther, living in the vicinity of marginalized indigenous groups.²⁶ (5) Hogan's prose fiction constantly situates us in-between myth and science as it interweaves cosmogonies of place with a keen naturalist's observation of the local ecosystem. (6) Events often conflate mythical time and space, and fictional time and space. In *Power* for example, the main plot braids the old stories of Sisa the Panther-Woman with the present-day story of the Florida panther and Ama—herself a contemporary shaman-like panther-woman. (7) Events can take place somewhere in-between dream and reality. Ama, for instance, receives her mission to kill the starving panther as the latter first visits her in her dreams. In addition, the porosity between what happens in Omishto's waking life and in her dreams is constantly foregrounded.

Via a similar process, *Solar Storms* recounts the initiation process of Angel, who undergoes healing on a personal level while being reintegrated into a multispecies, organic world of interdependence. Discovering her gift as a "plant dreamer," she finds out that she can help retrieve knowledge of herbal medicine. In line with shamanic tradition, once again, it is via some intuitive access to her ecological unconscious and the knowledge of the land mapped out in her dreams that Angel can locate the plants needed for traditional, herbal medicine: "But there was a place inside the human that spoke with land, that entered dreaming, in the way that people in the north found direction in their dreams. They dreamed charts of land and currents of water. They dreamed where food animals lived. These dreams they called hunger maps and when they followed those maps, they found their prey. It was the language animals and humans had in common. People found their cures in the same way."²⁷ In all three cases, the interlacing of the realms of dreams, myths, and of conscious environmental knowledge underlines how interwoven the stuff of our dreams and the stuff of our waking lives are and how, as such, they might influence one another, whether consciously or unconsciously, depending on how much access to one's unconscious one possesses.

In most postcolonial novels, the tangible and the world of the mind are not clearly bounded and separate. They are porous realms, where intellectual knowledge, physical sensations, and infra-linguistic intuition constantly inform one another. Outside of Hogan's ecotonal settings, Leslie Marmon Silko's *Ceremony* exhibits the same cluster of liminal qualities: the main character, Tayo is a mixed blood torn between, on the one hand, the Cartesian worldview and modern medicine which are imposed on him at school and in the hospital and, on the other hand, the old stories and practices passed on by his indigenous elders. As the novel deploys a stream of consciousness technique, readers can assess how Tayo relentlessly oscillates between the rational reading grids that are forced onto him by the establishment—the school, doctors, and the army—and the extensive knowledge acquired through his senses, via intuition, and through his dreams.

Throughout the narrative of his initiatory journey, Tayo's stream of consciousness imprints a way of dwelling wherein, taking my cue from Paula Gunn, "the earth *is* being, as all creatures are also being: aware, palpable, intelligent, alive."[28] His healing takes place at the end of a long initiation journey collapsing myth, individual and collective history, and where material reality is entangled within both corporeal and discursive practices. In Allen's terms, "Tayo lives the stories—those ancient and those new. He understands through the process of making the stories manifest in his actions and in his understanding, for the stories [. . .] are the communication device of the land and the people."[29] Western notions of time, space, and identity are reshuffled as Tayo's experiences defy all the boundaries erected by modern science. To him, mythical, past, and faraway beings manifest themselves in the here and now. Like Hogan's characters, Tayo's individual recovery hinges on liminal experiences situating him in-between the world of myth and his actual present, between elemental, animal, and human realms, with the intervention of shapeshifters moving in-between the spirit world and the physical world. Helpers include a mountain-woman and mountain-lion-man. Mediated as they are through the points of view of young hybrid protagonists, such ecopoetic liminal realistic narratives gradually initiate readers into a threshold world where the modern categories, concepts, and classifications that dictate how we normally conceive of space, time, identity, reality, and the living world can get reconfigured.

If they may at times seem destabilizing, even the dreaming and the shamanism pervading Hogan's novels find sense when read through the lens of animistic or individual and collective totemic bonds. Thus, Hogan's fiction makes way for the sensitivity of indigenous peoples who "continually receive direct, unmediated revelation from a sacred landscape and the genii loci that populate that landscape."[30] Drawing from Caroline Durand Rous's work on the reinvention of totemism in contemporary North American indigenous literature, I have shown how the ecopoetic liminality and magical realism in Hogan's *Solar Storms* and *People of the Whale* actually abide by the logic of totemism.[31] Reading *Power* through a similar lens, it appears clearly that here too, the liminal realism at play hinges on the totemism of indigenous traditions.

First, the relationships between Ama, Omishto, their Taiga people, and the Florida panther originate in collective totemism: "And the cat. Ama says this is its place, its territory. She always watches for it. She has seen it and she believes in it. In the old way, she says, the cat is her relative. My relative, too, since we are all in the same clan."[32] Aligning with totemic worldviews, the panther is called "a relative" or an "ancestor." It is alternatively referred to as an "older sister," a "cousin," an "aunt," or a "grandmother."[33] When Ama and Omishto address the panther, they call her "Aunt," or "Grandmother." Such words express a reverential kind of kinship that does not so much designate a direct blood relative as it points to a complex relationship of evolutionary kinship and intimacy as co-dwellers of one and the same territory.[34] Moreover, the family title encapsulates identification processes and acknowledged personhood suggesting mutual guardianship. Ama is the cat's "protector"[35]—a role Omishto inherits after Ama vanishes. Hence the panther's apparition out of the wild at the end of the book, resealing the covenant between the Taiga people and the animal:

> Again, something is around. When I get closer, [. . .] I see it. Standing still, looking back at me, the golden cat, large and with the tawny fur loose and healthy, lean-muscled. I don't move. It could kill me, swallow me. It thinks the same of me. We stand motionless and look at each other in the near morning and then I say, "Nho shi holo." I mean no harm, Aunt, Grandmother. I think this is the mate of the one Ama killed. Or maybe, as Ama maintained, it is the same one returned, fully grown and beautiful, or the one that was born alongside me at my beginning. As I watch it, as the trees open their ears and eyes and listen and watch, a breeze moves through and with it the spell is broken. The panther turns and walks away, slowly. It is not at all in a hurry and I want to say, "Run, we are dangerous people, every last one of us."[36]

As with the first human-panther encounter, a great part of the deep, mutual recognition at play has to do with a reciprocal registration of one another's vulnerability, placing the two individuals, and the greater species they represent, on an equal footing.

This crucial encounter recalls the exchange of glances between Tayo and the mountain lion in Silko's *Ceremony*—a key episode that is "part of the cycle of restoration" the book pivots on.[37] In that moment, Tayo's sensitivity to the natural world around is expressed by the many references to his enmeshment with the phenomenal world: "His face was in the pine needles where he could smell all the trees, from roots deep in the damp earth to the moonlight blue branches, the highest tips swaying in the wind. The odors wrapped around him in a thin clear layer that sucked away the substance of his muscle and bone; his body became insubstantial [. . .]."[38] Internal focalization relays Tayo's mindfulness of different earthly textures, translated via synesthesia. The leitmotif of entanglement is moreover introduced from the beginning, with many images of ties and knots interwoven in the opening passage. In the first pages of the narrative proper, the lexical field revolves around this notion: things past and present in Tayo's mind are "tied together like colts in a single file [. . .], with the halter rope of one colt tied to the tail of the colt ahead of it, and the lead rope tied to the wide horn on Josiah's saddle"; his "memories [are] tangled with the present, tangled up like colored threads from old Grandma's wicker sewing basket." Tayo "[can] feel it inside his skull—the tension of little threads being pulled and how it [is] with tangled things, things tied together, and as he [tries] to pull them apart and rewind them into their places, they [snag] and [tangle] even more." His thoughts "[become] entangled," as he tries "to think of something that [isn't] unraveled or tied in knots to the past."[39] Like his elders' art of basket weaving, there is something in Silko's ecopoetic craft that must honor and take part in the harmonious patterns that can emerge from the world's entanglements.

Like Omishto in *Power*, Tayo has learned from his elders to stay attuned to the more-than-human world. His uncle Josiah has taught him to lovingly guide cattle and, throughout his healing quest, Tayo spends most of his time outdoors. On coming face to face with the mountain lion, Tayo plunges his eyes into the wild creature's in way evoking both transcorporeality and a kind of deeper, spiritual connection: "The eyes caught twin reflections of the moon; the glittering yellow light penetrated his chest and he inhaled suddenly."[40] After Tayo and the mountain lion have been gazing at each other fearlessly, literally seeing eye to eye—and thus sealing a totemic pact—the mountain lion takes off and Tayo performs old rituals of offering, summoning stories of reciprocity and mutual care between humans and animals.

In this sense, Hogan and Silko constantly negotiate what Joni Adamson terms "the middle place where culture emerges from nature."[41] Such ecopoetic moments recall other animal encounters between humans and their totem animals in *Solar Storms* (Agnes and the bear) and in *People of the Whale* (Thomas and the whale). Situating us for a moment at the threshold between animal and human ways of dwelling, these awareness-raising, boundary-crossing, and humbling encounters with the wild play a similar function of renewing the broken covenants between humans and animals.[42] They initiate a reclaiming of the regard and interconnectedness between human and other-than-human worlds, a reweaving of the sacred—because unbreakable—ecological ties between humans and their cotenants dwelling within an animate, enchanting landscape. Of the mountain lion's elusive ways, Tayo observes: "Relentless motion was the lion's greatest beauty, moving like mountain clouds with the wind, changing substance and color in rhythm with the contours of the mountain peaks: darks as lava rock, and suddenly as bright as a field of snow. When the mountain lion stopped in front of him, it was not hesitation but a chance for the moonlight to catch up with him."[43]

The enchanting quality of such encounters is signaled by the expression "the spell is broken" at the end of the above quotation from *Power*, right before the panther turns away from the teenager and goes its own way. The saying highlights the startling, extraordinary and uplifting effect of such rare occasions, the power of which is relayed in much nature writing as pure moments of enchantment. In Annie Dillard's short essay "Living Like Weasels," the narrator recounts coming face to face with a weasel: "I don't remember *what shattered the enchantment*. I think I blinked, I think I retrieved my brain from the weasel's brain, and tried to memorize what I was seeing, and the weasel felt the yank of separation, the careening splashing into real life and the urgent current of instinct. He vanished under the wild rose. I waited motionless, my mind suddenly full of data and my spirit with pleadings, but he didn't return."[44] Albeit written here from a naturalist viewpoint, Dillard's defamiliarizing encounter with the weasel suddenly melts the boundaries between the animal's mind and her own:

> I startled a weasel who startled me, and we exchanged a long glance. [. . .] I was looking down at a weasel, who was looking up at me. [. . .] The weasel was stunned into stillness as he was emerging from beneath an enormous shaggy wild-rose bush four feet away. I was stunned into stillness, twisted backward on the tree trunk. Our eyes locked, and someone threw away the key.
>
> Our look was as if two lovers, or deadly enemies, met unexpectedly on an overgrown path when each had been thinking of something else: a clearing blow to the gut. It was also a bright blow to the brain, or sudden beating of brains, with all the charge and intimate grate of rubbed balloons. It emptied our lungs. It felled the forest, moved the fields, and drained the pond; the world dismantled and tumbled into that black hole of eyes. If you and I looked at each other that way, our skulls would split and drop to our shoulders.[45]

We may recall Jane Bennett's claim that "[to] be enchanted [. . .] is to participate in a momentarily immobilizing encounter; it is to be transfixed, spellbound." Describing enchantment as a moment of total rapture lifting us from our habitual stance, Bennett circumscribes "a condition of exhilaration or acute sensory activity." This resonates with Damasio's earlier definition of intuition. Characterized as an "odd combination of somatic effects," to be enchanted in Bennett's terms is to "be simultaneously transfixed in wonder and transported by sense, to be caught up and carried away."[46]

Like Omishto with the Florida panther, Dillard's narrator recounts having entered a liminal contact zone where the two realms may merge: "I tell you I've been in that weasel's brain for sixty seconds, and he was in mine. Brains are private places, muttering through unique and secret tapes—but the weasel and I both plugged into another tape simultaneously, for a sweet and shocking time."[47] Dillard's experience substantiates Bennett's approach of cross-species encounters, in particular her main contention that "enchantment can aid the project of cultivating a stance of presumptive generosity (i.e., of rendering oneself more open to the surprise of other selves and bodies and more willing and able to enter into productive assemblages with them)."[48] In essence, it also backs up Yves-Charles Grandjeat's claim about the "*dislocation*," the "cognitive rift" and perceptual unleashing which such wild encounters can trigger. Thus, argues Grandjeat, "[common] ground can then be found," with different species dwelling together at the wondrous intersection between human and other-than-human worlds.[49]

Nevertheless, a fundamental difference between Hogan or Silko's animistic worldviews and Dillard's lies in the latter's feeling that one cannot grasp the thoughts and language of animals: "Can I help it if [that tape] was blank? / What goes on in his brain the rest of the time? What does a weasel think about? He won't say. His journal is tracks in clay, a spray of feathers, mouse blood and bone: uncollected, unconnected, loose-leaf, and blown."[50] Where humans and other-than-humans seem to be able to meet and converse in postcolonial ecopoetic liminal realism, Dillard's fantasy of living like a weasel is one where she envisions "[living] under the wild rose wild as weasels, mute and uncomprehending." In Dillard's daydreaming about "living like weasels," as goes the title of her essay, images of oneness and entanglement once again resurface: "I could live two days in the den, curled, leaning on mouse fur, sniffing bird bones, blinking, licking, breathing musk, my hair tangled in the roots of grasses. Down is a good place to go, where the mind is single. Down is out, out of your ever-loving mind and back to your careless senses."[51] It is striking that Dillard opposes human to animal senses, as if the two were incompatible because of our self-consciousness. Yet, if from her standpoint, humans and animals cannot communicate, Dillard nevertheless offers a searing glimpse of some kind of access to infra-linguistic communication: "I remember muteness as a prolonged and giddy fast, where every moment is a feast of utterance received. Time and events are merely poured, unremarked, and ingested directly, like blood pulsed into my gut through a jugular vein."[52] Dillard here gestures to the short-circuiting of the logocentric channels of the brain which can take place via our instinctive, unmediated responses to the biosemiotic textures of the more-than-human world, in such situations where we might feel deprived of reliance on human forms of speech.[53]

In comparison, not only is Ama and Omishto's relationship with animals mediated by the traditional, mythical view that long ago, animals and humans spoke the same language, their own animal ways of being and knowing are furthermore constantly brought to the fore. To read the land, Ama and Omishto follow clues using their animal senses: "From tracks, [Ama] can tell anything. She can read the tracks of all the animals. She has a different intelligence than the rest of us."[54] Omishto confides "I think [Ama] could read the minds of animals too."[55] Characterized as a shaman, Ama lives at the threshold between human and animal realms, in a middle place where the world of dreams, myths, and our ecological unconscious merge with reality. As they are tracing a deer and, on its trail, the Florida panther who has previously visited Ama in a dream, beckoning her, we are told that Ama "looks like some kind of creature. She looks not human."[56] Observing her elder, Omishto witnesses the greater, animal responses to the land which Ama has developed:

> Ama keeps an eye on the tracks of the deer. *I can't see them, but she can*; even though they have been washed by the storm tide, *she can see and smell* them *as if by magic*. She *knows* where an animal has walked. I used to call her Bloodhound, to tease her, but it is *her gift* and try as I do to learn it, I never see anything but fresh-made tracks or newly-born twigs. *She has a sense I lack.* Now she seems to *catch the scent of the deer*; *she picks it out from the smell of fish* that were thrown out of water by the hurricane, *the wet smell of the storm on the ground*. [. . .] Ama stands still and *senses the land, feeling her way into* the brush or saw grass around her, into the oak hammock.[57]

Ama moves like a creature following its wild intuition: "She walks as if pulled by something, drawn as if she has no choice, the way we are drawn and held by gravity."[58] Half-human guided by her totemic logic, and half-animal responding to her gut instinct, Ama's gait and behavior recall the ways wild creatures are driven by the will to survive. In this sense, Ama embodies a tight relationship with the land, born from her felt interdependence with it. This reading of Ama's hypersensitiveness to the signs of other-than-human existences corroborate studies showing that human perception is dependent on the environment in which humanity has evolved. "The part perceived is just enough to safeguard our personal survival and reproduction,"[59] says biologist E. O. Wilson.

Ama's behavior fits with Dillard's contemplation of what it might be like to live as wild as weasels: "The thing is to stalk your calling in a certain skilled and supple way, to locate the most tender and live spot and plug into that pulse. This is yielding, not fighting. A weasel doesn't 'attack' anything; a weasel lives as he's meant to, yielding at every moment to the perfect freedom of single necessity."[60] The killing of the panther itself is presented in Hogan's novel as motivated by a force beyond human control—a necessity binding the panther's and the humans' fates together: "[Can't] a human decide what to do and what not to do? Can't Ama turn back now and can't I turn around and go home to my mother and sister who might even at this moment be homeless from the storm? I can't and I don't. And neither can Ama and I know this. We are carried in something larger."[61] Abiding by the logic of shamanism, it is inferred throughout the novel that Ama's and Omishto's destinies have long been singled out. Their high potential—whether at school or in the wild—manifests itself constantly. Omishto stands out for how brilliant a student she is, and, like Ama, she appears gifted with hypersensitivity, hyperesthesia, and hyperempathy. She feels more, sees, hears, and smells more strongly than most, and is constantly suffering for the land, feeling sick for the animals and for the trees when the latter are wounded, starving, or endangered. Called "the One who Watches,"[62] she can "read people" and intuit their moods and thoughts. "With Ama," she says, she has "learned to read sky and water, land." Her constant reading is motivated by a quest for meaning: "this reading [. . .] I do [. . .] as if to find something true that lies beneath all the rest."[63] In a nutshell, both Ama and Omishto live more intensely. They possess both greater animal and greater human intelligences, setting them apart as shamans.

Throughout her initiation into the threshold world where the panther's realm fuses with the human world, the adolescent's usual senses of time, space, and identity are confused. In that moment, mythical space-time catches up with actualized space-time as the two suddenly overlap within their clan's stories:

> I look back at the tracks we've left in the wet ground, as if we've grown from them, as if they created us and we grew upward, rose up as if from the footprints of our ancestors, to become the flesh of a woman and a girl. [. . .] Then, soon, I recognize that we are in the place where the stones look like backbone. Some people say this is God's back but we call it the Taiga Birthplace. In this place, the place of second creation, you'd think all this is true and that we walk toward the feet of that spirit, toward what meets firm and solid with the ground. It could be another world, another time, except that in the distance I hear a radio, and it sounds so strange to me. [. . .] Ama doesn't hear it, though, she only hears the deer walk.[64]

Omishto marks how she and Ama have learned different habits of attention, and consequently, how they inhabit different worlds. Although both are from the same clan, their experiences of the world are so different that they possess different *Umwelten*—they inhabit different perceptive worlds, with differently trained sensory-motor response-abilities: "Listen to [the deer's] hooves," [Ama] says, and I wonder how, always, she puts this world away as if it never happened and how she hears the little feet of the deer. [. . . And] it's this I like about her, that even though there are only twenty years between us, we live in different worlds. We do not even hear the same sounds. I try but I can't hear the sounds of animals walking and she doesn't hear a radio."[65] The teenager gradually learns to develop a similar hyperesthetic talent, so that by the end of the novel, she too can sense how a "place itself feels alive. Here, the land itself seems to have a sound, the soft brush of a breeze."[66] Omishto is expanding her own, human *Umwelt*, and in her wake, we too, as readers, may be acquiring a wider range of perceptions.

E. O. Wilson emphasizes how we "are primarily audiovisual, one of the few animals on the planet, along with birds and a smattering of insects and other invertebrates, that depend on sight and sound to find their way." Even then, our sensory capacities are limited. Sight-wise, "the only particle to which we respond is the photon. Still more restrictive," Wilson goes on, "our photoreceptors detect no more than razor-thin slices of the electromagnetic spectrum. Our vision begins at the low-frequency end with red (and does not even extend earlier to infrared) and it ends short of ultraviolet at the high-frequency end."[67] Regarding sounds, when compared with the "auditory geniuses of the animal world" such as those of echolocating bats, of moths, or of rumbling elephants, Wilson concludes that "we are close to deaf."[68] Far from giving in as Dillard does earlier to the anthropocentric illusion that, since we do not understand them, animals must be mute, Wilson highlights how "elephants rumble in complex conversation too low in frequency for our ears."[69]

As for smell, relying on the paucity of our vocabulary for scents in English, Wilson draws the conclusion that "humans by comparison with the rest of life are virtually anosmic. Every place, natural or cultivated, is alive with pheromones, chemicals used to communicate among members of the same species, and allomones, used by organisms to detect other species as potential predators, prey or symbiotic partners."[70] Living in the wild, and thus more vulnerable to predators and more dependent on hunting, fishing, and foraging, Hogan's trackers seem to have developed a greater olfactory sense for what Wilson refers to as "odorscapes"—"the thunderous round-the-clock chorus of olfactory signals, its varying mixes forming a riot of airborne scent."[71] Wilson mulls over the "phantom worlds in which we actually exist." "Surely," he ponders, "we cannot picture the living world and keep it safe without understanding the soundscapes and odorscapes by which it is organized."[72] Considering how Traditional Ecological Knowledge and environmental sciences such as ethology and biosemiotics can infuse *ecopoiesis*, it may well be that literature can play a role in amplifying our sensory capacities in response to the multiple textures of the world, thus refining our habits of attention.

From that standpoint, the translations of languages, stories, and worldviews belonging to indigenous peoples may prove essential, multiplying our capacities to perceive the multispecies vibrancy of the biocultural living world.

In *Power*, the tales surrounding Ama's past suggest that she was also initiated as a shaman. It is believed that she had to fend for herself in the swamps when she was twelve—the age of puberty, rather typical of initiation rites—and that she was later taken in by her elders living by tradition. Initially, Omishto exhibits caution as regards the myths circulating amongst her people, specifically those rubbing up against modern rationality. The earlier parts of her narrative teem with symptoms of internalized racism, mostly as Omishto discredits the old, animistic stories: "Some of the superstitious ones believed she'd been taken by the little people to learn the medicine ways. That is what the old people say happens in the woods at times, the little people take a person away and teach them things, and when that person returns they know medicine. But this is just another old belief and I don't give superstition even an ounce of weight."[73]

Omishto's initiation in Ama's steps gradually releases her from the stronghold of school-learned worldviews. By the end of the book, she embraces her culture and people, as their vision best corresponds to her own intuitive feelings about dwelling. But at first, Omishto's point of view is marked by profound ambivalence—thus making room for the Western hesitation that, according to Wendy Faris, is characteristic of magical realism when it comes to interpreting "supernatural" events. She is visibly both attracted to and repelled by the old stories and ways. This shows in the rumors she relays concerning Ama's elusive, potentially shape-shifting nature—another attribute traditionally lent shamans:

> Other people were afraid Ama'd been killed by that [. . .] bear [. . .] and that what returned was not really Ama but only looked like her, like a spirit that had changed bodies the way they used to do when people could turn into animals and animals could transform themselves into a human shape. And some people said she'd done that, she'd met and married a panther, and now she was an animal come back to observe us to see if our manner of worldly conduct toward them was right and kind, and because of this, those people [. . .] were afraid of her strength. [. . .] If I had to believe any one of these superstitious stories, it would be this last one, but only because it's the most interesting.[74]

Ultimately, Ama embodies and renews the old stories. She becomes first one of the storytellers of the Panther Clan—albeit one that remains on the fringe of human groups—and then the main character of the panther-woman story unfolding in the present.

Ama's liminal status is thus engrained in her characterization. We learn that even though she was partly raised by Janie Soto, the Panther Clan's leader, "Ama said the old ways are not enough to get us through this time and she was called to something else. To living halfway between the modern world and the ancient one."[75] Dwelling in a humble abode all alone somewhere in the swamps on the outskirts of both the tribal people's village and the more modern integrated and mixed-blood community where Omishto lives, Ama possesses all the attributes of the shaman. As David Abram assesses, such exceptional individuals "rarely dwell at the heart of their village; rather, their dwellings are commonly at the spatial periphery of the community or, more often, out beyond the edges of the village." Ama's different intelligence adheres with Abram's take on the shaman's role and place in society: "the magician's intelligence is not encompassed within the society; its place is at the edge of the community, mediating between the human community and the larger community of beings upon which the village depends for its nourishment and sustenance."[76] To a certain extent, and from an extradiegetic standpoint, the ecopoet herself writes as a shaman, trafficking in-between her readers and the more-than-human world, mediating different voices and points of view, and striving to restore the broken bonds between species. Like that of a shaman, Hogan's concern is with "[this] larger community [which] includes, along with the humans, the multiple nonhuman entities that constitute the local landscape, from the diverse plants and the myriad animals—birds mammals, fish, reptiles, insects—that inhabit or migrate through the region, to the particular winds and weather patterns that inform the local geography, as well as the various landforms—forests, rivers, caves, mountains—that lend their specific character to the surrounding earth."[77]

Tightly fitting the features of the shaman which Philippe Descola also details,[78] Ama's ability to communicate with the panther via her dreams and in her waking life abides by her function as mediator between animals and humans—a function that Omishto gradually inherits together with Ama's marginal dwelling place. At the heart of her voyage, Omishto is taken into a world obeying different laws from those learned in school in terms of time, space, dwelling, and personhood:

> Ama once said that space is full and time is empty; I think now I understand this. We are surrounded by matter, but time disappears from us. Or maybe, as Ama says, there are other worlds beside us all the time and every now and then we cross over and enter one, and every so often, too, one passes over and enters ours.
>
> Looking up, there is a hole in the sky, the way the old stories say about the hole pecked by a bird, a hole through which our older sister, the panther, Sisa is what

we call her in Taiga language, entered this world. And the anhinga birds with wings draping down have just come down through that hole; I see several of them as they sit in the trees and sun themselves. They don't move when we pass beneath the angle of their wings. We are that determined; we are that in nature. They can read us.[79]

In line with Eduardo Kohn's "ecology of selves," Omishto's narrative constantly foregrounds the reciprocity of the notion of personhood between humans and other-than-humans, whether trees or animals.

One of Omishto's defining traits is her empathy for nonhumans, expressing the kind of care most Westerners usually direct toward their fellow humans. This mutual recognition is tellingly enacted now and again as either of the two characters, the shaman and her initiand, "look out into the darkness and see [the panther's] eyes," and "[exchange] glances" with her and with other animals.[80] As they catch up with the "vulnerable and beautiful and bare cat" Ama is about to sacrifice—and spare from starving—the panther "raises its head and seems to look right at [Omishto], its eyes turning to light, round and glinting, its body all animal and lean muscle, its face so thin."[81] Highly aware in this suspended moment that either could be the other's predator or prey, their equal footing and shared vulnerability are expressed by the unexpected behavior on the part of the cat: "The cat looks back at us. It doesn't run. In the darkness its eyes shine and that is what I see. Eyes. It seems to look right through us. It sees though us. Then, at ease, as if it is certain we will follow, it moves slowly away. It is calling us forward. I can see this in the way it looks back at us from time to time, and in the fact that it is calm. [. . .] It is sure of us, and that we will follow."[82] Goading the women, the panther acts friendly and trusting, almost humanlike. Conversely, Ama's gait and posture become increasingly animalized: "At times, [the cat] disappears from view and I don't know how Ama traces it. [. . . It] appears every so often to lead us forward, as if it knows what it is doing. [. . .] And I think Ama is deadly more than I could know, stalking, crouching, slipping, dangerous and hungry under the brush and limbs."[83] Hogan's liminal characters thus prompt us to follow in their tracks and reconsider the pluriverse from a multispecies point of view, informed at once by animal, vegetal, and elemental ways of dwelling.

Taking my cue from Victor Turner's elaboration on the concept of communitas, I would refer here to Martin Buber's take on "community," which aptly describes the biocentric worldview expressed via Hogan's liminal realism: "Community is the being no longer side by side (and, one might add, above and below) but *with* one another of a multitude of persons. And this multitude, though it moves toward one goal, yet experiences everywhere a turning to, a dynamic facing of, the others, a flowing from *I* to *thou*."[84] Tied to a community sharing a "specific territorial locus,"[85] Hogan's stories of kinship make it evident that Turner's structure and communitas beg to be reframed from a biocentric perspective. Already, Turner pinpointed "the spontaneous, immediate, concrete nature of communitas, as opposed to the norm-governed, institutionalized, abstract nature of social structure." "Structure," he went on, "has cognitive quality; as Lévi-Strauss has perceived, it is essentially a set of classifications, a model for thinking about culture and nature and ordering public life."[86] Turner's anthropocentric thinking yet contains the seeds that germinate in ecopoetic liminal realism, which encourages us to reenter "into vital relations" with other lifeforms across a wider spectrum, rather than simply with "other men":

> Bergson saw in the words and writings of prophets and great artists the creation of an "open morality," which was itself an expression of what he called the *élan vital*, or evolutionary "life-force." Prophets and artists tend to be liminal and marginal people, "edgemen," who strive with a passionate sincerity to rid themselves of the clichés associated with status incumbency and role-playing and to enter into vital relations with other men in fact or imagination. In their productions, we may catch glimpses of that unused evolutionary potential in mankind which has not yet been externalized and fixed in structure.[87]

Making Turner's take on communitas more inclusive, the liminal experiences of ecopoetic characters can be summed up as "a transformative experience that goes to the root of each person's being and finds in that root something profoundly communal and shared."[88]

In Hogan's writing, the beauty and vulnerability of the living world palpitates at the heart of what her characters discover to be "profoundly communal and shared." Moreover, what Turner intuited about the power of artistic production casts light onto the eco-onto-epistemo-logical potential of liminal realism:

> Liminality, marginality, and structural inferiority are conditions in which are frequently generated myths, symbols, rituals, philosophical systems, and works of art. These cultural forms provide men with a set of templates or models which are, at one level, periodical reclassifications of reality and man's relationship to society, nature and culture. But they are more than classifications, since they incite men to action as well as to thought. Each of these productions has a multivocal character, having many meanings, and each is capable of moving people at many psycho-biological levels simultaneously."[89]

The literary corpus I have chosen for this book indeed proves that *ecopoiesis* activates the magical power of literature to induce the vision of a "spontaneous communitas." Liminal realism moreover corroborates Turner's claim that "[spontaneous] communitas is nature in dialogue with structure."[90] To put in a nutshell, tapping into our ecological unconscious and our embodied knowledge, ecopoetic liminal realism forms an expression of that evolutionary potential that resists the social and epistemological structures of modernity.

Hogan's reweaving of modern worldviews with animistic, totemic, and shamanic practices of dwelling is one essential aspect of the novel that Lydia R. Cooper's analysis of the novel has missed. As a result, Cooper concludes rather beside the point that Ama's "violent actions damage her own humanity and sense of kinship with nature and with other people."[91] Cooper claims that Omishto condemns Ama for the brutal killing of the panther, overlooking both the dramatic irony at play throughout the novel and the nondualistic, totemic logic underlying the killing in the first place. Because of our privileged access, via internal focalization, to Omishto's feelings and thoughts, we may experience by proxy her position in-between two cultures and two radically different ontologies. On the one hand, there are the treaty rights and the totemic culture whereby Ama's killing of the panther is meant "to renew the world."[92] It means to regenerate the old story whence her dream might have arisen, a story according to which "a woman went into the cypress to kill the cat so that the world might return to balance." Because the entire ecosystem tying together the many naturcultures that it shelters is out of kilter, and because, in her lawyer's words, Ama "believes in balance in the universe,"[93] Ama's act, like in the old story, seeks to generate a negative feedback loop, bringing both the indigenous and the nonindigenous groups back into the same biocentric worldview revolving around the now endangered panther. On the other hand, there are the white environmentalists defending the cat. "And I agree with them and with treaty rights, too," Omishto confides in her struggle to find moral sense. "How can there be two truths that contradict each other? And me. I am on both sides now; that is the worst. What Ama did was wrong, and it was right."[94] Only from this uncomfortable, liminal standpoint can we understand the oxymoronic ethics whereby Ama's act "is both grace and doom, right and wrong."[95]

The panther's killing is wrong for the beautiful wild cat that might have lived. It is right because it was carried out both out of mercy for the starving animal and as a sacrifice that would renew tradition. Benefiting from firsthand access to what Omishto sees and knows, we have stood witness to the whole hunting and killing scene. Thus, as opposed to the other characters in the story, we are aware that Ama has in fact scrupulously respected tribal customs from the beginning to the end, hoping to free the cat's spirit so that it could return in a different body: "I see [. . .] that she made it a bed of leaves in a circle of twine, that she offered it tobacco and food, that she did the correct thing [. . .]. She followed the old traditions of caring for the hunted cat, the prey, of giving it the proper respect. She even offered it pollen and corn, so its soul could eat before it left."[96] From our vantage point, we know more than the jury in the state court and the elders in the tribal court who sentence Ama to banishment—a sentence that, in cutting her off from her community and land, we are told, is equivalent to death. We also figure out Ama and Janie Soto's connivance early on, whereas it takes dazzled Omishto longer to come round and put together all the pieces to solve the puzzle of the missing panther skin. As a result, since we can identify before the teenager does the panther's skin laid out on Janie's lap during the elders' trial—the missing skin she implicitly came to pick up from Ama's the night of the killing—and because we have witnessed the killing at close range, we should fully grasp the true motive and the extent of Ama's sacrifice. We can see how deeply Ama hurts from the cat's pitiful condition, as it appears starving and scrawny, with "broken teeth," damaged fur, its long, scarred face—no longer "beautiful and large and powerful."[97]

As with the scene in *Solar Storms* where Agnes kills a bear to save it from its miserable, caged life,[98] we can appreciate Ama's identification with and sympathy for their totem animal. The same goes for Omishto: "I see that this is what has become of us, of all three of us here. We are diminished and endangered."[99] Paradoxically, this act of killing is both merciful toward the cat, and reempowering for Omishto and the Taiga people, whose treaty rights and worldviews are then acknowledged in court. Through dramatic irony, we, like Omishto, must be privy to the secret surrounding the panther's sickly, miserable state which triggered its sacrifice in the first place. If we know that revealing the killing as an empathetic gesture toward the suffering animal might protect Ama—"the woman who [most] loved and worshipped the cat"[100]—we are also granted access to the reason for this secret and for Ama's ensuing sacrifice. First, and again following a totemic logic, it is one way in which to restore balance. Something is given on both sides, animal and human. Second, Ama's banishment is the price to pay to allow her people to keep on cultivating hope. Rubbing against simplistic dualistic ethics of right and wrong, if the panther's sorry state is the one thing Ama forces Omishto to lie about in both courts, it is out of generosity, wisdom, and grandeur: it is simply because she knows that her people would be devastated to find out the truth about their sister-animal, that in these troubled times when their own naturcultures are threatened and could go extinct, they need to go on believing in the cat's power and resilience, the better to keep holding on to their own dignity and resilience.

Furthermore, we understand that Ama, in her sacrifice, has also planned for her own replacement. She introduces Omishto to the panther before shooting it, so that it will know who to return to—who will take over as the one watching over the world: "Ama watches, listens, and waits. And then Ama says, 'Omishto, come stand beside me,' and I do. The cat looks up and she shows me to the cat, and what she does is, she introduces me to it, it to me. She says my name as she looks at me, as if I am both an offering and a friend."[101] Hence the many allusions throughout to passages, to new beginnings and changes of skin, until the moment when, at the end of her liminal process, and feeling renewed, Omishto is fully reborn. Following a journey in and out of liminality that parallels those of Angel in *Solar Storms* and Thomas in *People of the Whale*,[102] this marks her aggregation within her Taiga people, where she embraces her role as the next mediator, the next shaman, and, most importantly, the next storyteller. "Stories are for people what water is for plants," the narrator concludes by the end of the book, in a very metafictional comment. By taking part in the dynamic regeneration of the old stories and rituals, Omishto's role as both character and storyteller dancing between two cultures is to sustain the people, to remind us that the earth is animate, that it is delicately interwoven with humans and with the stories we tell, and to help us nourish hope for reconciliation and healing.

Hogan's complex use of liminal realism thus renews for her readers the old sense of kinship between humans and panthers. This rekindled relationship of reciprocity involves deep connection—a spiritual, corporeal, and ecological connection that exists as both species dwell as keystone predators within the same ecosystem. We are reminded that as top predators, we might encroach upon each other's territory and food, even prey upon one another, or we might offer each other mutual respect, if not protection, and thus keep the entire biotic community in balance. Restoring the old relationships of interdependence with the land and between its co-dwellers, Omishto's story simultaneously restores our sense of ourselves as guardians and translators of the voices of the earth. Inspired by Hogan's characters, we are called upon throughout to heed the songs and voices of the earth: "[Janie Hide] hears the voices of the animals. And it's the whole of creation she has the duty to stand by, to speak for, to arrange."[103] As my next chapter explores, also braiding animistic and totemic ontologies in their otherwise naturalist ecopoetic novels, non-indigenous writers too may be deploying liminal realism as a strategy "to stand by, to speak for," and "to arrange" the land and its many co-dwellers.

NOTES

1. "Scheherazade," 167.
2. Hogan, *People*, 19.
3. Ibid, 19.
4. Peterson and Hogan, *Sightings*, 47.
5. Ibid, xviii.

6. German ethologist Jakob von Uexküll proposed the term "Umwelt" to circumscribe the milieu inhabited by an animal, which is defined by the animal's specific sensory-motor capacities (*Milieu*).
7. Peterson and Hogan, *Sightings*, 1.
8. Ibid, 2–3.
9. Descola, *Par-delà*.
10. Hogan, *Power*, 2.
11. Ibid, 3.
12. Ibid, 4.
13. Ibid, 25.
14. Ibid, 26.
15. Ibid, 26–27.
16. Ibid, 28. This typically fits Ingold's definition of animism, while being combined with a totemic worldview of kinship between the panther and the clan.
17. Ibid, 29.
18. Ibid, 178.
19. Ibid, 178–79.
20. Meillon, "Writing," 227, quoting from Hogan, *Solar*, 171.
21. Hogan, *Solar*, 171.
22. Forbes, *Columbus*, 182.
23. Watts, *Book*, 21, 79.
24. Allen, *Sacred*, 60.
25. Hogan, *Power*, 120.
26. See Meillon, "Writing," 209–10.
27. Hogan, *Solar*, 170.
28. Allen, *Sacred*, 119.
29. Ibid, 120.
30. Smith and Fiore, "Landscape," 50.
31. Durand-Rous, "Totem," and "Transmission"; Meillon, "Writing."
32. Hogan, *Power, 3.*
33. See Hogan, *Power*, 54, 55, 135, 233.
34. See Chadwick Allen's illuminating study of the blood/land/memory complex in indigenous cultures. Allen examines the reclaiming of ancestors central to the rebuilding of identity and worldviews for Maori and Native American literary activists.
35. Hogan, *Power*, 133.
36. Ibid, 233.
37. Silko, *Ceremony*, 196.
38. Ibid, 195.
39. Ibid, 6–7.
40. Ibid, 195.
41. Adamson, *Middle*, xix.
42. Meillon, "Writing."
43. Silko, *Ceremony*, 195–96.
44. Dillard, *Teaching*, 67–68, emphasis mine.
45. Ibid, 67–68.

46. Bennett, *Enchantment*, 5.
47. Dillard, *Teaching*, 68.
48. Bennett, *Enchantment*, 131.
49. Grandjeat, "Wondering," 4, 39.
50. Dillard, *Teaching*, 68.
51. Ibid, 69.
52. Ibid, 69.
53. For more on Dillard's ecopoetic prose, see Nathalie Cochoy's luminous contributions: "Imprint," "Réalisme," and "Dillard."
54. Hogan, *Power*, 133.
55. Ibid, 110.
56. Ibid, 52.
57. Ibid, 53–54 emphasis added.
58. Ibid, 54.
59. Wilson, *Origins*, 59.
60. Dillard, *Teaching*, 69–70.
61. Hogan, *Power*, 62.
62. Ibid, 5.
63. Ibid, 162.
64. Ibid, 54.
65. Ibid, 54–55.
66. Ibid, 159.
67. Wilson, *Origins*, 60.
68. Ibid, 61.
69. Ibid.
70. Ibid, 61–62.
71. Ibid, 62.
72. Ibid, 63.
73. Hogan, *Power*, 22.
74. Ibid, 22.
75. Ibid, 22–23.
76. Ibid, 6.
77. Ibid, 6–7.
78. Descola, *Par-delà*, 244–45.
79. Hogan, *Power*, 55.
80. Ibid, 57. See also p. 60.
81. Ibid, 63.
82. Ibid, 64.
83. Ibid, 64.
84. Buber, *Between*, 51, quoted in Turner, *Ritual*, 127.
85. Turner, *Ritual*, 126.
86. Ibid, 127.
87. Ibid, 128.
88. Ibid, 138.
89. Ibid, 128–129.

90. Ibid, 140.
91. Cooper, "Woman," 144.
92. Hogan, *Power*, 125.
93. Ibid, 135.
94. Ibid, 115.
95. Ibid, 62.
96. Ibid, 70.
97. Ibid, 69.
98. See Meillon, "Writing," 225.
99. Hogan, *Power*, 69.
100. Ibid, 69.
101. Ibid, 65.
102. See Meillon, "Writing."
103. Hogan, *Power*, 184.

Chapter Seven

Post-Pastoral Thought-Experiments with Totemic and Animistic Liminality

In this and the next chapter, I explore the resurfacing of animistic and totemic relationships between humans and nonhumans in non-indigenous contemporary literature, paying close attention to how this might tie in with liminal realism. To situate the emergence of such liminal realism in fiction published at the turn of the century, I first investigate Jean Giono's *Joy of Man's Desiring*, dating back to 1935. It can be hypothesized that Giono's seminal post-pastoral, ecopoetic prose has influenced both the literary mode of liminal realism and the return in later fiction to the notion of a "song of the world"—the phrase Giono picked for the title of his 1934 novel, *Le Chant du monde*, which likely harks back to cosmic poet Walt Whitman's "A Song of the Rolling Earth." In this chapter, I first focus on the liminal realism germinating in the French novelist's prose fiction, while I will address his rehabilitation of the topos of the song of the earth in Part Three. Having explored the central role of the buck at the heart of the community imagined by Giono, and the shamanic aspects of his Bobi protagonist, I will further examine the totemic function of the same forest animal in passages from Ann Pancake's *Strange As this Weather Has Been* (2007).

JEAN GIONO'S *JOY OF MAN'S DESIRING*

In her astute reading of the post-pastoral in Giono's oeuvre, Ginna Stamm draws from Terry Gifford's elaboration of the genre, which could also apply to most of the works in my corpus:

> The specificity of the post-pastoral as enumerated by Gifford are [sic] the following: "awe in attention to the natural world" [. . .], "recognition of a

creative-destructive universe" [. . .], a realization that "the inner is also the workings of the outer" [. . .], "awareness of both nature as culture and of culture as nature" [. . .], and the acknowledgments that "with consciousness comes conscience" [. . .] and that "the exploitation of the planet is the same mindset as the exploitation of women and minorities" [. . .]. Overall, these can be summarized as a sense of vulnerability to a world that is as threatening as it is beautiful, a sense of the union of the inside and outside world, and a coming to consciousness of the wider implications of exploitation of the environment. The post-pastoral acknowledges all the ramifications of a natural world that is "immanent" (152) to its inhabitants.[1]

The above characteristics in many ways apply to works such as Barbara Kingsolver's and Ann Pancake's Appalachian novels *Prodigal Summer*, and *Strange As this Weather Has Been*, Starhawk's Californian utopia *The Fifth Sacred Thing*, or again Powers' haunting human-tree saga, *The Overstory*. While Stamm has aptly demonstrated Giono's stamp over the post-pastoral genre, and while Jacques Noiray has elaborated a detailed analysis of the political utopia at the heart of Giono's *Joy*, I would like to delve into the totemic, liminal, and shamanic aspects of Giono's ecopoetic prose fiction that may establish it as a precursor of liminal realism.

The blurb promoting the English translation of Giono's novel presents it as "[a] true forebear of magical realism." Yet, with Faris's contribution in mind, it is hard to see what might be deemed an irreducible element of magic in that novel. A more sustainable approach might be found in pursuing Noiray's hunch that Giono's buck functions as a "totemic animal"—a claim Noiray makes out of the blue without elaborating any further.[2] Albeit not identifying them as such, the critic has nevertheless spotted many of the aspects that tie in with Giono's embracing of a liminal form of realism.

First, Noiray marks the "secrecy" of the territory serving as the setting for Giono's story, which, in the critic's own terms, is "radically separated from the rest of the human world by a barrier of cliffs and forests." Rather than concluding, like Noiray, that the effect is a nearly "fantastical" one,[3] I would venture that the indeterminacy and social separateness of the Grémone plateau provides the ideal geographical remoteness for the establishment of Giono's liminal experiment. Away from any town or village with socially regulated space, culture, and roles, the "individual and social revolution" Noiray comments upon in his predominantly political reading of the novel is to my mind first and foremost an ecopoet(h)ic one. Indeed, as Bobi gathers the farming families and couples of the Grémone plateau around the buck, they come to form the deer clan, whereby all of them start relating to the animal and engaging in relationships of reciprocity. In taking care of the buck—freeing him from bondage in the circus, providing him with the doe companions he needs, and returning him to the wild—the peasants initiate a pact of mutual care and a rewilding of their sensibility, of their ways of dwelling and expressing themselves which forms the novel's main project. The buck thus becomes a rewilding, tutelary animal for the entire Grémone plateau clan.

The only magic lies in the power of Bobi's poetic mind to rehabilitate his fellow humans' sensitivity to the animate world around: "Everything around them understood, from the tiniest plant to the tallest [oak], and the animals, and no doubt even the stars, and the earth here under his feet with its clods and its mat of undergrowth and its slender water courses. Everything understood and responded. They alone remained hard and impervious in spite of their good will. They must have lost the fine heritage of man to be so poor, to feel so despoiled and weak and incapable of understanding the world."[4] The initiation process starts with Jourdan and Marthe's encounter with Bobi, whose unfailing poetry brings them to feel the sap of the earth flowing through them, connecting them with the other languages and the "magic joy" of the forest.[5]

Bobi's nonconformist way of speaking gradually revitalizes the others' until-then-numbed capacity to dwell sensitively. Jourdan starts heeding the "long [incomprehensible] word" muttered by the wind. He can feel the "adolescent sap" flowing into a young birch "full of enthusiasm"—an enthusiasm for life that spreads from the world of plants to that of humans.[6] Contemplating a budding alder, they are reincorporated into the multisensorial flesh of the world, sniffing away the sweet, sugary scent that makes them say, "'Smell that odour'/ 'It's almost a taste.' [. . .]/ 'It smells like sugar.'"[7] Feeling suddenly wide awake in the middle of the night, Jourdan and Marthe receive an epiphany at the heart of the forest innerving all their feelers at once. Surrounded by many varieties of trees exploding with flowers in their blooming season, the characters' experience of the forest evokes a natural temple, or a cathedral of the senses, with interlaced odorscapes, soundscapes, and visual landscapes:

> The odours flowed fresh. It smelt like sugar, the meadow, resin, the mountain, water, sap, birch syrup, bilberry jam, raspberry jelly in which the leaves have been left, linden tea, new woodwork, shoemaker's wax, new cloth. There were odours that walked and they were so strong that the leaves bent with their passage. And thus they left behind them long furrows of shadows. All the rooms of the forest, all the corridors, the pillars, and the vaults, were silently alight and waiting.[8]

Giono is likely tapping into Baudelaire's famous poem describing nature as "a temple where live pillars sometimes let out confused words," where "perfumes, colors and sound" reverberate one another like "long echoes."[9] A great part of this Godless revelation is tied to making humans attuned to the polyphonic song of the earth: "On all sides could be seen the magic depths of the house of the world. / They were waiting for the wind. / Then it came. And the forest began to sing for the first time of the year."[10]

It thus dawns on Giono's human characters that all lifeforms have a way of expressing themselves in a multisensorial, biosemiotic landscape. Taking my cue from Wendy Wheeler's theory of biosemiotics, I would say that Giono's ecopoetic prose "places humans back in nature as part of a richly communicative global web teeming with meanings and purposes."[11] If Giono's characters are not equipped with the kind of technological instruments allowing for nowadays' bioacoustic analyses of a specific soundscape, they are nevertheless reawakened to their fine auditory and olfactory perception skills. This lends them what Murray Schaffer calls "clairaudience" in the domain of hearing,[12] which might be stretched to a larger form of "clairsentience" involving all their other, enmeshed senses, working in unison to form the flesh of the world. Focusing on characters whose stance to the world is one of openness, such numinous moments are by no means supernatural or irrational. Rather, they are tied to the sudden, vivid realization that, to take up David Abram's words, "[all] things have the capacity for speech—all beings have the capacity to communicate something of themselves to other beings. Indeed, what is *perception* if not the gregarious, communicative power of things, wherein ostensibly 'inert' objects radiate out of themselves, conveying their shapes, hues, and rhythms to other beings and to us, influencing and informing our breathing bodies though we stand apart from those things?"[13] From that perspective, it may be that, beyond the sounds produced by earth beings, even an odor, a shape, or a texture can be approached as a form of speech, a partaking in the "song of the earth."

The stag Bobi introduces on that same enchanting night is a liminal wonder in itself, "half beast, half tree," with its broad, woody antlers.[14] Converselyto Bobi's strange first name—Bobi in French sounds closer to a nickname, or a name one might give to a pet—his tamed creature is introduced with a human name, "Antoine." This grants the stag a nearly human kind of personhood: ""I was obliged to look for one that was almost human in order to make a proper mixture,' says Bobi.[15] Meanwhile, the oblique reference to Saint Antoine—prayed to in the Catholic tradition as the Patron Saint who can direct attention to that which one has lost—points to the deer's near miraculous gift to the humans, bringing them back the animal sensitivity they had forsaken: "And look, both of you, see how a thing that is pure and unfettered lights up the darkness. You see his eyes are the same colour as the buds, and see how our sight doesn't help us at all when we are in the pitch-dark amongst the wild things, as if our lids were weighted down with dead stones, because we have lost the joy of the seasons and innocent gentleness. See how his eyes shine!"[16] Bobi here serves as a spokesperson for Giono himself, calling for reenchantment through the reawakening of our senses, starting with the fresh, open look humans can cast onto the world. It is a look that has the power to illuminate, but which tends to get switched off in the disenchanting world of modernity.[17] Working toward reenchantment, Bobi relies on the deer, with his carnal aliveness, for his capacity to reopen his fellow humans" eyes, and, in addition, to stir the pleasure of feeling through their nostrils, their ears, and their pores all at once: "The deer lay down beside them. He stretched his head between their feet. His hot breathing warmed Marthe's legs and rose higher beneath her skirts, living, like the rhythmical throbbing of blood. [. . .] Marthe was pleased because the deer was sniffing her. [. . .] The forest was singing. A creature called from the depths of the woods. The voice rolled along its echoes. The deer raised his head. They could hear the scraping sound of his tongue on his lips."[18] What the tutelary animal brings these humans is altogether a change of heart affecting their whole stance: "'I went to get him and he will help us,'" Bobi explains. "'And now we are all three of us in the forest with him. And already we have lost the hearts of the inhabitants in the valleys. It is night, it is springtime. The forest is singing.'"[19] Starting with Bobi, the animal serves as an *Umwelt* amplifier, regenerating by proxy in the readers the look of wonder and the sustaining feeling of being alive, immersed in an animate landscape:

"What do you see" he said. "What is it you smell? [. . .] Yes, the earth, what do you smell in the earth?"

The deer uttered a long, deep word that made his throat and lips tremble. [. . .] He looked to see if Bobi was following.

"I'm coming," said Bobi.

He followed.

All at once he noticed that the great April sky also was abloom with thick clouds, dappled and dancing like an orchard in the wind. He had felt the warmth in his flesh, and the sun, risen and already hot, was no longer simply sunshine but a nutriment which was filtered through his skin into his heart, more nourishing than flour and saliva mixed.[20]

Bobi's liminal way of dwelling is informed by the animal's sentience, reawakening primal ways of feeling and knowing: "He felt almost deer, almost beast. He knew in advance through animal knowledge that where the deer was leading him was deer interest and interest for his, Bobi's, heart, nourished suddenly by the golden flour of the sun."[21] In the above quotes, Giono apprehends the spiritual nourishment which humans may derive from a very material embracing of greater-than-human nature. Throughout, the characters' most intense joy and sense of fullness simply come from a humble material merging with the natural world through the senses, bringing about a deep sense of plenitude, of belonging, and of dwelling in harmony within a meaningful world of multispecies entanglements.[22]

Moving in-between the perceptive realm of humans and that of the buck, Bobi the outsider takes on the role of a shaman. In his practice of contortionism, he makes himself a kind of shapeshifter, becoming in turn a "man-frog, with thighs on shoulders," or a "man-snake, with arms and legs knotted about him."[23] Bobi's physical flexibility proves to be the outward indicator of his inner elasticity. It lends him no true metamorphic power, but simply the ability to partially switch perspectives with the buck to enlarge the scope of his own world. His amplified bodymind plugs into other-than-human ways of dwelling. Very much like a shaman, Bobi claims to speak the same language as the deer. As is also the case for the shamanic characters in Linda Hogan's novel, this capacity is not so much presented as having to do with the supernatural, but with a form of companionship and communication mediated through the empathetic look they lay upon one another: "'I've known [Antoine] for a long time. We have worked together. He used to love to watch me. I used to love to look at his eyes. We have often talked to each other about our mutual desires.'"[24] The first lesson Bobi and the buck teach the others is to stand quietly at attention, to make oneself all ears: "'Let's listen,' said Jourdan. / And all four were silent for a long time. The forest was singing its most solemn song."[25]

Along the same lines, the first concrete change Bobi urges Marthe and Jourdan to make is to replant hawthorn. Although it may not at first seem as useful as the cultivation of wheat, Bobi draws their attention to other-than-human perspectives, and to their human responsibility in the design of the soundscapes of the world:

> "You should plant hawthorns," said the man. "Some hawthorn hedges around the buildings and hawthorn thickets at the corners of the fields. It is very useful. [. . .] You need hawthorn hedges to bound the fields, not as fences, but you take too much land in your ploughing. Leave a little for the others. [. . .] One thing only, to make you understand. If you understand that, you understand all. With the hawthorn there are birds. Ah!"[26].

Bobi's take hits a spot deep inside Joséphine the bird-lover, as the source of joy she finds in the presence of birds intersects with the pleasure the birds themselves may find in the bushes near the house, and with the subsequent occasion for humans to revel in birdsong. In that matter, Giono's novel might have set the tone for Rachel Carson's introductory chapter in *Silent Spring*.

Indeed, after the birds have returned in Marthe and Jourdan's field, Bobi tells them, in a way that potentially foreshadows Carson's dysphoric "silent spring": "'The birds do not belong to us. There above the pasture, a whole nation of birds has gathered, and by that we are assured the sky will not close in over us all winter. That is really something. [. . .] For,' said Bobi, 'the sky closes in over country where all the birds are dead.'"[27]

Seventy-five years before David Abram's books on ecophenomenology, Giono's project was basically the same ecopoetic one:

> Thus do we shelter ourselves from the harrowing vulnerability of bodied existence. But by the same gesture we also insulate ourselves from the deepest wellsprings of joy. We cut ourselves off from the necessary nourishment of contact and interchange with other shapes of life, from antlered and loop-tailed and amber-eyed beings whose resplendent weirdness loosens our imaginations [. . .]. For too long we've closed ourselves to the participatory life of our senses, inured ourselves to the felt intelligence of our muscled flesh and its manyfold solidarities.[28]

Giono might have influenced the philosopher to the point where it may be hard to distinguish at times one's prose fiction from the other's essays. Abram indeed adheres with Giono's project of a liminal reenchantment:

> Owning up to being an animal, a creature of earth. Tuning our animal senses to the sensible terrain: blending our skin with the rain-rippled surface of rivers, mingling our ears with the thunder and thrumming of frogs, and our eyes with the molten sky. Feeling the polyrhythmic pulse of this place—this huge windswept body of water and stone. This vexed being in whose flesh we're entangled.
>
> Becoming earth. Become animal. Becoming, in this manner, fully human.[29]

Abram often sounds directly inspired by Giono's writing and Bobi's shamanic character. His book resounds for instance with this dialogue between Bobi and the nearby farmer who rejects Bobi's ecopoet(h)ic project. The farmer speaks in favor of a Communist vision of human "intervention," one relying on human reason and on land management via powerful technology:

" . . . [Your] joy is not a solid one."

"It is based on simplicity, on purity, on the common things in the world."

"It is animal."

"We are animals."

"Yes, tragic animals. We make tools."

"Let us become once more . . . "

"[To become] is [to go] forward. Never backward."

"Is it for that that water barks in all the valleys, that the wind sways the forests, that the grass ripples, and that, unceasingly, the white clouds cross the sky like ships?"[30]

The quest relayed by Bobi is one of rebecoming animal. Hence the long list of animal joys that Giono's characters gradually learn to channel again through their somatic experience of the world:

> In the midst of the winter calm, the men rediscovered ancient joys in all the secret resources of their bodies. Joys in their armpits, in their elbows, their knees, and in that part of the shoulders that rises from the neck of one's shirt and that is more sensitive to cold than to heat. To all these joys could be given names of animals. There were dog joys in the spine: to stretch, to feel the heat rising along the loins like the growth of a plant with glowing branches and broad leaves of soft, caressing warmth that folded gently over the breast to shelter the heart and lungs like a silk worm cocoon. There were fox joys in the legs that walked frozen paths [. . .]. The suppleness of the thigh, the cold that seized the skin and a centimetre of flesh around the leg, while from deep inside the body flowed a burning blood that descended to the legs, and thigh, and knee, calf, and foot began to live with an ardent force. That was no longer a part of man, but it was a part of the world, like mountains, torrents, clouds, or the great wind. [. . .] There were bird joys in breathing the glacial air [. . .].

Giono's writing thus paved the way early on for Abram's "work of recuperation"[31] of both "the natural magic of perception"[32] and the sensuous language to translate it. "'[Poetry] is an initiating force, and a great force, dynamite that lifts and tears our rock,'"[33] Bobi insists. Even the farmer who condones the human mastery and control of nature admits: "'I have always thought so [. . .]. We need poets. You have touched me in a spot that each day I cover with earth. I have felt it alive down there, at your appeal, like water beneath the divining rod.'"[34] Taking up Giono's reenchantment project nearly a century later, Abram writes "This is a book about becoming a two-legged animal, entirely a part of the animate world whose life swells within and unfolds all around us. It seeks a new way of speaking, one that enacts our interbeing with the earth rather than blinding us to it. A language that stirs a new humility in relation to other earthborn beings [. . .]. A style of speech that opens our senses to the sensuous in all its multiform strangeness."[35]

Giono's avatar in Bobi the ecopoet is recognizable in the idiosyncratic way the latter expresses himself. As Noiray notes, Bobi's style is replete with figures of speech and images. Noiray quotes the "Orion Queen Anne's lace" leitmotif that becomes emblematic of Bobi's language—it is "the sign of a utopic language that no longer speaks to reason, but to dream and imagination, a language that no longer makes you understand, but that makes you see."[36] There is something oxymoronic about Bobi's magical formula, at once visionary and grounding, synthesizing earth and sky, flowers growing in the soil, grounded right beneath our feet, and thus made accessible as one might pick and eat them, and out-of-reach stars constellating the immensity of the sky, forever away in both space and time and which yet sustain our poetic, mystical imagination. As Henry Suhamy remarks, "Poetic language forces an idiom to express the irrational, so that it becomes the idiom of the irrational. [. . .] The concept of irrationality besides is inconvenient as it starts with a privative prefix, triggering mistrust. It can be replaced with other notions, with more positive connotations, such as sensibility, empathy, imagination, vision."[37]

Together with the repoetization of their lives, the characters undergo a reconnection to the ecological intuition available to them through their dreams, another aspect of Giono's creative craft in favor of liminal, rather than magical, realism. Dreaming of a white bull, one of the characters explains how this otherworldly encounter has brought him deep happiness. He goes on to recount these nocturnal visits from his "dream bull," followed by pleasing apparitions of a goat, a horse, then of his mare with "ten flame-colored colts," and, finally, of a sow.[38] These dreams translate how regular contact with the buck and his does has opened a door giving an outlet for the character's ecological unconscious. Hence the animal dreamer's regained awareness of his deep desire to be more tightly involved with the lives of beautiful animals: "'I should like to have,' he [says], 'a fine bull and a fine cow, a fine ram, a fine ewe, a fine she-ass for my donkey who is already fine, and to have a beautiful stallion from the bellies of my mares. My hands itch to create fine animals.'"[39] Like Bobi, the character named Jacquou, a livestock farmer, develops a form of interspecies communication via the liminal zone his dreams open onto: "Jacquou [. . . was] going on to talk about his dreams, about his marvellous animals, his bulls almost too fine for this world, his cows with their cream-colored udders, that even talked at times, until one wondered how Jacquou could have imagined them. [. . ..] He talked of them as if it were all true, as if these animals were already there."[40] At this point, the world of their dreams and that of their working lives have become enmeshed, revitalizing their essential ties with animal lives, sustaining them in spiritual and affective ways that go far beyond the usual payback one might expect from farming in a more capitalistic mindset: "But Randoulet understood because he knew what it was to be linked to beasts as with iron bolts and nails. And it was useless to tell him that sometimes one went without eating so as not to do without some other nourishment that one considers at the moment much more necessary."[41]

Meanwhile, in a like rejuvenating manner, Jourdan has become a plant-dreamer, talking about his "dreaming green."[42] Following his green dreams, where all sorts of plant species are intermingled, Jourdan has decided to put his newly discovered, green thumb to concrete use. Achieving the design revealed to him in his night visions, he begins to sow less wheat for selling, making time instead for "extraordinary seeds" of grasses and flowers.[43] As it turns out, Jourdan's gratuitous cultivation of beautiful, fragrant flowers will attract animals and humans alike to a site of wonder and repose for all. Thus, blurring the lines between the world of dreams and reality, the events are not so much "fantastic" as they make way for the characters' ecological unconscious to feed into their mindfulness. Guided by their oneiric visions of animals and flowers, the farmers start yielding to their resurfacing desire for connection with the more-then-human world. They reclaim their innate capacity to care for the earth and its co-dwellers, thus reckoning how they feel called to simply acknowledge and enjoy their participation in the natural world. Having refreshed their response-abilities to the natural world, they gradually come to take responsibility for their own culture and happiness by trusting the existential revelations granted through their dreams, through poetry, and through contemplation of the more-than-human world.

As opposed to Noiray's reading of the ending of Giono's novel as marking the failure of Bobi's utopian project, I would argue that the character's death coincides with the shamanic interpretation I have ventured. As a shaman, Bobi must remain marginal, on the fringe of the community. When he gets involved with passionate Joséphine however, who turns clinging and possessive, it becomes clear that Bobi is not meant to settle in as a full-time resident of the community he has inspired. Moreover, as he is ostentatiously drawn also to young, wild Aurore—an attraction that is reciprocal—Bobi seems torn between his animal drives and the tacit social and moral codes forbidding the satisfaction of his every carnal yearning. As demonstrated in Part One, his death lends itself to an ecofeminist reading. Essentially, as Bobi is struck by lightning in a moment of ultimate epiphany, Giono's poetic language once again highlights the character's liminality. Like his totem animal, the last vision we get of Bobi indeed transforms him into a being of light, partly animal, partly human, and partly tree: "The flashes sprang from everywhere like the force of a forest. [. . .] The sweat steamed from his naked body. Suddenly, he was warned like a bird by a sparkling under his tongue. [. . .] The lightning planted a golden tree between his shoulders."[44] On top of the poetry of this passage, turning Bobi into a luminous forest being, his nakedness at that point gestures to the symbolical value of his death as, paradoxically, a sort of cosmic rebirth. Bobi is returned to his earth mother where he will decompose, enrich the forest soil, and be born again via the continuum of life and death processes on which a forest thrives. As he is struck by lightning, the vision is that of a searing moment, revealing what a powerful conductor for earthly energies his character embodies. His extraordinary death thus brings the final touch to his characterization as mythical prophet.

In addition, beyond his death, the word Bobi has spread of simple joys, of dwelling poetically and cultivating hope stays with the community, as visible in the closing pages of the novel: "That evening [Joséphine] felt no grief. She had never received so much from the world. It seemed to her that everything helped her. She drew strength from all; from the evening, the air, the colours, the fragrance and the murmuring of the forest. It seemed to her that Bobi now had a hundred ways of being with her."[45] If anything, Bobi's spectacular passing away anoints him as mythical being and mediator between worlds, an eternal storyteller whose ecopoetic parables will be passed on. This is an aspect of the terrestrial prophet making him potentially live on forever, as he does in any case in Giono's timeless prose.

ANN PANCAKE'S *STRANGE AS THIS WEATHER HAS BEEN*

In its powerful ecofeminist indictment of Mountain Top Removal (MTR) in the Appalachian range, Ann Pancake's *Strange as This Weather Has Been* takes up Giono's post-pastoral reenchantment project. Yet, contrarily to Giono's novel, facing at once the ruins from MTR, from blindly exploitative capitalistic systems, and from climate change, the small rural community at the heart of Pancake's story does not rediscover their physical and spiritual attachment to the land thanks to an outsider's intervention or via a utopian project. Rather, the novel stages three generations of people who are deeply attached to the native mountains the defense of which becomes their main struggle.

Living modest lives, the locals take up available jobs such as mining and waitressing, while coming from a tradition of roaming, gathering, and hunting in their backyard forests and hills. Although not from an indigenous background in the sense of Native American peoples, Pancake's characters nonetheless exemplify a process that Potawatomi writer and biologist Robin Wall Kimmerer identifies as "becoming indigenous to a place." These characters indeed testify to a way of "living as if your children's future mattered," a "becoming indigenous" that means "to take care of the land as if our lives, both material and spiritual, depended on it."[46] What I would like to demonstrate here is that Pancake's incursions into what could be approached as magical realism are better interpreted in the light of Giono's avant-garde use of liminal realism, with his poetic rehabilitation of animistic and totemic worldviews in an otherwise naturalistic context. My contention about Pancake's ecopoetic use of liminal realism is that it taps into the oral liveliness of Appalachian English while re-rooting culture within nature, corroborating Kimmerer's vision of "becoming native" to a place: "[To] become native to this place, if we are to survive here, and our neighbors too, our work is to speak the grammar of animacy, so that we might truly be at home."[47]

The polyphonic novel comprises interlaced chapters relaying six different voices, three of which are mediated by a third-person heterodiegetic narrator using the stream of consciousness technique, while the other three come in the first person, conveying a confidential impression. In interviews, Ann Pancake has frequently explained that what turned into a novel had started as short stories. The overall vision produced is a kaleidoscopic one, shifting in and out of various consciousnesses, each with its own frame of mind, age group, and type of sensibility. Those different experiences of the same place compose a rich palette of colors and perceptions. Via the conversational tone of the chapters staging homodiegetic narrators in relationships of great intimacy with the land and its many dwellers, readers are spurred along the way to "think like a mountain"—as goes Aldo Leopold's famous saying.[48] The complex narrative strategy facilitates the inclusion of various points of view, voices, and types of discourse that together layer the "invisible landscape."

The ecofeminist and political dimensions of this dialogic novel are mostly conveyed by the chapters told by the two female homodiegetic narrators—Lace, the activist, and her daughter Bant, a teenager whose development turns the novel into an initiation narrative. In addition, the history of the true-life Buffalo Creek Disaster is relayed in chapters told by a heterodiegetic narrator using stream of consciousness via the educated character of Avery, whose firsthand perspective is established as reliable, both as a survivor of the episode, and as a result of his scholarly research into the facts and the ensuing legal issues.[49] Two of the other perspectives in the novel come from children, two of Bant's three little brothers, Dane and Corey. Their inner worlds are similarly accessible via the stream of consciousness technique, with much free indirect speech liberating their idiosyncratic voices and individualized sensitivities into the narrative thickness. And finally, liminal realism is injected into the one and only chapter told in the first person by Bant's Uncle Mogey, with an orality of tone suggesting an interview, and coming at the very center of the novel. This essential chapter intertwines animistic and modern worldviews as it relays Mogey's epistemological conflict, torn as he is between "what they teach at church and what [he knows] from the woods."[50]

Post-Pastoral Thought-Experiments with Totemic and Animistic Liminality 181

If it sporadically veers into liminal realism, Pancake's novel first and foremost effectively engages with concrete realities apprehended in a rational and scientific way. This is apparent in the way the 1972 Buffalo Creek catastrophe is covered throughout. Avery, the focalizing agent endowed with a business degree, serves to synthesize the research into the accident, exposing the industrial liability, the legal injustices, and the true cost for the poorly compensated survivors. Moreover, as the victims of this and other MTR-related traumas strive to comprehend the despoiling of their land and communities, they unravel the industrial processes and their destructive impact over humans' bodies and psyches, over their families and communities, over the fauna and flora of the local ecosystem, over the life of entire mountains. The mostly realistic novel thus embarks us on a consciousness-changing journey, calling for awareness of the actual ongoing destruction of a mountain range older than the Himalayas.

Mogey's chapter plays a crucial part in the dialogism of Pancake's novel, set as it is in conservative rural Appalachia, where Christian faith is as pregnant in people's minds as coal mining is in their lives. In fact, defenders of MTR frequently and openly invoke God's will, via literal anthropocentric readings of the Old Testament. Hence the very realistic debate voiced by Mogey in the novel, reflecting an ideological battle that might have been relayed by a true inhabitant of West Virginia:

> Although I have been a Christian all my life I have never felt in church a feeling anyplace near where I get in the woods. This worried me for a long time. Even when I prayed in church, I couldn't make much come, where woods, I had only to walk in them. To walk in woods was a prayer. But I knew it was wrong. Some kind of paganism or idolatry, I didn't know what you'd call it, but I knew it must be a sin. I used to feel so guilty about it I finally talked to the pastor one time. [. . .] So I told Pastor Dick my concern, and he said, "Mogey, God gave man the earth and its natural resources for our own use. We are its caretakers, and we have dominion over it . . . " And he went on like that, saying stuff I'd been hearing since I was little.[51]

As Mogey confesses his inner struggle, the narrative hinges on his problematic allegiance to his church, clashing with the naturculture derived from direct contact with his homeplace. Deep down, Mogey cannot adhere to teachings that extract humans from the earth: "But part of me knew, even back then, that's not what it is. I knew we wasn't separate from it like that."[52]

As it turns out, Mogey represents an inspiring older figure for Bant, providing her with guidance as she walks forward into adulthood. A patient tracker and forager, Mogey receives what he gathers from the forest as gifts: "I've always loved looking for stuff in the woods. That feel you get when you sudden-spy, as you're moving, the deep green leaf of ramp. The crinkle of a morel. Presents the woods give you just for paying attention, that's how I saw it."[53] In this sense, Mogey embodies a relationship of attentiveness and reciprocity with the land that resonates with Robin Wall Kimmerer's musings over the Potawatomi conception of wild strawberries as nature's gifts: "Even now, [. . .] finding a patch of wild strawberries still touches me with a sensation of surprise, a feeling of unworthiness and gratitude for the generosity and kindness that comes with an unexpected gift all wrapped in red and green. [. . .] After fifty years they still raise the question of how to respond to their generosity."[54] As with Mogey, berry-picking in the wild has given rise to a wholesome land ethic, respectful of its bounty as much as of its mystery. Kimmerer writes: "Strawberries first shaped my view of a world full of gifts simply scattered at your feet. A gift comes to you through no action of your own, free, having moved toward you without your beckoning. It is not a reward; you cannot earn it, or call it to you, or even deserve it. And yet it appears. Your only role is to be open-eyed and present. Gifts exist in a realm of humility and mystery—as with random acts of kindness, we do not know their source."[55]

While Mogey and his wife carefully cultivate their vegetable garden with nonendemic tomato plants, the seeds of which they first sprout in Styrofoam cups, they also gather and hunt from the unmanned garden growing in the hollows nearby: berries, and roots, nuts and greens, mushrooms–all sorts of wild, seasonal gifts gathered from their bountiful mountains. The grandmother—Mogey and Lace's mother, herself a teacher of inhabitation wisdom and practices—repeatedly tells Bant: "'You can live off of these mountains.'"[56] Yet, in their reaping, these characters know how to gather thoughtfully, not to litter and pollute. Thinking like a mountain, they know and teach their children not to kill any of the predators (such as the snakes) that keep everything in balance all the way up and down the mountain's food chain. In short, in contrast with the extractive type of land use carried out by MTR, itself driven by a market economy, these characters are aware of their interdependence with the health of the mountains that they subsist from. The system they live by thus comes closer to a garden economy.

Pancake's characters derive from their *oikos* much more than just food for the body or financial profit. Their mountain woods provide a terrestrial kind of spirituality and a profound, earthly well-being. Bant, Lace, and Mogey cultivate a bodymind relationship with the mountain that one might call "awareness" or "love"—an attachment that is spiritual and corporeal at once: "I got drafted [. . .] and saw other mountains, and now I know people not from here probably don't understand our feeling for these hills. Our love for land is not spectacular. Our mountains are not like Western ones. [. . .] In the West, the mountains are mostly horizon. We *live* in our mountains. It's not just the tops, but the sides that hold us."⁵⁷ Whatever word might best suit this organic relationship, it is one of deep intra-actions, one that keeps humans alive, making them feel connected, sound, blissful, and whole.

Mogey's chapter instills a sense of mystery via his preternatural childhood encounter with a deer, following which their relationship gradually acquires totemic value. The story contains what Faris would call "irreducible elements of magic," aligning with magical realism. It moreover possesses specifically initiatory, or liminal, qualities. Taking place right before Thanksgiving—a symbolically loaded date, as the national holiday was originally derived from reciprocal indigenous practices of thankfulness for the land's offerings—when Mogey is ten years old, an seemingly failed hunt turns into a numinous experience of natural grace, forever connecting him to the elusive buck. Mogey's apprehension of the world has ostensibly been shaped by years of alert animal and plant tracking. Consequently, his sensitivity to animal presences is finer than what the eye can see: "That buck came out after the last drive. I don't mean he was driven to us. He was not, he came out on his own [. . .] and I felt him before I seen him. The way a big animal throws something off himself, something he carries around himself that you can feel without seeing."⁵⁸

The narrator details the animal's bizarre behavior, apparently unafraid of hunters, as if intentionally coming to meet them. The animal thus represents yet another mysterious offering from the wild: "The buck held himself still, like he should not have, an animal that old knows better, and I stared at him, wondering at that stillness."⁵⁹ Mogey uses the personal pronoun "he," enhancing the recognition of personhood in the animal. As Mogey takes in the animal's greatness, paying attention to all the physical details of his amazing presence, his older brother Robby shoots his rifle, triggering even more incomprehensible phenomena:

> When he fired on him, that buck didn't show in any way he moved that he'd been hit by a gun. The shot knocked him off the ledge more like a punch than a bullet. And after he fell, he didn't just crash and come to rest on the next outcrop like he was supposed to. No. He went rolling. It was the third strange thing he done that day, after showing himself to boys with a gun and then standing still, practically posing for a shot. I've seen nothing like it since, big buck hooping down that mountain end over end, antlers over backside, whiteside, the antlers, then the white rump, coming up over and over again.⁶⁰

Throughout his description of the event, Mogey insists on the material reality of the buck's body, attested by the two boys' clear vision and hearing: "Me and Robby leaned out, each of us hanging off a tree, and we watched him roll what had to have been well over five hundred feet, and that buck never hung up on a thing. Not a bush, not a ledge, not a rock. He just never hung up, like he should have. After a while, he disappeared out of sight, although we could still hear the thumping and even rattle of dirt and rock, and then, after a little bit, we lost the sound of him too."[61] Next to this miraculous fall, the boys logically go searching for the body which they assume should be lying somewhere below them. But they can find no trace of a corpse: "Not only no body, but no blood, and no tore-up leaves or brush, and no knocked loose rocks."[62] Just as the buck strangely appeared to the rifle-toting boys in the first place, it has now inexplicably disappeared into thin air. Beyond the outlandish phenomenon, it is understood that ever since that first moment of nonverbal interspecies communication between him and his totemic animal, Mogey himself has "never hung up" the link then established between them.

This encounter with the deer brings Mogey various moments of epiphany, thus marking its liminal power. After the buck inexplicably vanishes, Mogey experiences his first mystical communion with the nonhuman world in a natural shelter provided by rocks on his mountain slope: "It was a spot where the shelf between the Ribs and the creek broadened a little. Turned out, although I couldn't see that yet, that it made enough space for a little sunk-down place like a room, and it seems even more like a room because there was rocks all around it. Somehow a rock fall had come and made like this room."[63] Half enclosed, half open, this naturally designed cove makes for a perfectly isolated, small place at the heart of the woods, a kind of natural chapel to meditate in. Not surprisingly, this is where Mogey receives his first revelation about dwelling:

The buck was not there in body. But something else was.

I stepped into that little room, I stopped and looked around me. And something layered down over my self. At first it seemed to wrap me. But then it was somehow in the center of me, starting there, and then it washed on out through all of my parts. It was the feel of a warm bath with current in it, a mild electric, it prickled my skin, every inch of my skin it touched. And the thing was, once it had currented all the way through me and reached my very ends, it kept going.

It melted my edges. It blended me; I don't know how else to say it, right on out into the woods. It took me beyond myself and kept going, so I wasn't no longer holed up in my body, hidden [. . .]. It made me feel bigger in myself, and it made me feel more here [. . .]. And with it came total sureness. And with the total sureness came peace.[64]

The evanescent buck brings Mogey enlightenment, guiding the boy to experience a dissolution of his ego. Mogey is suddenly swayed from the illusive sense of existing primarily as an individual self, separate from the rest of the world. The buck leads Mogey in his tracks to experience a moment of revelation as to his being at one with the world—a liminal experience that, as he is then aggregated back into his daily life, Mogey stores within himself, forever tucked in his innermost secret psyche.

Stronger than the creed preached in church, Mogey's epiphany from then on rules his spirituality and ontology. His experience with the buck has opened new doors to his perception, powerfully liberating his ecological unconscious at once. No matter the religious ideology and law meant to mold his consciousness and control his superego, his intuitively awakened knowledge prevails:

> I tried for a long time to pull the two together, what I knew from church and what I knew from mountains. Of course, it would only be right if I could keep the church part ruling the woods part. So when I'd first walk into the woods, I'd say to myself, "Look here what God's give us." But just about as fast as I could have that thought, this second one would come from deeper: "This is God." And then, from under that thought, from deeper yet, another thought would come, saying, "I go here. This is where I go." And last of all, the most certain thought, but also the most dangerous: "This is me. This, all this, is me."[65]

The encounter with the buck triggers Mogey's ecological unconscious, which, as may be the case for Pancake's readers, had until then been repressed by the teachings of his culture.

Forever connected to Mogey, the buck later appears repeatedly in his dreams. Making the realms of waking life and of dreaming porous, this human-animal tie makes Mogey a kind of postmodern shaman initiated in the wild. At first, the animal resurfaces in Mogey's unconscious in forms where it metonymically embodies the mountain dying from MTR:

> I used to dream a good bit about that buck. [. . .] I'd dream we'd come up on the sunken place, me and Robby together, and the buck would be there, but the feeling would not. The buck has a broke back, but he's still alive, trying hard to get up, him hoofing in the sloppy wet leaves for a grip. His big rack is dragging at him, pulling on his head, and there comes in me a tear in my chest, like cloth tearing, such pity do I feel. The rocks lie in a circle, making the room where the buck struggles, and Robby is afraid to shoot him for fear a bullet might ricochet off a rock and hit one of us. But I crouch down behind a big beech, I press my cheek to it, those trees that look like they're wearing human skin, and Robby takes aim from behind another. I hear the explosion and then the echo off the mountain across the creek. But when I peak around to see the body, there ain't nothing there.[66]

Through paronomasia, Robby's name can sound like "robbing." As a matter of fact, the character metonymically symbolizes the dominant extractive attitude of mainstream Western culture. His ruthless shooting at the animal, instead of welcoming its awesome presence or feeling empathy for its agony, like his brother does, is a metaphor for the industrial explosives detonated by the MTR engineers to carve out the mountain tops and then get to the coal inside the mountain ribs. Thus, Robby's worry in Mogey's dream that "the bullet might ricochet off a rock and hit one of us" possibly expresses Mogey's underlying anxiety that killing off what is left of the agonizing mountain might in turn trigger human demise.

In another recurring nightmare, the dead deer obliquely evokes the state of the nearby mountains, killed off by coal gutting:

> Me and Robby come up upon him not fresh dead, but a day or two dead, his body twisted unnatural and his coat matted with rain. His coat matted in a way it would never get live. He's shrunk up, how much littler he looks dead, and collapsed around his ribs like somebody has gutted him already. I look at the rack. The rack looks huge. It looks aliver than him. And something has already ate his eyes, even though you would have thought the cold rain would have kept them back in their holes. There again comes in my chest the pity-feel of cloth-tearing. And then, after the pity feeling, once more he's just not there. [67]

Here the diminished deer reads as a metonymy for its larger woods habitat. The grotesque, monstrous aspect of the deer in Mogey's dreams points to the "unnatural" deforestation of the mountain, sheered of its trees and collapsing under the regular blasting of its ridge tops.

As Mogey discovers traces of a nearby slurry impoundment, he gets sicker and sicker with migraines. His dreams and headaches then symptomize his growing awareness of the ongoing ecocide. Via liminal realism, the buck in his night visions reveals greater signs of spoliation, the taking over of the mountains by draglines and tractors where the deer used to roam, and the industrial dumping:

> As the headaches get worse, the dreams do, too. Looking back now, I believe it started, these new ones, with me dreaming animals with metal for teeth. A couple of times, I dreamed just that, normal deer with metal pressed in their gums. Then I dreamed I shot a buck and went to gut him, and I found he had a plastic bag for a belly. After that, I dreamed I was out walking and finding glass scat. I dreamed leaves falling as ash.[68]

His dreams gradually betray an even greater angst, with apocalyptic environmental scenarios that may well be tied to larger climate change issues–a global perspective implied in the title *Strange as this Weather Has Been*:

Then those dreams passed too, and I stopped dreaming animals, I stopped dreaming woods at all. Instead, I dream that the world tilts, and I see crowds of people can't keep their footing, and they all fall and slide into a corner. Or I dream I'm out in my yard, and everything just stops. It's like a clock running down, one where you don't notice the ticking until it stops, but then it does stop, and I feel the universe dead quiet in its halt. And now, finally, I've got to where I dream without pictures at all. It's just a dream of sound. There is nothing to the dream but an alarm going off, a horn with a beat to it: *Mwaaa. Mwaaa. Mwaaa. Mwaaa.* I don't need no Daniel to interpret that dream.[69]

Between the disappearance of animals and woods, the prescience of a global tipping point that could tilt humans off the face of the earth, and the ticking motif, Mogey's unconscious stages some of the imminent end-of-the world scenarios that scientists have been ringing the alarm to. In Mogey's nightmare, it is significantly the crying out of the deer that rings the alarm ("*Mwaaa. Mwaaa. Mwaaa. Mwaaa.*"),[70] forcing him to listen to the voices of the earth speaking to him through animals, through the land, and through his own dreams.

As Mogey strives to make sense of his distressing dreams of a silenced, dead world, he gradually learns to rely on his intuition. Thus, he allows himself to trust his regained ecological consciousness rather than the Christian mythology of Man's separateness from and dominion over the earth:

There is what my reason tells me. There is what my church tells me. There is what my dreams tell me. There is what the land tells me. I'm coming to accept that I'll never bring all those things together before I die. But on my strongest days, I can tell myself without guilt or fear, it is not paganism or idolatry or sacrilege or sin. It's just what I know. And what they tell me, these things I finally let myself trust, is what you're doing to this land is not only murder. It is suicide.[71]

While other characters such as Lace or Avery provide the historiographic and scientific accounts of MTR and start unravelling the links with climate change, Mogey's poignant chapter calls onto the readers' own ecological unconscious, poetically stirring a form of awareness brought about through embodied knowledge and intuition rather than cool reason.

What Mogey ultimately listens to is the voice of the earth, speaking to him through the buck, through his visionary dreams, through his corporeal sensing of his own embeddedness into the earth: "And I could make myself feel how I was part of the land just by letting down something inside me. I got practiced that way. A letting down at will."[72] Mogey thus becomes a visionary, or shaman. In touch with the living world, he can open to the numinous rapture that comes from paying attention, from making oneself porous to the more-than-human world. Furthering the liminal, mystical strand of Mogey's narrative, the single chapter recounted in his voice ends on a powerful note, as he recalls another searing moment of green epiphany:

> . . . I dreamed I was in a little grassy clearing. [. . .] Then, while I was standing there quiet and glad, an old doe walked up to me. She stepped right up to me, and I looked back into her brown eyes, and she said, "That's what it's like inside my head."
>
> Then she shelled her head open. It just fell open into halves. And as she did it, there spilled out of it and over me this light a color green I'd never seen before.
>
> The light from her head carried in it the feeling I'd had in the little room where the buck wasn't. [. . . This] time, when I blended beyond myself with the sureness, the peace, the sureness and peace kept growing. Bigger and beyond anything I'd ever felt, it swelled and spread, I swelled and spread, until there was not anything else. No woods and no doe and no light and no me. Until there was all. It was all. Not nothing. Not something. Just all.
>
> I guess you'd call it the peace that passeth understanding. I guess you'd say it'd come by grace.[73]

Because of the confidential tone adopted in this chapter ("I can guarantee you I've never talked before about any of this out loud. The buck, the dreams, the feel in the woods."[74]) and the lack of explanation a controlling, omniscient narrator might have provided, the reader is called upon to make her own sense of this mind-blowing chapter. As apparent in most of the above quotes, Mogey's speech is overall so vernacular in its search for an authentic expression of Appalachian orality that it often strays from the prescribed rules of normative English. The unexpected use of ancient conjugation in this last passage ("the peace that passeth understanding") takes up Modern English Elizabethan translations of the Bible in Philippians 4:7. Foregrounding heteroglossia, Mogey's discourse thus deconstructs the ideologies framing his learning, meanwhile writing a new local, sylvan gospel.

Mogey's chapter irradiates the rest of the novel with a mindfulness of other-than-human nature and a sense of interconnectedness resembling a kind of Appalachian Zen. It prompts readers to question the modern ontology whereby humans would be made of a different, more intelligent and inspired stuff than the rest of terrestrial matter. As it is, Pancake speaks in an interview of having spent one year in Japan, where she was struck with the stark attention to everyday beauty and to the present moment which people derived from Zen Buddhism—the philosophy of which, the writer explains, has much influenced her.[75] At the same time, Mogey's chapter potentially offers a new version of Emerson's transcendentalism:

> In the presence of nature a wild delight runs through man, in spite of real sorrows. [. . .] In the woods, we return to reason and faith. Standing on the bare ground—my head bathed by the blithe air and uplifted into infinite space—all mean egotism vanishes. I become a transparent eyeball; I am nothing; I see all; the currents of the Universal Being circulate through me; I am part or parcel of God.[76]

Yet, laced at the very heart of Pancake's book (page 180 out of 360 pages), the above passage recited in Mogey's voice pulses with a glow of light revealing not so much a transcendent reality, but, rather, the immanence at the core of all matter, the inseparability of matter and mind, and a form of continuity between the material agency of all life forms.

Mogey's vivid synesthetic and green epiphany channels through the rest of the book the voice of the earth and its numinous light. Blending the blue and gold colors that dominate Christian symbolism of the divine, Mogey's profane revelation of dwelling within an encompassing sentient world mixes the two colors into the green of the forest—a color here evoking the natural beauty of forest light, bathed in chlorophyll. Breaking away from the Christian tradition, Mogey's enchanting vision is yet permeated with a sense of a sacred, ecological continuum. It relays the mountain's glorious energy, to be found at the heart of all agential beings and things which are inextricably tied to one another. Meanwhile, it communicates a sense of grace and delight by proxy to the reader. Consequently, seen through the lens of Mogey's ecopoetic mysticism, what seems sacrilegious is no longer the turning away from an alleged anointed human exceptionalism, but precisely the human lack of humility and reverence for the sacred entanglements of lifeforms that together compose the vibrant life of a mountain forest, and which soundly sustain human spirituality. Pancake's ecopoetic use of magic suggests that what she is really going for in her post-pastoral novel is liminal realism—a form of realism partly tapping into totemism and operating at a crossroads between Modern and animistic ontologies and epistemologies, the better to reignite our awareness of living in an animate world made of intersecting, interdependent, and intra-active multispecies forms of sentience and agencies.

Writing elsewhere about how to creatively respond to the world's unravelling, Ann Pancake corroborates my analyses of liminal realism as a reweaving of consciousness with our ecological unconscious: "I believe we artists must open ourselves wider to how art performs politically *beyond* bearing witness, because I've concluded that the only solution to the current mess is a radical transformation of how people think and perceive and value. In other words, we must have a revolutionizing of people's interiors."[77] Emphasizing how, in a cultural context where "imagination is impoverished and misdirected," literature possesses the specific "power to exercise, develop, and revitalize the imagination," Pancake stresses the role of literature to provide "new vision and ideas."[78] She emphasizes that literature "can reunite an individual's conscious and unconscious" and argues that "many of our contemporary ills are caused or exacerbated by our culture's rending the conscious from the unconscious, then elevating the conscious—the intellect, rationality—to the complete neglect, if not outright derisions of the unconscious. This is disastrous," Pancake explains, "not only because such psychic amputation cripples people, contributing to feelings of emptiness, insatiability, depression, and anxiety, but also because within that castoff unconscious—in intuition, in dreams—dwell ideas, solutions, and utterly fresh ways of perceiving and understanding [. . .]."[79] Hence the room made by Mogey's chapter for the "boundless [. . .] realm" of the unconscious. It is a realm that, in her Pancake's words, is "explosive with energy and light," "eons ahead of [the] intellect," "worlds larger in vision than [the] rational mind."[80] Further sustaining my reading of *ecopoiesis* as a revealing of those parts of the world that are inaccessible through human sight and rationality, Pancake elaborates:

> . . . refining this notion of artists' reintegration of the conscious and the unconscious, I'll propose that artists are also translators between the visible and invisible worlds, intermediaries between the profane and the sacred. Only by desacralizing the world, over centuries, have we given ourselves permission to destroy it. [. . .] Literature resacralizes by illuminating the profound within the apparently mundane, by restoring reverence and wonder for the everyday, and by heightening our attentiveness and enlarging our compassion.[81]

Moving on, I will now turn toward contemporary novels whose greater use of liminal realism increasingly flirts with the supernatural ascribed to magical realism. As I will show, a possible ecopoetic reading of the liminality and magic in these novels again reveals an underlying logic proper to shamanism and totemism. To this end, the next chapter deals with Starhawk's *The Fifth Sacred Thing* (1993), and finally, with Richard Powers's *The Overstory* (2018).

NOTES

1. Stamm, "Post-pastoral," 5.
2. Noiray, "Utopie," 67.
3. Ibid, 65.

4. Giono, *Joy*, 93. I have corrected the translation of "chêne" by "oak" instead of "ash" in the published translation.
5. Giono, *Joy*, 93.
6. Ibid, 92.
7. Ibid, 93.
8. Ibid, 95.
9. Baudelaire, *Fleurs,* translation mine.
10. Ibid, 95. As I demonstrate in part 3, more than a mere topos, the song of the earth is materialized in Giono's densely poetic, sensuous prose. This translates in his tendency toward incantation—as in the above long quote—and in his reliance on the onomatopoeic and imitative aural dimensions of language to foreground the sonorous texture of *ecopoiesis*, with its capacity to offer poetic echoes of the earth.
11. Wheeler, "Biosemiotic," 270.
12. Schafer, *Paysage*, 33.
13. Abram, *Becoming*, 172.
14. Giono, *Joy*, 96.
15. Ibid, 96.
16. Ibid, 96–97.
17. While Giono's novel indicts the dehumanizing exploitation of peasants that is a byproduct of Capitalism, French Nobel Prize writer Jean-Marie Gustave Le Clézio would later underline in his avant-garde experimental novel *Les Géants* (1973) a similar effect of Capitalism in the hyper-consumerist urban world of supermarkets and malls: "Man's face had been conceived with those two burning suns, that cast their light onto the world. Much fiercer than suns, much brighter even. X-rays that could pierce holes through metal armoring, that could seek insight into anything, unbounded. Eyes, or headlights: and the light they carried was the light of thought. Darkness could then be parted, and everyone was their own prophet. But the light has gone out, and nowadays, our eyes can no longer see through to the other end of the universe. There are no more suns. Faces are grey. They are turned toward one another and, where the eyes should be, instead of the headlights, there are only two glaucous balls of dim color" (*Géants*, translation mine).
18. Giono, *Joy*, 97–98.
19. Ibid, 98.
20. Ibid, 121.
21. Ibid, 121.
22. By "dwelling in harmony," I do not mean to naively imply a world with no chaos, suffering, or violence, but one where human suffering, death, and decay find sense as they form part of the larger, chaotic living world. Moreover, as I will get back to in part 3, the notion of a harmonious earth can serve to refer to the polyphonic texture of the living world, with different species each singing into its own acoustic niche, thus producing harmonies that together form the intricate song of the world.
23. Giono, *Joy*, 343. Here I would have translated "the toad-man, with his legs coming over his shoulders; and the snake-man, with his arms and legs coiled around himself."
24. Ibid, 98.

25. Ibid, 98.
26. Ibid, 20–21.
27. Ibid, 316.
28. Abram, *Becoming*, 7.
29. Ibid, 3.
30. Giono, *Joy*, 244.
31. Abram, *Becoming*, 9.
32. Ibid, 8.
33. Giono, *Joy*, 245.
34. Ibid, 243–44.
35. Abram, *Becoming,* 3.
36. Noiray, "Utopie," 70.
37. Suhamy, *Stylistique*, 216, 17, translation mine.
38. Giono, *Joy*, 338–39.
39. Ibid, 341.
40. Ibid, 391.
41. Ibid, 391–92.
42. Ibid, 339.
43. I am translating from the French "des graines extraordinaires," rather than quoting the English translation "unusual seeds," Giono, *Joie*, 340.
44. Giono, *Joy*, 459.
45. Ibid, 463–64.
46. Kimmerer, *Braiding*, 9.
47. Kimmerer, *Braiding*, 58.
48. Leopold, *Almanach*, 137–41.
49. On February 26, 1972, a sediment pond filled with coal slurry burst near Buffalo Creek, in Logan County, West Virginia, creating a flood that killed 125 people, injured 1,121, and left four thousand homeless. In the subsequent legal procedures, Pittson Coal Company, who had built the faulty impoundment and dams, described the accident as an "act of God."
50. Pancake, *Strange*, 174.
51. Ibid, 168.
52. Ibid, 168.
53. Ibid, 170.
54. Kimmerer, *Braiding*, 23.
55. Ibid, 23–24.
56. Ibid, 38.
57. Ibid, 173.
58. Ibid, 169.
59. Ibid, 169. Tim Ingold has written interesting pages on "totemic" and "animic" relationships and views of hunting in indigenous societies. He connects ethological understanding of deer behavior meant to throw off predators with the way deer behavior is interpreted in various ontologies. See "Totemism, animism and the depiction of animals," in *Perception*.
60. Ibid, 170.

61. Ibid, 170.
62. Ibid, 170.
63. Ibid, 172.
64. Ibid, 173–74.
65. Ibid, 173.
66. Ibid, 173–74.
67. Ibid, 174.
68. Ibid, 178.
69. Ibid, 178–79.
70. This is not without echoing Bobi's last cry to Mother Earth in Giono's novel, tackled in Part One.
71. Pancake, *Strange*, 179.
72. Ibid, 179.
73. Ibid, 180
74. Ibid, 178.
75. Gipe, "Straddling," 178–81.
76. Emerson, "Nature," 6.
77. Pancake, "Creative," 412.
78. Ibid, 412.
79. Ibid, 412–13.
80. Ibid, 413.
81. Ibid, 413.

Chapter Eight

Liminal Realism and Interspecies Thought-Experiments in Contemporary Fiction

ECOFEMINIST WITCHES, OR SHAMANS, IN STARHAWK'S CLI-FI NOVEL, *THE FIFTH SACRED THING*

Quite radical a resort to liminal realism is to be found in writing by Starhawk—the ecofeminist, neopagan, and Californian, self-declared "witch." Her novel *The Fifth Sacred Thing* works toward an ecopoetic reenchantment of the world via the imaginative freedom granted by the cli-fi genre. Set in 2048 in times of drought, Starhawk's novel oscillates between dystopia and utopia. While the background is one of dictatorship in dire times of drought, with a totalitarian white patriarchal and capitalistic régime established in the city of Los Angeles, the story focuses on a resistant community. The latter forms a multicultural, ecological, and horizontally governed society residing in the City of San Francisco, living in respect of biocultural diversity and harmony.

As Claire Perrin has noted, the cli-fi genre serves as a laboratory for imagining ways of reacting to climate change. Whereas Yannick Rumpala asserts that apocalyptic novels serve the following four functions, (1) a pedagogical, critical disenchantment working as warning; (2) catharsis and consolation; (3) habituation, allowing us to get used to our changing world; and (4) empowerment,[1] I would venture that, while dealing with climate change and dystopic scenarios, novels such as Starhawk's paradoxically effectuate a form of ecopoet(h)ic reenchantment of the world. As broached in part 1, the syncretic, ecofeminist vision and practices transpiring through Starhawk's fictional community promote a dynamic reinvented spirituality which the neopagan activist and writer expands upon in her nonfiction.[2] In the face of violence, injustice, and destruction, her novel undertakes a reenchantment project by embracing a form of liminal realism that does not wholeheartedly reject modern science or technology—to the contrary, the community has developed advanced communication means and a sustainable transportation system of gondolas based on renewable energies—but creatively reinvests animistic and totemic worldviews. Starhawk's liminal realism expresses a hankering for the restoration of partnership models honoring the interdependence between humans and nonhumans. As I hope to demonstrate, the novel stages a form of interspecies communication that rests on studies of insect behavior while poetically imagining entanglements between human and other-than-human worlds which both serve the plot and expand our notion of dwelling.

Starhawk casts a set of extraordinary characters including witches endowed with various psychic powers. Madrone, the midwife and healer, can through some sort of deep "psychic vision"[3] connect to other people's vital energies and cure them. As the dictatorial government in the south unleashes scientifically engineered viruses onto the population, Madrone must tend to the sick and try to identify the virus to hinder the pandemic. Trained in midwifery, gynecology, and herbal as well as Chinese medicine, much of her practice is "'what you might call laying on of hands,'"[4] by which she essentially channels into one's *ch'i*. Through some sort of empathic resonance, Madrone can acquire insight into one's marrow or plug herself into one's immune system, where she can detect the presence of a virus or cancer. From there, she can even change the energy patterns activating the body's responses to a virus. Via an intuitive, synesthetic reading of matter and energy fields that is half-way between feeling, seeing, and hearing, Madrone can analyze a virus strand, "each with their characteristic signature in the energy realms."[5] In a world where energy vibrates and potentially translates into sound, her vision is linked to a deep kind of listening she can activate at will, hearing for instance the acoustic signature of the deadly virus—a high-pitched whine, "like one small mosquito in a large room," below or above the other frequencies commonly available to humans on the audio spectrum.[6] Thus, Madrone negotiates the liminal realm between visible and invisible matter, between the sensitive and the subliminal. Like shamans, she seems to access these liminal zones through senses presented as finer than in most ordinary humans, allowing her to develop greater intuition and to access her unconscious.

Resounding with physicist's Rupert Sheldrake's theory of morphic resonance,[7] Madrone explains that she has halted one of the epidemics by intentionally contracting the virus herself, the better to use her powers and enter another dimension, going after "'the oversoul, or the morphogenetic field.'" One of her co-doctors explains: "'In the *ch'i* worlds, something like a virus is a collective entity. What we *see* of it is a symbolic representation of actual form-generating forces. [. . .] So what happens to its *ch'i* image reverberates in the physical world."[8] Capable of moving through dimensions other than that of the visible "upper world," Madrone in her healing travels "[down] to where sounds and smells [disappear], down to the level where everything [is] energy, *ch'i*, and, below that, through the place where fear and pain and the light of spirits moving across the veil [give] way to something even deeper. The level of *cause*."[9] By changing the pattern of a virus in her psychic journeys, Madrone triggers a change in the behavior of the virus, out there in the physical world, responding to the new formation habit spurred by Madrone.

Madrone moreover becomes "a Dreamer." As explained to her by the "Deep Listener" Lily, one of the elders, being a Dreamer means accessing liminal zones: "'There is the world of physical form [. . .]. What we know and can touch. And there are the realms of the *ch'i* world, realms of energies and spirits that infuse and underlie the physical. The division between the worlds is never absolute. Always there is bleed-through. So, a Dreamer stands on the boundary. Did you know the German word for *Witch*, *Hexe*, comes from *haggibutzu*, she who sits on the hedge?'"[10] From the point of view of threshold ecopoetics, it is interesting that in her creative thinking, Starhawk should align much of her witches' magic powers with an extraordinary finetuning to the world's vibrancy translating into sound. Beyond the esoteric dimension they move in, these characters gesture to the rich auditory texture of the material world, much of which eludes humans and yet is averted by science.

Mixing developmental biology with Chinese medicine and giving the whole an exciting sci-fi or cli-fi twist, Starhawk's characters braid diverse epistemologies and ontologies, opening onto a realm where matter expresses itself in ways that her postmodern witches have learned to decipher. Madrone's poetic *in*-sight thus makes her inclined to believe that the virus was engineered in the first place:

"When I *saw* it, it was like something constructed from metal and bolts and bicycle locks."

"Can you diagram it?" Aviva asked. "That might help Flore work up a computer model; figure out the morphic field equations. Because if new attacks come, they're likely to be constructed along similar lines."

"I can't diagram the molecular structure," Madrone said. "I wish I could. But that's not how it works. The mind translates patterns of energy and *ch'i* and molecular structure into symbols, things. I could draw them, but it'd look pretty silly."[11]

Madrone claims to have "changed [. . .] the morphic field that generated the disease."[12] These postmodern witches ecopoetically braid physics with pre-scientific animistic intuitions about the entanglements at play in the world involving observation, intention, and discourse: "Modern physics," Starhawk elsewhere points out, "no longer speaks of separate, discrete atoms of dead matter, but of waves of energy, probabilities, patterns that change when they are observed; it recognizes what Shamans and Witches have always known: that matter and energy are not separate forces, but different forms of the same thing."[13] Throughout the novel, the omniscient narrator never expresses any skepticism as to the exciting, magical powers of these futuristic witches who can quite poetically "[translate] patterns of energy and *ch'i* and molecular structure into symbols." Having become a "Dreamer," Madrone is moreover able to communicate with people *in absentia*, using telepathy. None of those dialogues are subjected to doubt by the narrator, who relays the conversations between the characters with the use of direct speech and quotation marks. Madrone's characterization as a shaman is complete as she learns to shapeshift. In addition to proposing a picture of the world making room for postmodern ecofeminist shamans, these tantalizing powers play a great part in the construction of the plot. They provide formidable moments of tension and relief as the characters use their magic to wrench themselves out of rigged situations or to save one another.

Starhawk's fictional mode abides by Amaryll Chanady's insistence that contrary to the fantastic, which encodes hesitation as it underscores "the simultaneous presence of two conflicting codes in the text," magical realism relies on "authorial reticence," which "naturalizes the supernatural," so that, as a result, the magic "does not disconcert the reader." As Wendy Faris puts it, "because the magical realist narrator accepts the antinomy and promotes the same acceptance in the reader, the antinomy is resolved."[14] Beyond Madrone's healing powers that seem too good to be true—especially in our Covid-ridden times—the irreducible elements of magic are legion in the novel. The elder and storyteller named Maya for instance is constantly engaged in dialogues with the dead which are transcribed onto the page with the usual quotation marks, as with any other ordinary dialogue. As for Lucy, one of the "Deep Listeners," she "[maintains] a constant protective vigil in the spirit world, alert for threats to the people."[15] Yet, for all the references to a "spirit world" of the dead, most of her characters' liminal experiences are counterbalanced by descriptions grounding them in a phenomenal world in touch with concrete reality. My contention is that the thought-experiments cooked up by Starhawk in her ecofeminist cauldron bring us to explore our ecological unconscious and to envision liminal states whereby the separations between matter and mind, between self and other, and between human and other-than-human worlds are erased in a way that ecopoetically changes the landscape of consciousness.

Following the logic of shamanism,[16] Madrone and her lover, Bird, have been initiated into their powers since early childhood. Madrone followed her witch mother into the jungle, learning to gather medicinal herbs and to listen hard: "'She told me not to be afraid of snakes, to sit quietly and listen to the animals and plants and try to understand what they were telling us."[17] Developing a keen hearing at an early age, Madrone has acquired the habit of listening for other-than-human voices. Initiated fully at sixteen, spending her "watershed year" living with Bird in the forests, Madrone fasted for three nights and days and went on to her "vision quest."[18] Similarly, Maya spent time alone in the mountains as a young woman, surviving with very little and gradually opening up her mind to the more-than-human presences around her: "'The rocks are very beautiful up there,'" she recounts, "'a clean granite, gray-white with dark flecks and little sparkles of quartz. After I had been alone for a while, they began to speak to me. Everything came alive and had its own voice, and I could hear it. [. . .] Without knowing the word for it, I became a witch."[19] Maya thus turned into a storyteller from hearing earth's voices.

Starhawk's liminal characters move between human and other-than-human worlds in different ways. Bird, escaping prison after ten years of sequestration and abuse, can read the flying of a crow guiding him by mapping out the topography of the land before him through some sort of echo-localization: "A black crow became Bird's guide. He would see it fly up before them, to reveal a way across a ravine, or hear it call, beckoning them down a certain path at a crossroads. [. . . .] The crow led them through the dunes that bordered the hill country and flapped down in an abandoned garden by a flat marshy lake where they were able to gather grapes and self-seeded tomatoes."[20] As crystallized in his first name, Bird has learned to become bird and see like a hawk,[21] either by dreaming, or through "spirit-traveling"—a flying of the mind to other places through visualization techniques. He first breaks free from within his cell: "Alone in the dark, he'd begun to fly. He had always been good at spirit travel; now he had infinite time to explore and few outside distractions. He went to his power place in the mountains; it was winter there, and in his astral body the crystalline structure of the snow became a labyrinth of rainbow chambers where we could wonder for hours or days."[22]

Bird's initiatory vision quest revealed his power as a musician and singer, turning "his hand on fire with melodies and rhythms that [have] snaked their ways into his dreams."[23] Describing his "redwood country" to a character from the South who has never beheld a sequoia, Bird explains of the majestic trees, "'They're like . . . guardians. When you're around them you feel protected. Watched over. They collect fog in their branches, way above your head. People say the spirits of your beloved dead hang out there.'"[24] Living in-between the world of sky and the deep world of soil, the redwoods, like Bird and his fellow witches, stand between different realms. Moreover, Bird's reverence for the trees evokes a kind of totemic relationship. He is so enmeshed with his homeland and co-dwellers that the mere thought of redwoods intuitively brings back all his somatic memories of the live land, including the song he has echo-vocalized in honor of the giant tree. Bird thus works as an instrument for the Redwoods, translating their song and olfactory signature into human music:

> He felt good, remembering the redwoods of Mount Tamalpais, the damp earth smell when you were down in a grove of them, the soft, rough, ridged bark, and the bay laurel trees raising their graceful limbs in between and wafting their pungent perfume around you. A tune he had forgotten came back to him, and he sang it for them.
>
> "That's the redwood song," he said. "It kind of sounds like they are."
>
> The music sang itself inside his head, and his hands ached for an instrument to chase it with.[25]

After a prolonged period of amnesia, this passage where Bird reawakens to his gift as an earth musician highlights Starhawk's conception of art as a form of surrendering to the vital forces of the earth running through one's being, activating the mind and body at once. Art, then, is a channeling of one's ecological unconscious into the making of *ecopoiesis*.

Metafictionally weaving into her characters' discourse her own vision of the narrative agency of stories, Starhawk pinpoints the empowering function of myths. Bird for example has been raised on a diet of shamanic stories: "[His] mother, Brigid, had put him to sleep with tales of shape-shifters and magicians. Now he recognized that she had been teaching him magic, telling him history in a form that kept him eager for more and more. 'Once there were some children who could turn themselves into birds.'"[26] This in turn induces his night visions. In his dreams, he "[becomes] a hawk, soaring over the hills to the north" and scrutinizing the landscape from a bird's elevated viewpoint. Reversely, late-blooming Maya remembers being brought up in a traditional Jewish family where she was "raised [her] whole life not to feel, not to trust [her] intuition, not to notice if [she saw] an aura or [felt] the energy move."[27]

Starhawk's witches highlight how the world of dreams, like that of myths, whispers unconscious truths to us which can shape a vision.[28] This vision, when made manifest, can influence our awareness of reality and, in turn, the political stances and concrete choices we make that will then shape reality: "'In an ordinary dream,'" Lily explains to Madrone—and by proxy to the reader—"'the spirit world speaks to us. But a Dreamer can speak back, can make shapes and patterns in that world that later take form in this. [. . .] There are many ways to *dream*. Some do it at night, with their eyes closed, some open-eyed in the light of day. Some, like Maya, tell stories that become the dreams of many.'"[29] Adhering to shamanic practices, Madrone can cure her sick fellows by modifying patterns in her sleep. She also learns the art of lucid dreaming, whereby active meditation and visualization are equated with a form of "flying."[30] Thus attuned to her unconscious, she pursues the vision she has had of going down south to help resistant groups in need of a healer.

Taking her liminal thought-experiments one step further, Starhawk thinks up a becoming-bee for Madrone that will ensure the happy resolution of the novel's suspenseful plot. As she travels through the desert, following her calling, Madrone is initiated into a beehive by witches that have learned the art of human-bee communication. In one of the most enchanting episodes of the novel, Madrone is introduced to this human-bee community and their interspecies relationship of reciprocity. The bees are acknowledged as the group's "little sisters" and "friends." They feed the human people with honey and tend to the wounded in swarms. In this group, some of the women have integrated the hive altogether: "a woman was wearing a cloud of bees like a cloak. Their buzzing was a sustained hum, like a chant. The air seemed to vibrate in harmony, and Madrone felt it move through her body like a sudden rush of intoxication. She smelled something on the wind, like the distilled essence of wild blossoms: honey. The woman seemed to be wearing no other covering but the bees. They crawled over her body like a second skin."[31] Obviously cognizant of scientific studies on bee societies and communication, Starhawk has imagined a world where liminal characters could enter bee consciousness and communicate on a par with them, as if dwelling partly in bee skin: "'The sisters work with us to heal the wounds, but we who are bonded to them cannot come near the sick. They have a horror of illness. In the hive, they kill the sick bees. The wounded, too—but over the years we have been able to teach them to work with us on injuries, as long as they don't get infected. It was difficult to train them. We had to enter into the hive mind and become part of it. But it has also become part of us.'"[32]

Madrone's initiation into the hive mixes all the traditional ingredients of a rite of passage. She is taken away from the group by the bee shaman, "the Melissa," to a remote place in the wild, in a "rocky outcropping where long ago a river had scoured a small hole in a bank, a dome-shaped cave just large enough for one person to lie curled up inside."[33] Activating the symbolism of a ritual rebirth, the place is described as feeling "warm and [. . .] safe, like a womb."[34] As Turner has observed of rites of passage, "liminality is frequently likened to death, to being in the womb, to invisibility, to darkness, [. . .] to the wilderness [. . .]."³⁵ Madrone is here undressed, covered in honey, and fed a mysterious, consciousness-changing liquid of "fermented honey and something else," making "everything around her [disappear]." Her modified consciousness is once again described as "flying," and the experience grants her enhanced consciousness of the world of smells: "Lilac was not a name but a realm of the air that called her into wild places where her whole body throbbed with delight. Sage was a universe, pungent, bracing." While immersed in the hive, she can hear the Melissa communicating in a kind of bee language: "'Let go,' the Melissa buzzed and hummed and murmured." In addition, her bodily form seems to have metamorphosed: "Madrone wanted to clutch her human form but she no longer had hands to grip with, only wings that beat incessantly, gossamer propellers to carry her away [. . .]."³⁶

While the rational mind could fall back on hypnosis and drugs to explain the induced voyage outside of Madrone's human mind and form into those of a bee, the point of this magical flight experience lies in her augmented *Umwelt* at the end of the process, prompting us in her wake to consider the world from the perspective of bees: "She was moving in the dark safe hive, where body brushed against scented body, learning from movement and smell what the hive knew, the paths through the air to the nectar flow, the health of the brood, the golden warmth of the sun. And under it, the queen's smell."³⁷ Responding from then on to the pheromonal world inhabited by bees, the woman becomes a drone—a transspecies, transexual metamorphosis foreshadowed by her name, "Ma*drone*": "Within her own body flowed rivers of scent and taste, and suddenly she knew them in a way even she, a healer, never had before—knew the scents her sweat could produce and what each signified and how they could be messages and conversations and offerings."³⁸ Scent, it is implied, takes part in the song of the world.

Once Madrone is aggregated back into the society of humans, her perception has been transmogrified, integrating the sensory communication capacities of bees: "Bees hummed lazily around her; their sound was now like music to her, operas and symphonies and oratorios, and at the same time like a crowd of gossiping friends, telling her everything she needed to know. [. . .] She felt a sense of vertigo, almost a double vision. She could see through multifaceted insect eyes more easily than she could look at things straight on in her old human way."³⁹ By proxy, as Madrone's "double vision" from then on shifts back and forth between a human's and a bee's, readers must consider the odorous and gustatory textures of the more-than-human world which carry biosemiotic messages eluding human vision and language: "[Words] seemed awkward, clumsy, unnecessary when a molecule of scent could convey the same thing. [. . . Words] seemed primitive and inadequate compared to the delicate subtleties of taste and smell."⁴⁰ On the level of the plot, this human-bee association saves the day as Madrone will call upon the bees to help Bird and their entire community out of trouble. Meanwhile, liminal realism goads us to reconsider the "boundaries between the bee mind and [one's] ordinary mind." For Madrone, these boundaries have become "permanently blurred": "The route was a succession of plant smells, fragrance after fragrance, wafting down from the well-watered gardens above. She was intoxicated by carob and night-blooming jasmine and other sweet traces she could not name. Lights whirled in kaleidoscope patterns, as if prisms were fixed before her eyes. [. . .] Her eyes darted around with the insect alertness that had become second nature."[41] In the wake of the characters' fictional shapeshifting, the reader is thus embarked on a perception-shifting trip. Besides, this ecopoetic process adheres to Philippe Descola's conclusions as to the nature and function of metamorphosis in animistic shamanic practices:

> Metamorphosis, then, is not an unveiling or a disguise, but the summit of a relationship wherein each person, by modifying the observation point imposed by her original physicality, strives to coincide with the perspective he or she believes the other to envision herself from. Thus displacing the angle of approach where one tries to walk "in the shoes" of the other by embracing the intentionality one lends her, the human no longer sees the animal as she does ordinarily, but the way the latter sees herself, as a human. Moreover, the shaman is perceived not as she usually sees herself, but rather as she wishes to be seen, as an animal. More than a metamorphosis, in sum, what is at stake is an anamorphosis.[42]

It may be concluded at this point that liminal realism consists in triggering interspecies anamorphoses, spurring humans to shift perspectives onto the world and develop an other-than-human *Umwelt*.[43]

While Starhawk inspiringly imagines new forms of human-bee relationships, her experimental writing may remind us that there are indeed many ways to "enter into the hive mind," like the Melissas—one of which being through the scientific studies on bee societies and communication which Starhawk toys with.[44] Entomological studies provide insight into the complexities of bee language through their vibration dances and pheromonal emissions, together with the intricate workings of their communities, each functioning like a superorganism. As a matter of fact, even the incredible ending of Starhawk's novel—when Madrone calls to the bees for help in the human conflict being waged—is somewhat realistic, in that it reflects actual phenomena that have been observed among bees. Indeed, in their division of labor, bees in a hive include guardians and soldiers who cooperate in case of danger. When sensing a potential attack on the brood, the guardian bees emit a banana-tasting pheromone through their mandibles which the soldier bees receive as an alert message summoning them to go sting the source of danger. To a certain extent then, sending out scents that call upon the bees for help, Madrone's gesture makes perfect sense according to the science on bees. Meanwhile, it reactivates totemic bonds of guardianship and cooperation between human and animal populations.

Starhawk's ecofeminist novel brings to the fore the parallels between the collective intelligence of bees and her utopian ecofeminist community based on partnership models of power-with, rather than those of power-over at the heart of modern patriarchal thinking. We may thus be reminded of our actual human-bee associations through apiculture, a very old human-bee partnership bringing about much pleasure and many benefits to the human health via the bees' production of honey. Furthermore, if Starhawk's scenario retains a sense of the impossible, it obliquely says something of the ecosystemic services rendered by pollinator bees and which many anthropic activities both depend on and impact. As bee populations around the world have plummeted in great part because of pesticide-spraying, climate change, and intensive monocultures, Starhawk ecopoetically encapsulates some of the tangible ways in which human and bee existences are, in fact, very much entangled.

Essentially, her ecopoetic thought experiment makes headway into the possibility of imagining how to abide by a multispecies swarm intelligence. It can be concluded that as she constantly interweaves scientific visions of reality with ecopoetic ones, Starhawk means to affect the codependent evolution of consciousness and reality, an interweaving of mind and matter which she phrases in the following conundrum: "Consciousness shapes reality/ Reality shapes consciousness."[45] Encouraging readers to envision a world beyond species boundaries, where there is more to see, feel, hear, smell, and interact with than most of us do in our ordinary lives, Starhawk's liminal realism abides by her "discipline of magic": "We need the discipline of magic, of consciousness-change, in order to hear and understand what the earth is saying to us."[46] Reflecting on how to heal from the sense of alienation from nature at the root of our ecological crisis, Starhawk speaks for a practice of attention to other-than-human languages: "Everything around us is always speaking. We can heal only by first learning to hear, to understand, and in time, to respond. As we do, the world becomes richer, a more complex and vibrant place."[47]

As with the other writers in my corpus, Starhawk's use of liminal realism offers that "vision across boundaries" which, according to Paul Shepard is required by ecological thinking. Liminal realism thus rehabilitates the "relatedness of the self," the "delicate interpenetration" between the individual and the larger world. It corrects a distorted vision whereby, as Shepard puts it, [we] are hidden from ourselves by habits of perception," encouraged to think of ourselves as "an isolated sack, a thing, a contained self." Reaching across the "ego boundary" and across the human-nonhuman frontier, liminal realism sheds light on "the beauty and complexity of nature" as "continuous with ourselves." Liminal realism thus proves essential to renovate our perception as it "reveals the self ennobled and extended."[48]

SHAMANISM AND FOREST ECOPOETICS IN RICHARD POWERS' POSTMODERNIST *THE OVERSTORY*

In Richard Powers's 2018 novel, *The Overstory*—which won the 2019 Pulitzer Prize for Fiction—interspecies communication is established between humans and trees in ways that pivot on liminality while staying grounded overall in a postmodern ontology. Also coherent with Rupert Sheldrake's theory of morphic resonance, the story casts a set of characters across the United States in the early 1990s who simultaneously awaken to the wonders of trees, some of them beginning to hear tree voices. My claim is that in his postmodernist resort to liminal realism, thereby interweaving various approaches to reenchantment, Powers casts light onto the ecopoetic complementarity between our scientific and poetic minds in their conjoined capacity to renovate our sense of awe and to restore healthier notions of what dwelling ecologically might mean.

Allowing for various types of reader identification, Powers's characterization involves widely different mindsets. His tree-hearers and tree-rights crusaders include an artist born into an Iowa family of farmers (Nick), a tree biologist and researcher (Patricia), an American-Asian mystic influenced by Chinese Buddhism (Mimi), a down-to-earth Air Force veteran who has become a tree-planter and seasonal ranch worker (Douglas), a college pot-smoker and drop-out who turns into an environmentalist leader (Olivia), an intellectual property lawyer and his stenographer wife (Ray and Dorothy), a genius Gujarati-American computer programmer who starts his own ecopoetic video game company (Neelay), and a psychology student and later academic obsessed with individual and collective behavioral patterns and cognitive biases (Adam). Two of the main strands of Powers's dialogism are rooted, on the one hand, in the historiographic reconstruction of the 1990s Californian Timber Wars—by the end of which 95 percent of coastal redwoods had been clear-cut—and, on the other hand, in the latest science on forest life and tree communication. Recent findings underly the novel, such as those popularized by Professor of Forest ecology Suzanne Simard,[49] by Peter Wohlleben's widely distributed *The Hidden Life of Trees: What They Feel, How They Communicate: Discoveries from a Secret World*, by (2016), or the less acclaimed, *The Secret Life of Trees: How they Live and Why They Matter* (2005), by Colin Tudge. As the first dialogic thread has provided inspiration for the group of five characters who become radical environmental activists dedicated to saving *Sequoia sempervirens*, the second one underpins Patricia's character and research, making her a literary mouthpiece for the science writers. In addition, the novel contains many references to some of the most influential nature writing from Emerson, Thoreau, Muir, and Leopold, either in the form of epigraphs, or via the readings of the protagonists. Powers's postmodernist text thus recycles these early writers' foundational revelations as to our various intra-actions with the other-than-human world, including those sustaining our spiritual lives.[50]

Providing an inkling of the onto-epistemological braiding at play, Powers designs a novelistic threshold made of three epigraphs, respectively by Ralph Waldo Emerson, James Lovelock, and the late aboriginal William Neidje—the last speaker of the now extinguished *gaagudju* language. Already, and despite the different ontological frameworks underpinning them, each of the three quotes sheds its own light on the agency of trees and interspecies communication. In a famous passage from Emerson's essay on "Nature," the transcendentalist philosopher and poet wrote: "The greatest delight which the field and woods minister, is the suggestion of an occult relation between man and the vegetable. I am not alone and unacknowledged. They nod to me, and I to them." As the association of epigraphs heralds, the novel about to open seeks somehow to conciliate Emerson's ecopoetics with Lovelock's science-based Gaia hypothesis, grounded as it is in the very material agency of the biosphere, and, here, of trees: "Earth may be alive, not as the ancients saw her—a sentient Goddess with a purpose and foresight—but alive like a tree. A tree that quietly exists, [. . .] endlessly conversing with the sunlight and the soil. Using sunlight and water and nutrient minerals to grow and change." Finally, both takes on the world chime with Bill Neidjie's expression of an indigenous animistic worldview: "Tree . . . he watching you. You look at tree, he listen to you. He got no finger, he can't speak. But that leaf . . . he pumping, growing, growing in the night. While you sleeping you dream something. Tree and grass same thing." Accordingly, if some of Powers's characters' approaches to trees are predominantly naturalistic, i.e., based on careful empirical observation and rational thinking, they are nonetheless interlaced with mystical, aesthetic, and affective responses to trees, thus forming an intricate knot of human-tree relationships. Examining the liminal realism crafted in this novel, I am interested in the way a writer with some training in physics and computer science, in music as well as in literature,[51] opts for an ecopoetics of reenchantment that braids science with animistic and totemic worldviews.

While all the characters in the novel exhibit an unusually acute form of sensitivity to the existence of trees, Patricia the scientist and Olivia the college drop-out and voice-hearer provide two essential polar opposites.

As such, they offer complementary entries into a form of liminal realism challenging the notion of a separation between the world of trees and that of humans. Albeit via different routes, both characters function as shamans mediating between the enmeshed realms of humans and plants. Furthermore, each accesses the world of trees through marginal kinds of perception, challenging the boundaries of what is ordinarily accessible to humans on auditive, visible, and olfactory spectrums. Completing the picture of a human and tree communitas, Mimi's revelation sheds a different light on green enlightenment. Her revelation is compatible with her cultural background, that is, with Zen Buddhism and its theory of the interconnectedness of all things. As for the male characters' fascination with trees, it springs from different sources—environmental art for Nick, an accident and coding for Neelay, and the discovery of the magnitude of deforestation in Douglas's case, which gets him dedicated to tree planting and radical activism. Finally, Ray and Dorothy's love of trees originates in their backyard, where for the first years of their marriage, they planted a tree with each passing anniversary. After Ray has suffered a stroke leaving him paralyzed and fully dependent, Dorothy spends much of her time reading to him while he contemplates the view over their garden. Their captivation with trees is sparked by their reading of Patricia's best-selling nonfiction writing on how trees sense and communicate, how they form partnerships and underground networks that enrich the soil and engineer much of a healthy ecosystem. Quite metafictionally, Patricia's book on the lives of trees is systematically presented as a powerful eye-opener. In Ray and Dorothy's case, it encourages them to rewild both their garden and their imagination—a central point that has curiously been overlooked in the existing critique of the novel, and which I will get back to later.

Although aligning with modernity in her overall method, Patricia nonetheless displays many of the attributes of the shaman. She has never entirely given up the animistic vision prevalent in childhood, which, back then, fed into her imaginative playing with a fabulous world "made of twigs," very much "alive" to her: "squirrels made of sweetgum balls, dragons from the pods of Kentucky coffee trees, fairies donning acorn caps, and an angel whose pinecone body needs only two holly leaves for wings."[52] As long as her child's mind and anamorphic eye creatively animated her little human-tree theater, she projected the capacity for speech onto her "twig creatures," identifying with them: "All her twig creatures can talk, though most, like Patty, have no need of words. She herself said nothing past the age of three. Her two older brothers interpreted her secret language for their frightened parents, who began to think she must be mentally deficient."[53] Interestingly, because of "a deformation of her inner ear," Patricia finds herself in a position similar to that of plants—because they cannot comprehend her cryptic language, most people around her give up on her, treating her as an inferior with diminished capacity to think and express herself: "The clinic fitted her with fist-sized hearing aids [. . .]. When her own speech started to flow at last, it hid her thoughts behind a slurry hard for the uninitiated to comprehend. It didn't help that her face was sloped and ursine. The neighbor's kids ran from her, this thing only borderline human. Acorn people are so much more forgiving."[54] Owing to her experience as an outcast, Patricia is separated early on from the wider society including her human peers—a condition widening the scope of her imagination, freeing her in part from the restrictions imposed by common sense, and setting her up to become a shaman.

Patricia's childhood experience indeed singles her out with the necessary attributes to become a mediator between trees and humans. Questioning the dominant anthropocentric take on notions of language, communication, and agency, Patricia and her father are from the start attuned to biophony and geophony. They quietly sit outdoors and "listen to what other people [call] silence."[55] As a result, they have learned to pay attention to what nonhuman lifeforms might be saying. Heeding nonhuman songs and languages, they work for example on figuring out how these might translate in meaningful ways for humans, as for example when Patricia listens to cricket songs and remembers the "old formula" her father has taught her, "one that converts cricket chirps per minute into degrees Fahrenheit."[56] Aware that the snowy tree crickets' song translates changes in temperatures,[57] Patricia and her father make the reader sensitive to the different ways in which beings emit meaningful signals in relationship with their environment. Paying close attention to the biosemiotic textures of the world, these two characters may inspire us as they strive to grasp how the world is structured in acoustic niches where different species express signs of their own health: "For sixty years, the nighttime orchestra all around [Patricia] has been playing one of those folk dances that keep speeding up until all the players tumble in a heap."[58] Listening for what the crickets are telling us, whether they mean to or not, about climate change, Patricia negotiates a kind of interspecies communication. Cognizant early on of most people's narrow-minded take on the notions of personhood and language, she herself forms a deep connection to the different ways in which trees offer signs of their own intelligence and communication skills—how despite their nonhuman ways, they too possess an intricate, wonderful form of sentience and thrive on myriad mutualisms that include humans.

Patricia's father—an agricultural extension agent with comprehensive knowledge of botany—gets both the child's fascination with plants, her "woodlands world," and "her every thickened word." If anything, he seems to recognize a greater potential in his differently gifted daughter, his "little plant-girl, Patty" living in-between human and other-than-human worlds.[59] The hybrid world Patricia inhabits soon becomes explainable through science: "acorn animism turns bit by bit into its offspring, botany. She becomes her father's star and only pupil for the simple reason that she alone, of all the family, sees what he knows: plants are willful and crafty and after something, just like people."[60] This vision of botany as descended from animism echoes Robin Wall Kimmerer's effort to reconcile the worldview that suffuses Traditional Ecological Knowledge with current, mainstream science: "Because we can't speak the same language [as animals and plants]," she writes, "our work as scientists is to piece the story together as best as we can. We can't ask [them] directly what they need, so we ask with experiments and listen carefully to their answers. [. . .] We measure and record and analyze in ways that might seem lifeless but to us are the conduits to understanding the inscrutable lives of species not our own." Underscoring the shamanic function of scientists in our postmodern world, Kimmerer argues: "These are just ways we have of crossing the species boundary, of slipping off our human skins and wearing fins or feathers or foliage, trying to know others as fully as we can."[61] Kindling her passion for the world of trees, Patricia's father brings her to contemplate all the riddles trees have in store for humans, meanwhile pointing out the myopic view of modern science: "'We know so little about how trees grow. Almost nothing about how they bloom and branch and shed and heal themselves. We've learned a little about a few of them, in isolation. But nothing is less isolated or more social than a tree.'"[62] Reducing the lives of trees to what can be observed in laboratories while studying specimens that have been uprooted from their natural *oikos*, scientists long missed out on the underground and airborne routes that trees use to communicate and intra-act with their kindred species as well as with humans, animals, insects, and fungi.

Decentering human attention, a dialogue between Patricia and her father metafictionally expresses Powers's use of ecoliterature as a corrective lens, filling the gaps of an anthropocentric education, social system, and even literature, where trees generally get too little attention: "He teaches her to tell a shellbark from a shagbark hickory. No one else at her school can even tell a hickory from a hop hornbeam. The fact strikes her as bizarre. 'Kids in my class think a black walnut looks just like a white ash. Are they *blind*?' / 'Plant-blind.'"[63] Patricia's imagination is moreover unleashed by Ovid's *The Metamorphoses*, filled as it is with tales of shapeshifters:

> She loves best the stories where people change into trees. Daphne, transformed into a bay laurel just before Apollo can catch and harm her. The women killers of Orpheus, held fast by earth, watching their toes turn into roots and their legs into woody trunks. She reads of the boy Cyparissus, whom Apollo converts into a cypress so that he might grieve forever for his slain pet deer [. . . , and of] Myrrha, changed into a myrtle after creeping into her father's bed. And she cries at that steadfast couple, Baucis and Philemon, spending the centuries as oak and linden [. . .].[64]

Allowing her to envision the possibility that trees and humans may form part of one and the same realm, reading Ovid leads Patricia to ponder the continuity between humans and trees, the two kinds having co-evolved from the same wild stuff: "The fables seem to be less about people turning into other living things than about other living things somehow reabsorbing, at the moment of greatest danger, the wildness inside people that never really went away."[65]

Owing to her open, imaginative stance, Patricia is not blindfolded by the hierarchical classification of the living world prevailing in her scientific community, nor by the dualisms that have made it impossible for forest management services to just let a forest be: "*Improve forest health*. As if forests were waiting all these four hundred million years for us newcomers to come cure them. Science in the service of willful blindness: How could so many smart people have missed the obvious? A person has only to look, to see that dead logs are far more alive than living ones. But the senses never have much chance, against the power of doctrine."[66] With an oxymoronic vision making room for the alliance between life and death processes in the complex ecosystem of a forest, Patricia's nondualist epistemology is constructed empirically, from a liminal standpoint making room for fables *and* science, for intuition *and* reason, for data *and* metaphor. Her ontology of ecological continuity splits neither mind away from matter, nor life from death, nor our capacity to reason from our capacity to feel through our various sensors. Moreover, as she strives to translate scientific data into accessible human concepts and tales, like Simard and Wohlleben who poetically extend the concepts of motherhood, families, partnerships, and societies to trees, forests, and mycorrhizal associations, Patricia conjugates an acute analytical mind with a gift for speaking in metaphors.

Essentially, Patricia's father makes her sensitive to phenomena that are *a priori* invisible to the human eye: "He teaches her how to see a tree, the living sheath of cells underneath every square inch of bark doing things no man has yet figured out."[67] The man invites Patricia to perceive the biosemiotic messages emitted by trees. Her incorporation of tree presences later percolates into her research, suffused by "the stink of humus filling up her nose with relentless musky life."[68] As she gets into graduate school studying forestry and teaching botany, she spends prolonged time doing ground research in the woods, something feeling to her like "an animist's heaven."[69] Having kept her childhood sense of wonder intact, Patricia's scientific enquiries are determined by her intuition that trees must be "social creatures" forming "mass mixed communities [which] must have evolved ways to synchronize with each other."[70] Following her hunch against her field's hegemonic common sense, Patricia nevertheless keeps in line with the scientific method and logic.

She sets out to demonstrate that trees do possess a form of intelligence and language, which she translates in both scientific and ordinary terms:

> ... something else in the data makes her flesh pucker: trees a little way off, untouched by the invading swarms, ramp up their own defenses when their neighbor is attacked. Something *alerts* them. They get wind of the disaster and they prepare: she controls for everything she can, and the results are always the same. Only one conclusion makes any sense: The wounded trees send out alarms that other trees smell. Her maples are *signaling*. They're linked together in an airborne network, sharing an immune system across acres of woodlands. [. . .] [71]

Drawing the only possible conclusion from her data, Patricia casts light on how trees literally "get wind" of an attack: "*The biochemical behavior of individual trees may make sense only when we see them as members of a community.*"[72]

Patricia Westerford thus serves to diffract the wide-ranging, scientific communication work carried out by the likes of Suzanne Simard, Colin Tudge, and Peter Wohlleben—the latter whose initials the character shares. Meanwhile, Patricia also reads as a fictional avatar for prescient women scientists such as Simard, or before her and in a different field, Rachel Carson, both of whose work was first widely criticized by their male-dominated scientific communities of peers, unwilling as they were to give credence to a female fellow intuiting and demonstrating what had not yet been reasonably or widely accepted by modern science. As such, Patricia's posture long remains marginal: "'My whole life, I've been an outsider. [. . .] We found that trees could communicate, over the air and through their roots. Common sense hooted us down. We found that trees take care of each other. Collective science dismissed the idea.'"[73] The novel thus foregrounds how subjective and fragile notions of truth and reason may be even from a scientific point of view.

Like Simard's, Patricia's research challenges anthropocentric notions of communication. First, she seizes the odorscapes in which we bathe, which convey non-verbal signs as wafts of pheromones travel by air. Patricia points to biosemiotic messages that may remain subliminal for humans. Walking around a Californian campus with great tree diversity, she considers the likelihood that "[every] student must be drunk on the air's intoxicants without even knowing it."[74] Second, alluding to the current research in the chemical signals and sounds which plants emit below human earshot—signals which are involved in tree communication and, possibly, in interspecies communication—Patricia ventures ideas that humanist philosophy and scientific rigor tend to prohibit, although they are in fact being looked into worldwide by specialists of plant sentience and communication:[75] "Trees know when we're close by. The chemistry of their roots and the perfumes their leaves pump out change when we're near. . . . When you feel good after a walk in the woods, it may be that certain species are bribing you. [. . .] Trees have long been trying to reach us. But they speak on frequencies too low for people to hear."[76] Probing throughout how "trees talk to one another, over the air and underground," Patricia's sharp response to tree smells can make us wonder to what extent humans living in contact with plants and trees could be unconsciously responding to such emissions: "The wind wafts through the window, smelling of compost and cedar. The scent triggers an old, deep longing that seems to have no purpose. The woods are calling, and she must go."[77] It may be wondered how far Patricia's scientific "calling" to study the forest superorganism might be accounted for by her sensitive attunement to the world of trees with which our bodyminds have co-evolved: "Our brains evolved to solve the forest. We've shaped and been shaped by forests for longer than we've been Homo sapiens."[78] Thus, Powers's liminal realism explores the possibility that the odorscapes modern humans have evolved to filter out of our consciousness might yet provide a potential space for interspecies communication.

Patricia cues us to toy with the idea that humans, like some insects, might be receptive to some of the signals sent out by plants: "[We] discovered that trees sense the presence of other nearby life. [. . .] That trees feed their young and synchronize their masts and bank resources and warn kin and send out signals to wasps to come and save them from attacks."[79] Hinting at his own creative process, Powers here gestures to the starting point of his thought-experiment whereby humans are called upon by trees to save our partnerships with them. His liminal realism, we come to understand, is rooted in actual science on tree communication. In this respect, Powers' ecopoetics of reenchantment follows a logic similar to Starhawk's. Both craft liminal *ecopoeisis* that stretch the imagination from science to fictional interspecies communication and partnerships, which then play a crucial role in their plots.

Hence the "[signals that] flood [Patricia's] muscles, finer than any words," stopping her from committing suicide after her work is castigated by her peers, who start shunning her as "the woman who thinks that trees are intelligent"[80]: "She listens to the forest, to the chatter that has always sustained her. [. . .] Something stops her. [. . .] *Not this. Come with. Fear nothing.*"[81] When Patricia's groundbreaking discoveries are rejected by her colleagues and when, after six months spent "at the bottom of a well," she decides to take her own life, she is stopped at the last minute by the irruption in her head of non-audible voices. These could express either her own unconscious life-drive, or, following a liminal realistic logic, her tuning in with tree language. The use of italics here is ambiguous and destabilizing, translating a shift in speakers which is hard to locate in the absence of quotation. Patricia being alone, it seems logical that the italics should translate a voice that is heard internally. The unusual syntax and somewhat cryptic language, with a telegraphic style, evoke a nonhuman language, which coalesces with the "signals [flooding] her muscles, finer than any words."

In fact, the italics, the polyphony, and the vocabulary echo the strange incipit of this first part of the novel, focusing on an anonymous woman receiving the gift of communication with trees—a woman which we can later identify as Buddha-like, enlightened Mimi:

> [In] a park above the western city after dusk, the air is raining messages. / A woman sits on the ground, leaning against a pine. Its bark presses hard against her back, as hard as life. Its needles scent the air and a force hums in the heart of the wood. Her ear tunes down to the lowest frequencies. The tree is saying things, in words before words. [. . .] *Something in the air's scent commands the woman [. . .]. A chorus of living wood sings to the woman*: If your mind were only a slightly greener thing, we'd drown you in meaning."[82]

Thus, as Patricia seems to be developing an extrasensory gift allowing her to decode tree language—most of which results from contact with the trees and empirical observation—readers might hesitate briefly between going along with her incredible character or wondering about her sanity. Cueing us toward accepting the former, the greening of her mind and her role as translator between humans and trees is then after expressed in terms oscillating between traditional nature writing and science writing. Besides, the possibility that the italics may ascribe direct discourse to trees whenever a character plugs into tree language is confirmed when Olivia, in the following chapter, starts hearing tree voices beckoning her to their rescue. Moreover, Powers's thought experiment implicitly relies on the underlying hypothesis that Patricia's inner ear deformation may paradoxically have turned into a gift. Thus, it can be inferred, she may be more prone than her fellow humans to pick up vibrations on frequencies too low for most other humans. She can register for example the "humming" of the many seeds collected in her seedbank, which, we are told, are "singing something—she'd swear it—just below earshot."[83] In *Thus Spoke the Plant*, scientist and shaman Monica Gagliano writes of humans' capacity to attune to "green symphonies." Differentiating her "plant-writing" from "ventriloquizing by assigning a voice to plants or speaking for them," Gagliano—whose discoveries have also encountered much initial resistance—asserts that "the human is not an interpreter who translates a mental representation in her head [. . .]. Rather, the human is a listener who filters out personal noise to hear plants speak, who engages in active dialog with these nonhuman intelligences." Thus, she concludes, "the human acts as a coauthor who physically delivers those conversations on the page," creating "stories [that] emerge out of a human-plant collaborative endeavor and mixed writing style."[84] As a matter of fact, Gagliano's nonfiction book also exhibits all the traits that I have established as characteristic of liminal realism, thus confirming that the paradigm shift I have identified is underway beyond the confines of ecopoetic fiction.

Patricia falls off the academic grid for several years. She evolves in isolation from her human peers, going deeper and deeper into the woods, a kind of modern-day, atheist Thoreau. Experiencing transcorporeal fusion with nonhuman nature, she camps out in the forest, "turned wildly around by the smell of inland oceans, sleeping on beds of thick lichen, sixteen inches of brown needle pillow, the living earth beneath her bag, its fluid influence rising up into the fiber of her and all the towering trunks that surround and watch over."[85] Like a shaman living in-between the world of humans and the world of trees, Patricia gets "busy learning a foreign language,"[86]—that of trees. Undergoing a liminal phase, separated as she is from her human peers, Patricia joins the multispecies community of sylvan life: "And day and all night long, her only people are the trees, and her only means of speaking for them are words, those organs of saprophytic latecomers that live off the energy green things make."[87] Powers thus gradually but surely instills a sense of wonder bringing readers to envision biotic communities. As early as 1949, Aldo Leopold heralded that "the evolution of a land ethic is an intellectual as well as an emotional process."[88] Leopold encouraged an ethic of love and respect for the land, while laying down the basic principles of ecology: "We abuse the land because we regard it as a commodity belonging to us. When we see the land as a community to which we belong, we may begin to use it with love and respect."[89]

Further paving the way for when other characters start hearing tree voices, the lexical field of wonder riddles even the more scientifically steered narrative. Taking her on tours of botanical marvels, Patricia's father draws her attention to "extraordinary things" that science has yet to explain.[90] In Patricia's teaching and writing, the language of magic applies to the tangible, ordinary wonders accomplished by trees: "It's a miracle, she tells her students, photosynthesis: a feat of chemical engineering underpinning creation's entire cathedral. All the razzmatazz of life on Earth is a free-rider on that mind-boggling magic act."[91] Without straying from rational, scientific culture, Patricia nonetheless imparts deep reverence for the marvelous genius at work in nature, much of which remains unfathomable: "She leads her charges into the inner sanctum of the mystery: Hundreds of chlorophyll molecules assemble into antennae complexes. Countless such antennae arrays form up into thylakoid discs. Stacks of these discs align in a single chloroplast. Up to a hundred such solar power factories power a single plant cell. Millions of cells may shape a single leaf. A million leaves rustles in a single glorious gingko."[92] Mingling scientific data and parlance with a sense of awe, liminal realism here places us in between animistic worldviews and naturalistic, empirical readings of a world that remains amazing in its life-sustaining complexities and agency. Turning into a translator for trees, "Plant-Patty," as she is later called by the narrator, reads as a fictional avatar for the writer—himself an ecopoetic mediator between science and poetics, between trees and humans.

Much of the story hinges on rehabilitating the old notion that human fates have always been intricately intertwined with those of trees. The ancient bonds of kinship at the heart of totemic practices are first revived through the rational lens of genetics and evolution. Hence the opening paragraph of Patricia's book, underscoring the scientific view of kinship between humans and trees, repeated like a leitmotif throughout the book: "You and the tree in your background come from a common ancestor. A billion and a half years ago, the two of you parted ways. But even now, after an immense journey in separate directions, that tree and you still share a quarter of your genes."[93] Presenting trees as distant kin, Patricia's vision "blurs the line between those nearly identical molecules, chlorophyll and hemoglobin."[94] One might recall that red and greed are in fact two complementary opposites are on the chromatic circle. Moreover, both our blood's red cells and plants' photosynthetic chlorophyl are in themselves invisible wonders resulting from millions of years of coevolution between various lifeforms, allowing organisms of different kinds to live in constant interaction with their immediate environment—the one assimilating carbon dioxide thanks to chlorophyll, and the other distributing oxygen to all human cells.

Totemic connections are reformed in the novel as the ecowarriors adopt tree names via rituals designed to seal their belonging to the community of tree huggers pledging allegiance to the lives of trees. Thus, Olivia is rechristened Maidenhair—a common name for the Gingko biloba—while Mimi and Adam are respectively called Mulberry and Maple, in reference to trees they have grown up with and which seem to have presided over their destinies. In Adam's case more specifically, the totemic logic underlies his family history, his father having planted a different kind of tree in their backyard for every one of his children's births. When Adam's sister's tree gets killed by the Dutch elm disease caused by a fungus, Leigh herself disappears—most likely kidnapped and murdered by a stranger. Abiding by a similar logic, toward the end of the novel, as the ecowarrior group blow one of their missions to save trees from clear-cutting, Olivia's accidental death provides the human sacrifice balancing out the earlier felling of the gigantic redwood that the group failed to save—an ancient "mother-tree"[95] which Olivia had been summoned to protect and had christened "Mimas." However, like the stump of the thousand-year-old redwood, it is suggested that Olivia's connections with her fellow ecowarriors may continue to thrive underground. Through some sort of rhizomatic meshwork, Olivia's mythical tale still influences the other characters' growth—their choices, ethics, and spirituality, as well their continued environmental resistance.

Moreover, this totemic relationship tying humans to specific trees accounts for Powers's tweaking of real-life events—the long-term occupation in the late 1990s of a giant Californian redwood named Luna by Earth First activists, and particularly by Julia Butterfly Hill, who reported her sense of communicating with trees and of answering to tree sentience and to "the spirit of the forest."[96] Departing from the positive outcome of Hill's tree sit, the tragic fate of both Mimas and Olivia in Powers's fiction is determined by his mythopoeic rewriting of the history of Joan of Arc together with his reliance on a totemic worldview necessitating give-and-take equilibria—when Mimas is felled, his guardian in the human community must be sacrificed to restore balance.

That each of the characters in the book is specifically connected to one type of tree is signaled at the beginning of the first eight chapters, devoted in turn to each protagonist. Each chapter is introduced by the eponymous character's first and last names, followed by a delicate, realistic black and white drawing of tree foliage, sometimes with flowers and fruit. In time, each chapter turns out to include some sort of totemic intimate bond between its human protagonist and the specific tree whose singular existence is made manifest on the page by these visual close-ups. For example, Olivia's sudden awareness of her deep connection to trees seems triggered by her unconscious knowledge of her ties with the gingko thriving outside her college apartment. The drawing of the ginkgo presiding over the chapter introducing Olivia first calls attention to those ancient, adaptable, and synanthropic trees that now thrive alongside humans in cities.[97] Besides, the biography of each of the drawn tree turns out linked at one point or another to that of its human correspondent. Affirming the biocentric stance of the novel, trees moreover tend to get described at greater length than the characters. With each chapter focusing as much on trees' existences, names, and particularities as they do on humans, the novel whittles our curiosity for trees and restores our attention to them, cuing us to reconsider the many ways in which human and tree existences are intermingled.

Patricia's role as a postmodern shaman fits with her hermit life in the woods. Working for the Bureau of Land Management, she can "follow the rhythms of the land," and stay tuned to the trees, "listening to the wood put forth fresh cells."[98] Like many shamans, Patricia develops her own way of communicating with the dead: "Out in the woods, her father is with her again all day long. She asks him things, and the mere act of asking out loud helps her see."[99] In a like manner, after her husband dies, Patricia continues to discuss her research out loud with him—a habit allowing her to find her own answers in conversations with her *alter ego*. Patricia's husband describes her as being "practically a prophetess,"[100] laying down the "gospel of new forestry."[101] She equates her "life's work" with "listening to trees" and "telling people what [trees are] saying."[102] With such a postmodernist kind of animism—or, one might say, a kind of self-reflexive, liminal agential realism—ecological science replaces old religions, focusing our attention onto material wonders: "This gospel of new forestry is confirmed by the most wonderful findings: beards of lichen high up in the air, that grow only on the oldest trees and inject essential nitrogen back into the living system. Subterranean voles that feed on truffles and spread the spores of fungi across the forest floor. Fungi that infuse into the roots of trees in partnerships so tight it's hard to say where one organism leaves off and the other begins."[103] Quite ecopoetically, the word "religion" finds its roots in Latin, through *relegere*, meaning to read over, to reread, and/or *religare*, meaning to relate, to bind or tie. Thereby dovetailing with ecology—a science seeking to read how an organism is tied to other organisms, and how each interacts with the milieu in which it evolves—religion as a human concept may then well be reinvested from an ecopoet(h)ic and materialistic stance, as an invitation to translate the enchanting book of nature in search for what binds all lifeforms together.

Patricia's character fosters a secular kind of spirituality. Illumined by the revelations of science, her worldview encourages a resacralization of material agencies: "Love for trees pours out of her—the grace of them, their supple experimentation, the constant variety and surprise. These slow, deliberate creatures with their elaborate vocabularies, each distinctive, shaping each other, breeding birds, sinking carbon, purifying water, filtering poisons from the ground, stabilizing the microclimate. Join enough living things together, through the air and underground, and you wind up with something that has *intention*."[104] In that sense, Powers's ecopoet(h)ic enterprise chimes in with Gary Snyder's bridging of sciences and spirituality:

> The biological-ecological sciences have been laying out (implicitly) a spiritual dimension. We must find our way to seeing the mineral cycles, the water cycles, air cycles, nutrient cycles as sacramental—and we must incorporate that insight into our own personal spiritual quest and integrate it with all the wisdom teachings we have received from the nearer past. The expression of it is simple: feeling gratitude to it all; taking responsibility for your own acts: keeping contact with the sources of the energy that flow into your life (namely dirt, water, flesh).[105]

All in all, such ecological spirituality emanates from the various ways of dwelling encapsulated in the books of my corpus.

Patricia moreover reinterprets the nature writing of John Muir and H. D. Thoreau: "The particle of her private self rejoins everything it has been split off from—the plan of runaway green. *I only went out for a walk and finally concluded to stay out till sundown, for going out, I found, was really going in.*" Blurring the borders between self and environment, Patricia's retreat into the woods is an initiatory one whereby Powers revisits the transcendentalist writers' meditations, giving them a liminal realist twist: "She reads Thoreau over wood fires at night. *Shall I not have intelligence with the earth? Am I not partly leaves and vegetables myself?* And: *What is this Titan that has possession of me? Talk of mysteries! –Think of our life in nature,—daily to be shown matter, to come in contact with it,—rocks, trees, wind on our cheeks! the* solid *earth! the* actual *world! the* common *sense!* Contact! Contact! Who are we? where are we?"[106] As she becomes intent on hearing the forest's "underground [...] massed symphonic choruses,"[107] it is suggested that her bodymind might also be unconsciously decoding some of the auditory and olfactory information put out by trees, those "airborne semaphores"[108] which she breathes and which may potentially provide a subliminal source of *inspiration* for her foray into the "foreign language"[109] of trees.

The forest becomes a liminal place where commonly accepted classifications, logic, and languages are reshuffled, and where all the senses are at once interlaced in enmeshed soundscapes, odorscapes, and landscapes, producing an entangled synesthetic, multispecies flesh of the world: "Clicks and chatter disturb the cathedral hush. The air is so twilight green she feels like she's underwater. It rains particles—spore clouds, broken webs and mammal dander, skeletonized mites, bits of insect frass and bird feather."[110] Patricia bathes in a green world suffused with microscopic signs interwoven in the material fabric of the forest, some of which may be intangible to the human eye. As she "walks in silence," Patricia turns into a patient tracker, "watching for tracks in a place where at least one of the native languages uses the same word for *footprint* as *understanding*."[111] Possibly alluding here to historian Carlo Ginzburg's claim that, as Claire Cazajous-Augé has put it, "the epistemological method consisting in reading almost imperceptible clues, in order to uncover events one did not directly experience, emerged in a cynegetic context,"[112] Patricia's synesthetic forest reading furthermore gestures to the invisible layers of meaning that she primarily intuits before she and her colleagues can analyze them.

While Patricia's scientific method abides by naturalism—that is, supposing, in Descola's parlance, a discontinuity between the interiorities of humans and nonhumans—in the field, she behaves very much like an animist conversing with trees. Conscious that the fate of humans on earth "will depend on the inscrutable generosity of green things,"[113] as she comes across "an immense western red cedar" in the "prodigious forest," Patricia notices how it smells of incense, bears "a candelabra of boughs," and how "[a] grotto opens at ground level." Inspired by the "cathedral hush" of the forest, Patricia spontaneously improvises a thanksgiving ritual reminiscent of indigenous traditions: "Thank you for the baskets and the boxes. Thank you for the capes and hats and skirts. Thank you for the cradles. The beds. The diapers. Canoes. [. . .] The rot-proof shales and shingles. The kindling that will always light." With this address, Patricia reinitiates a relationship of reciprocity between humans and trees. Enlightened by her vast empirical knowledge of sylvan life, she goes further to ask for forgiveness on the part of humans for the overconsumption of wood: "'We're sorry. We didn't know how hard it is for you to grow back.'"[114] Negotiating with trees in this way, Patricia the shaman braids indigenous practices and animistic worldviews with scientific understanding. Powers thus makes headway into the possibility for a postmodern, interspecies communication that does not hinge on the belief in a spirit world but recognizes the agency of nonhumans in a way that is compatible with science. Moreover, combined as it is with a poetic mind, Patricia's predominantly scientific approach to tree sentience prepares readers to accept, in the following chapter, Olivia's telepathy with trees.

While Patricia the biologist was born partly hard of hearing, and consequently impervious to most humans' language and egotistical preoccupations—an impairment which, following Powers' creative logic, turns into a gift, allowing her to filter out much of human babbling, but not tree messages—Olivia is a college student who has spent most of her young life as a narcissistic, plant-blind person. And yet, she too is suddenly gifted with the extrasensory capacity to hear the voices and decipher the language of trees. For any reader needing to co-opt the marvelous, Olivia's mindset could easily be equated with a form of schizophrenia. However, rather than condemning the character to such a devastating psychiatric diagnosis, the narrator more ambiguously presents her as a voice-hearer endowed with superior attributes. Her gift makes her a mystically inspired leader in the protest movements against clear-cutting. Additionally, her strange power is by the end shared by other characters whose backgrounds and experiences are harder to interpret as cases of psychosis. Rather, all these characters are construed differently to form a kaleidoscopic vision conducive to liminal realism.

There are two ways of reading Olivia, depending on whether one's desire is to debunk the marvelous, or whether the reader is happy to go along with Powers's thought experiment. In accordance with the dominant view of auditory hallucinations in psychiatry, Olivia's belief that she can communicate with nonhuman "presences," or "beings of light" beckoning her to drop out of college and drive across the country to save the remaining California redwoods can be dismissed as a case of psychotic delirium. To make it worse by normative medical standards, Olivia on the night of her first epiphany is stoned, under the influence of alcohol, and traumatized by an electric shock that nearly killed her—three facts that could be held to have altered her functional state of mind. Yet, on telling her story to Nick, lucid Olivia herself coopts the irrational aspect of her story: "'I don't feel crazy. That's the weird thing. [. . .] Believe me, I know what I sound like. Resurrection. Bizarre coincidences. Messages from television sets in a discount warehouse store. Beings of light I can't see.'"[115] Nick's reaction to her story also serves to soothe readers' potential need for rationalism: "By every reliable measure, she's more than a little titled. But he wants her to stay like this, talking crazy theories in his kitchen [. . .]. 'You don't sound crazy,' he fibs."[116] The labelling of her condition as a case of schizophrenia is thus envisioned; and yet, readers must go along with the story of how Olivia becomes another inspiring tree-prophetess thanks, in fact, to her near-death experience—a common trope associated with liminality and rebirth. Leading the war against tree-cutters, Olivia turns into a modern-day Joan of Arc. While the latter was considered a heretic and burned on the pyre like a witch, Olivia dies a martyr in a botched ecoterrorist attack.

The characters' awareness of the incredible quality of Olivia's experience makes it more palatable to skeptical readers. Yet, the narrative voice relays her perception without explicitly introducing any doubt. Tree voices are translated by the omniscient narrator in italics: "The car is filled with beings of light. They're everywhere, unbearable beauty [. . .]. They pass into and through her body. [. . .] Her joy at their return spills over, and she starts to cry. / *You were worthless*, they hum. *But now you're not. You have been spared from death to do a most important thing.*"[117] Powers' story fits with Wendy Faris's study of the "unsettling doubts" afflicting readers of magical realism: "The question of belief is central here, this hesitation frequently stemming from the implicit clash of cultural systems within the narrative, which moves toward belief in extrasensory phenomena but narrates from the post-Enlightenment perspective and in the realistic mode that traditionally exclude them."[118] As it goes, rather than discrediting voice-hearing as delusional, there may be a consciousness-changing angle on the phenomenon if one approaches it as a form of lucid dreaming, expressing contents arising from the character's ecological unconscious. This is hinted at by Olivia, as she ventures: "'Maybe it's all my subconscious, finally paying attention to something other than me.'"[119]

On the night of Olivia's epiphany, the narrator underscores the presence outside her house of a ginkgo biloba which the young woman obliviously walks by daily. It is "a singular tree that once covered the earth—a living fossil, one of the oldest, strangest things that ever learned the secret of wood."

Some of its biological characteristics are described in detail, together with its identifying features such as its peculiar leaves, which "vary as much as human faces," or its "limbs," with their "extraordinary profile [. . . making] the tree unmistakable." Yet, despite having just spent an entire semester in the immediate companionship of the ginkgo, now "lit by streetlamp in front of her house," self-centered Olivia doesn't know it's there."[120] Olivia's character is nonetheless evidently tied to the gingko. As the oldest tree on earth with incredible adaptive skills, the gingko might stand as the emissary talking to Olivia in the name of all tree life, waking her up to the multispecies voices and songs of the world: *"The most wondrous products of four billion years of life need help.* / She laughs and opens her eyes, which fill with tears. *Confirmed. I hear you. Yes.* [. . .] The air all around sparks with connections. The presences light around her, singing new songs."[121]

It can be wondered if Powers was influenced in his creative process by the Hearing Voices Movement, which challenges the authority and diagnostic categories established by mainstream psychiatry. The Hearing Voices Movement provides an original outlook and recommends nonconventional treatment for auditory verbal hallucinations—experiences that typically get diagnosed and treated as schizophrenia. As Angela Woods underscores, "5000 years ago we were all voice-hearers. Socrates, Moses, Margery Kempe, Joan of Arc, Virginia Woolf and Ghandi are among the most famous figures to have been identified as voice-hearers, and to remind us that these experiences were until recently at least as strongly associated with spiritual enlightenment, saintliness, creativity and philosophical insight as with madness and disease."[122] Positively welcoming the perception of verbal messages and voices in the absence of any other identifiable human speaker, Olivia stands as a "voice-hearer."[123] After retracing the emergence in the Western world of the figure of the "voice-hearer" at the end of the 1980s, Woods recapitulates:

> . . . the figure of 'the voice-hearer' has identified herself, and been recognised, as someone for whom the experience of voice-hearing takes on a significant if not central role in the constitution of identity; as an expert, capable of founding new traditions of empowerment and self-help while also challenging the expertise of those working in the mental health professions; and as someone who has explicitly rejected the label 'schizophrenic.' The 'voice-hearer' draws her interlocutor into a view of the world whose founding gesture is less the negative affirmation of psychiatry, and more the assertion of a new language (even if in practice the latter cannot happen without substantial efforts towards the former).[124]

Indeed, by creating these liminal characters, Powers limns the existence of a tree language which humans might potentially be able to decipher and translate in comprehensible terms. As with shamans in animistic cultures, Powers's modern-day tree-hearers "[produce and perform] an identity in whose name [ecopoetic] meaning can be made, [. . .] and flourishing envisaged"[125] on both an individual and a collective, multispecies level.

As schizophrenia is first logically considered by Nick, after hearing the destabilizing story of Olivia's "calling," his research into mental health offers insight into the liminality of Olivia's experience: "He reads the encyclopedia article on mental disorders. The section on diagnosing schizophrenia contains this sentence: *Beliefs should not be considered delusional if they are in keeping with societal norms.*"[126] As a matter of fact, Olivia's experience is one that profoundly upsets modern conceptions of the human mind and our definition of human society: "[The beings of light] don't scold her for forgetting the message they gave her. They simply infuse her again. [. . .] They speak no words out loud. Nothing so crude as that. They aren't even *they*. They're part of her, kin in some way that isn't yet clear. Emissaries of creation—things she has seen and known in this world, experiences lost, bits of knowledge ignored, family branches lopped off that she must recover and revive. Dying has given her new eyes."[127]

Olivia's epiphany furthermore contains many of the elements traditionally associated with initiation rites. On that night, she is fasting and retreating in her bedroom where she indulges in alcohol and pot-smoking, blissfully modifying her state of consciousness. Moreover, she has accidentally wounded herself, and the confrontation with the bleeding gash in her ankle heightens her consciousness of her own vulnerability—something which, as she recognizes, all living beings have in common. Before identifying the voices of other-than-human beings, Olivia displays a fondness for hallucinatory doped up perceptions: "Soon the beautiful brainstorms come [. . .]. The universe is big, and she's allowed to fly around through the nearby galaxies for a while, zapping things for fun, if she doesn't abuse her powers or hurt anyone. She does so love this ride."[128] Her psychedelic trance opens onto a world where she perceives a kind of self-generated music making it hard for her to concentrate on her more logically organized thoughts: "Then the tunes start up, the inner ones. [. . .] Hard to hear anything, over the magic melodies of her own devising."[129] Finding an outlet for her stoned melomania, Olivia has creatively invented her own musical notation system. Evoking Wassily Kandisky's genius for painting music, Olivia translates her powerful synesthetic experiences of music: "Real musical notation reads to her like so much secret writing, but she has devised her own system for preserving the tunes that come to her while stepping out. Line color, thickness, and location all encode a record of the gift melodies. And the next day, after her buzz wears off, she can look at these scribbles and hear the music all over again. Like copping a contact buzz, for free."[130]

While it might first seem tempting to mock Olivia's revelation as a drug-induced, delusional state, the near-death episode brings the final touch to the liminality of her character, already gifted for stepping into other dimensions and for deciphering non-verbal languages.[131] Waking up "naked and comatose," Olivia is reborn. She has crossed over to a different way of dwelling, in touch with "large, powerful, but desperate shapes [beckoning] to her," and "[showing] her something, pleading with her."[132] Open to the possibility of a communitas including earth others, Olivia acquires a whole new sense of self: "For the first time, she realizes that *being alone* is a contradiction in terms. Even in a body's most private moments, something else joins in. Someone spoke to her when she was dead. Used her head as a screen for disembodied thoughts. She passed through a triangular tunnel of strobing color and emerged into a clearing. There, the presences—the only thing to call them—removed her blinders and let her look *through*."[133] Readers may resist the marvelous because of the drugs, the somewhat cliché, numinous experience of death, and the overall dry, even sardonic and unforgiving tone at times of the narrator—all of which may prod us into defiance. Yet, our rationalizing is increasingly hindered as Olivia's magic powers spread to other characters whose overall stances are less questionable. As a result, the cognitive bias initially cuing us to conform to the dominant view of auditory hallucinations in modern psychiatry is gradually curtailed, allowing us to immerse ourselves completely for the time of the reading within an alternative way of dwelling.

Powers's liminal ecopoetics constantly challenges our very notions of what may seem realistic and logical, or, to the contrary, irrational and supernatural. First, it is through botany that Patricia's father initiates her—and the reader in her wake—to the "miracles that green can devise." The father and daughter thus embody the humility and awe that many scientists derive from their field: "Other creatures—bigger, slower, older, more durable [than humans]—call the shots, make the weather, feed creation, and create the very air."[134] As Patricia scrutinizes "hidden things" in an aspen forest, examining the mycorrhizal underground networks linking all the "seemingly separate stems" observable above ground, she estimates the rhizome mass, "this great, joined, single creature that looks like a forest," to be around a million years old—a very natural reality that nevertheless seems extraordinary to the human mind: "The thing is outlandish, beyond her ability to wrap her head around. But then, as Dr. Westerford knows, the world's outlands are everywhere [. . .]."[135] When it comes to trees and forests—these liminal wonders both visible and invisible, extending and interacting both above and below ground—earthly enchantments are everywhere, still beckoning us humans to develop a green enough mind for them.

Even the stuff of the most natural phenomena may commonly seem "outlandish" or supernatural to the limited bodyminds of city-dwellers. It is in the

Brazilian old-growth Amazonian forest for instance that Patricia marvels at the complex order that a scientist using the right lens might discern within chaos: "Species clog every surface, reviving that dead metaphor at the heart of the word *bewilderment*."[136] Powers highlights the etymology of the verb "bewilder," which, as Roderick Frazier Nash underscores, "comes from 'be' attached to 'wildern.' The image is that of a man in an alien environment where the civilization that normally orders and controls his life is absent."[137] Reminding us that what is truly wild eludes human control and, in many ways—because of the hidden patterns that chaotic complex systems are constantly working on—human grasp, Powers's characters prove that one does not need to return to pre-scientific worldviews to perceive the truly wondrous aspects of the real world. Having opened his eyes to the miraculous engineering taken on by trees underground as well as aboveground, thanks to Patricia's book, Ray also experiences "pure astonishment at [the] performance unfolding just outside his window,"[138] a performance easily accessible to him as he spends long hours gazing onto the back garden that he and his wife Dorothy have decided to simply let go wild again.

In addition to turning topsy-turvy our habitual notions of what is natural and ordinary vs. extraordinary and magical, Powers subverts conventional takes on the rational and irrational. Olivia first turns the tables on mainstream notions of sanity: "'What's crazier? Believing there might be nearby presences we don't know about? Or cutting down the last few ancient redwoods on Earth for decking and shingles?'"[139] Beyond the collective folly of humans blindly destroying their *oikos*, Olivia furthermore pinpoints some of the conceptual hindrances for the modern mind when it comes to interspecies communication: "'What's crazier—plants speaking, or humans listening?'"[140] For, Powers reminds us, outside of Modern Western thinking, many indigenous cultures have long been open to the notion that trees possess sentience and agency. Hence the short exchange between Nick and a Native man helping him make forest land art: "'It amazes me how much [trees] say, when you let them. They're not that hard to hear.'/ The man chuckles. 'We've been trying to tell you that since 1492.'"[141]

Like Linda Hogan and Robin Wall Kimmerer, Powers redeems the ecological knowledge weaved into indigenous stories, practices, and rituals, which were narrow-mindedly looked down upon by Europeans with crude understanding, in fact, of the naturcultural environments they were "discovering." In Hogan's terms, "[ecology] and harmony within a working system were late sciences for the new arrivals, yet they are concepts that, in our time, are becoming alive again."[142] In his typically satirical tone, Gerald Vizenor on his part makes a similar statement, implicitly quoting Shakespeare's *As You Like It*. In the Elizabethan play, Duke Senior declares: "And this our life, exempt from public haunt / Finds tongues in trees, books in the running brooks, /

Sermons in stones, and good in everything."[143] Echoing this passage, one of Vizenor's characters in *Wordarrows* is a white scientist boasting about his findings: "we find tongues in trees, books in brooks, phrases from the mouths of fish, oral literatures on the wings of insects, sermons in stone."[144] At that, in a way resembling the exchange at the end of Powers' novel, Vizenor's indigenous protagonist retorts, "tribal people have known that since the beginning of the world."[145] In both cases, there is a reclaiming of ancient, ecopoetic understanding of the polyphonic book of nature. Powers has his scientist regretting humbly, "it may take centuries to learn as much about trees as people once knew."[146]

Speaking for indigenous cultures, botanist Robin Wall Kimmerer writes: "We have always known that the plants and animals have their own councils, and a common language. The trees, especially, we recognize as our teachers. [. . .] In the old times, the elders say, the trees talked to each other. They'd stand in their own council and draft a plan."[147] Disenchanting the world, modern scientists, Kimmerer bemoans, "decided long ago that plants were deaf and mute, located in isolation without communication. The possibility of conversation was summarily dismissed."[148] Kimmerer takes issues here with science's claim to objectivity, when it appears rather clearly in the case of plants that our biased epistemologies shape ontologies that are apprehended through distorted subjective lenses—here a zoocentric prism: "Science pretends to be purely rational, completely neutral, a system of knowledge-making in which the observation is independent of the observer. And yet, the conclusion was drawn that plants cannot communicate because they lack the mechanisms that *animals* use to speak. The potentials for plants were seen purely through the lens of animal capacity. Until quite recently no one seriously explored the possibility that plants might 'speak' to one another."[149]

Hinting at ancient cultures which Europeans have long derided, all of which recognized the vital interdependence between human and nonhuman worlds and the possibility for interspecies communication, Powers's call to "become indigenous again" is voiced and embodied by Olivia.[150] To put it in a nutshell, this amounts to rehabilitating an enchanted way of dwelling that, to take up Gary Snyder's poetic way of putting it, has regained "the sense of the magic system, the capacity to hear the song of Gaia,"[151] but also to translate the latter into ethical relationships of reciprocity and partnerships with the land and its many codwellers. Olivia and her co-hearers and tree-saviors thus play out as restorers of a kind of biotic communitas, where humans may (re)learn to heed what other-than-humans have to say and do, mindfully computing from there how to keep the various agencies of the world in balance. To put it in David Abram's terms, "[the] 'spirits' of an indigenous culture are primarily those modes of intelligence or awareness that do *not* possess a human form."[152] Possibly alluding to Abram's study of the ways in which oral languages have emerged from direct contact with the animate landscape, and how, in many ways, the aural aspects of a language still reverberate physical qualities of place, Powers writes of the anonymous indigenous character helping Nick at the end: "The man in the red plaid coat says a few words to the dog in a language so old it sounds like stones tossed in a brook, like needles in a breeze, humming."[153] As I will go further into in part 3, certain material aspects of language can indeed still produce poetic echoes of the earth and its many other-than-human voices. Moreover, at no point does the novel imply a cleavage between the characters' mystical and scientific takes on trees. On the contrary, they appear throughout as oxymoronically compatible, as the science on tree communication offers a materialistic reading grid for otherwise incomprehensible experiences of wellbeing and connectedness.

Mimi the female Buddha completes Powers's gallery of liminal beings. Her enlightenment once again takes place half-way between the mystical and the rational. With Chinese ancestry relating her to famous Buddhist arhats, and with a practical modern mind trained in engineering—she realizes as some point that she can name all the machines involved in clear-cutting, but hardly any of the trees—Mimi yearns to identify the pine tree just outside her office window and receives her epiphany in an urban setting. She revels daily in the view of the tree, embodying the soothing effect of tree and plant companionship which scientists have now demonstrated.[154] The way Mimi relates to the Ponderosa pine evokes its invisible biosemiotic messages: "The smell is on her before she reaches the trees—the scent of resin and wide western places. [. . .] The music of the trees, too, tuning the wind."[155] The more intensely Mimi sniffs out what the tree is putting out in the air, the more it makes her aware of their deeply ingrained connection through scent: "She remembers. Her nose slips into one of those dark fissures between the flat terracotta plates. She falls into the smell, a devastating whiff of two hundred million years ago. She can't imagine what such a perfume was ever meant to do. But it does something to her now. Mind control."[156] Inviting us to ponder the enchanting power of olfactory signals, and what functions those might serve in plant life, this passage underlines how much of a tree's appeal to humans may lie in its fragrance. Yet, the subtleties of the signals we pick up from trees can remain hard to name, both because of our limited capacity to consciously analyze smells and because of the dearth in our vocabulary when it comes to specific smells: "It's neither vanilla nor turpentine, but replete with highlights of each. A shot of spiritual butterscotch. A sprig of pineapple incense. It smells like nothing but itself, pungent and sublime. She breathes in, eyes closed, the tree's real name."[157] Not only does this passage underscore how the complex, yet specific fragrance of every kind of tree presents its olfactory signature, it moreover touches upon the transcorporeal spell that fragrant plants can cast upon our bodyminds. Powers thus ignites the possibility that human-tree partnerships have long been fueled by a nonverbal language the workings of which follow partly subliminal channels.

Functioning like Proust's madeleine, the smell of the tree brings back Mimi's childhood memories of fishing with her father during the camping trips in the wild which she so enjoyed. Far from being presented as mystical, the scene is described in phenomenological terms gesturing to the very material ways in which trees possess the intoxicating power to modify human sensations and consciousness: "She doses herself for a long time, like a hospice patient self-administering the morphine. Chemicals rush down her windpipe, through the bloodstream to her body's provinces, across the blood-brain barrier and into her thoughts. The smell grips her brain stem until she and the dead man are fishing side by side again, under the pine shade where the fish hide, in the soul's innermost national park."[158] Mimi's neurocognitive, corporeal interlacing with the pine wakes up her childhood memories of being in the wild as well as her ecological unconscious. Her anagnorisis is thus two-fold: on the one hand, she travels back to cherished moments of encounters with the wild; and on the other hand, she receives an existential revelation, the nature of which seems triggered by the tree's scent, activating a biophilic response developed over eons of tree-human companionship:[159]

> Blissed by memory and volatile organics, [. . .] Mimi leans back into the tree, falling one last time into that unnamable scent. Eyes shut [. . . something] comes over her. The light grows brighter; the smell deepens. Detachment floats her upward, buoyed by the tides of her childhood. She turns from the trunk with a profound sense of well-being.[160]

Mimi's green epiphany gradually turns her attention to trees' need to be defended.

Her Buddha-like character moreover conveys the audacious hypothesis that the numinous experience originally assigned to the famous mystic could correspond to a sudden greening of the human mind induced via smell, among other unconscious pathways to the brain. Toward the end of the novel, after she spends hours and hours on end fasting and sitting still in a lotus-like position under a pine in a city park, Mimi attains revelation. Her stance then evokes the mode of being which the poet Rainer Maria Rilke aspired to, which he referred to as the "open," where consciousness merges with nature. It may be wondered whether Powers might have sought inspiration for Mimi's enlightenment in Rilke's intense letting-go into the "open," which he experienced reclining against a tree in an urban public garden.[161] The scene again sparks the hypothesis that spiritual enlightenment may be induced by transcorporeality:

> Some slight change in the atmosphere, the humidity, and her mind becomes a greener thing. At midnight, on this hillside, [. . .] Mimi gets enlightened. The fear of suffering that is her birthright—the frantic need to steer—blows away on the wind, and something else wings down to replace it. Messages hum from out of the bark she leans against. Chemical semaphores home in over the air. Currents rise from the soil-gripping roots, relayed over great distances through fungal synapses linked up in a network the size of the planet.[162]

At one again with her *oikos*, Mimi the postmodern female Buddha gestures to a kind of spiritual enlightenment rooted in very material, time-old, and earthly entanglements.

Mimi thus offers a third kind of liminal character, one with mythical antecedents, who reopens the doors of perception onto a fleshy world interweaving matter and mind, humans and trees. Defending how an epistemological shift to the notion of transcorporeality "can become an ethical matter," Stacy Alaimo writes: "trans-corporeal subjects must also relinquish mastery as they find themselves inextricably part of the flux and flow of the world that others would presume to master. [. . . A] recognition of trans-corporeality entails a rather disconcerting sense of being immersed within incalculable, interconnected material agencies that erode our most sophisticated modes of understanding."[163] In a way that is reminiscent of Mogey's wild epiphany in Ann Pancake's novel, Mimi's realization brings about a sensation of total well-being grounded in a letting go of human logocentrism and a sensuous blending with the vibrancy of the greater-than-human world. As Mimi embraces her material connectedness to everything around her, she acquires a sense of cosmic belonging anchored in transcorporeality. Besides, Mimi's experiences confirm studies in transpersonal psychology, a branch of ecopsychology that examines how nature plays a central role in bringing "optimal mental health and psychological development, mystical and spiritual experiences, inner peace, compassion, trust, fully realized aliveness, and selfless service."[164] John Davis argues that wilderness experiences encourage "the sense that we are each unique and individual, and, at the same time, part of the larger whole." With the potential to dispel feelings of separation and meaningless, contact with the greater-than-human world, Davis shows, can "trigger the sense that the world is enchanted, alive, whole, and meaningful."[165]

Whereas Patricia serves to weave scientific discourse into the novel, Mimi casts a green light onto the mysticism at the heart of Buddhism. These various reenchantment angles thus work together, forming a kaleidoscopic multidimensional take on animacy. Through liminal realism, Powers designs an ecopoetic charm fostering an immersive, uplifting experience of the world of trees which may affect, at the end of the reading voyage, how readers conceive their own belonging to a multispecies communitas. Moreover, to any reader familiar with the well-being derived from living in conviviality with plants and trees, Powers's creative thought-experiments may ring a bell with the ways in which the presence of trees often does indeed trigger sensations that can ripple into a spiritual experience. Eventually, reading this novel might very well augment one's *Umwelt*, encouraging a finer perception of the ways in which our environments are perfused with tree signals, i.e., biosemiotic messages signaling a form of sentience both alien and fellow. In a way that may compare with Starhawk's earlier conception of a human-bee swarm intelligence, Powers' formidable epic eventually paves the way for us to imagine multispecies forms of ecological awareness and problem-solving.

Going back to the metafictional dimension of Powers' novel which I have touched upon earlier and which has been problematically overlooked by the critics elaborating on the writer's densely dialogic project, I would now like to turn to the *mise en abyme* of storytelling by dint of which Powers brings to the fore the role of ecopoetic fiction in the revision of our disenchanted ontologies. Toward the end, as Dorothy reads both fiction and nonfiction literature to her paralyzed husband, the story takes an unexpected turn, shedding new light on the puzzling story. Opening onto the possibility that Olivia's character might in fact be the offspring of the childless couple's imagination, the last passages focusing on them indeed reshuffle the narrative cards, suggesting that the omniscient narrative voice throughout might have been no other than Dorothy's, playing a what-if game with her husband from whence the rest of the story must have unfolded.

One of the blinders to the overlapping between Dorothy's character and the omniscient narrator lies in the fact that Ray and Dorothy are identified as the Brinkmans, while Olivia is called Vandergriff after the husband she recklessly married during her first years in college. Yet, Powers drops several clues paving the way for this final unravelling of how all the different narrative and discursive threads connect in rhizomatic hidden patterns not unlike those of underground mycorrhiza. For example, "[years] before and far to the northwest" of where Olivia starts following tree voices, the Brinkmans attend the opening night of *Who's Afraid of Virginia Woolf*. In the 1962 play by Edward Albee, it turns out that the infertile couple of protagonists have compensated their incapacity to bear children by giving life to an imaginary son, whom they regularly talk about and invent a life for. In addition to this first intertextual hint, the totemic trees linking Nick, Adam, Mimi, and Douglas respectively to a chestnut, a maple, a mulberry, and a Douglas fir turn out to be growing in the Brinkmans's garden, goading them to research the botanical properties and histories of each. In addition, before his stroke, Ray the intellectual property lawyer ponders the idea of extending legal personhood to trees and other nonhuman beings. Thus, it is inferred toward the very end of the book that the other characters' stories may have been invented altogether by Dorothy and Ray, interweaving fiction and reality, myth and science, and producing a mythical offspring—Olivia the tree-hearer—to fill in for the child they were unable to conceive, meanwhile creating an inspiring postmodern tale.

Considering this suddenly revealed postmodernist assemblage, Dorothy's name may obliquely evoke the *deus ex machina* dénouement of *The Wizard of Oz*, suggesting several similarities with the famous fairy tale that itself already contained elements pointing to the mechanisms of enchantment. Taking place just before the novel comes to a close, this revelation prepares readers for their reaggregation into the nonfictional world from which the liminal realism at the heart of the novel has temporarily uplifted them. At this crucial point, the metafictional vertigo induced by this unforeseen twist is likely to cue readers to meditate on the ways Powers' enchantment feat hinges on our willful letting go of the norms and categories ordering the world outside of fiction and poetry. As a result, we must assess the deep initiatory function of ecopoetic liminal realism, reflecting over how the experience of reading might function as both an esthetic and ethical prompter and onto-epistemological revealer. Simultaneously, we must reevaluate the links between the many intertwined stories, most of which now appear as intradiegetic fabulations.

As a matter of fact, Powers has weaved into his novel several intertextual and metafictional motifs that constantly point to his own enchanting devices. Paradoxically, this may both temper and strengthen his postmodernist *ecopoiesis* of reenchantment. Levelling the implausible capacity developed by some of the characters to communicate with trees, Ray for instance reads nonfiction that we may easily identify as serious propositions made in the extradiegetic world as regards legal tree representation. Thus, we are brought to meditate on the dialogic texture of Powers's tree novel, in actual conversation with much writing that sustains today's discussions on environmental issues, and therefore with very real philosophical, legal, and political ramifications.[166]

Readers must ponder whether fictional *ecopoeisis* may indeed provide an ideal liminal space to experiment with the notion of tree sentience and multispecies representation. Ray's reflections on the subject provide a sweeping vision of historical changes of mentality, many of which, we might remember, have been helped by great awareness-raising novels such as those written by Charles Dickens or Harriet Beecher Stowe, to quote just those two.[167] One of the quotes from Stone's essay offers a clue to what Powers may be aspiring to: "*What is it within us that gives us this need not just to satisfy basic biological wants, but to extend our wills over things, to objectify them, to make them ours, to manipulate them, to keep them at a psychic distance?*"[168] Via the endless possibilities offered by liminal realism, Powers's what-if game abolishes for a moment that psychic distance which Modernity has introduced between humans and nonhumans.

Quite tellingly, Adam the academic specializes in cognitive biases and in the bystander syndrome, which happens to be fully at work in our lack of reaction to the planet's demise. His research focuses on how dominant collective worldviews rule individual psychology. Bringing into the picture Foucault's history of madness and, potentially too, his study of biopolitics, Adam's character serves to debunk the modern notion of reason:

> But the question interests Adam. What did the dead Joan of Arc hear? Insight or delusion? Next week he'll tell his undergrads about Durkheim, Foucault, crypto-normativity: How *reason* is just another weapon of control. How the invention of the reasonable, the acceptable, the sane, even the human, is greener and more recent than humans suspect. [. . .] He hears himself answer: "Trees used to talk to people all the time. Sane people used to hear them."[169]

Laden with philosophical, political, and metafictional implications, this passage nails down the liberating power of liminal realism. The fictional mode can indeed explore realms beyond what is visible to the human eye or certified by modern science, gesturing to obliterated dimensions of the flesh of the world. It can challenge the premises on which rationality was built and reveal some of the ways in which most societies rely on epistemologies blindly striving to perform the impossible extraction of humans from our worldly entanglements.

Adapting Stephen Slemon's view of magical realism in a postcolonial context to tackle liminal realism in a posthumanist context, one might say:

> " . . . magical [liminal] realist texts comprise a positive and liberating [multispecies] engagement with the codes of imperial [and anthropocentric] history [. . .]. This process, they tell us, can transmute the 'shreds and fragments' of colonial [and environmental] violence and otherness into new 'codes of recognition' in which the dispossessed, the silenced, and the marginalized of our own dominating systems [i.e. earth others] can again find voice and enter into the dialogic continuity of [a biocentric] community and place.[170]

Because the reading of fiction relies on willing suspension of disbelief, Powers's exhilarating narrative designs a liminal experience shared by his readers who voluntarily embark in collective "effervescence," to take up Durkheim's key concept in his study of collective enchantment processes. Without returning to old religious views of a transcendent God or spirits, liminal realist literature nevertheless provides the stage for an ecopoetic reenchantment of the world, which banks on the emotional and psychic contagion triggered by the reading of fiction. Such ecopoetic fiction may thus play a great part in the spreading of social "effervescence," which Durkheim defines as a common passion, "an extraordinary degree of exaltation,"[171] whereby in this very case, readers might acquire an entirely new, reenchanted vision of the relationships between humans, trees, and forests.

As is made clear by the eventual shattering of the fictional illusion produced by Powers's multilayered and self-reflexive narrative, the point however is by no means a simplistic return to or appropriation of non-European animistic cultures, nor does it cue us to renounce postmodern epistemologies. Rather, via the enchanting potential of fiction, Powers conceives an initiatory experience whereby the findings of postmodern science *make sense* and come alive in an ecopoetic vision affecting readers in their esthetic, ethical, ontological, and political stances. Simultaneously, the novel highlights the storytelling at play in both science and literature, where wild nonhuman entities possess narrative agency as they become enmeshed with human observation and discursive practices. In light of Powers's tale, it is tempting to take up Thomas Pughe's paper on ecopoetics as a reinvention of nature: "Faced with the disenchantment of nature, understood in the philosophical sense of the Enlightenment period, ecopoetics fights back by insisting on its reenchantment, demanding a representation respectful of its wild character."[172]

Blurring the lines between the various diegetic levels conventionally separating the characters from the narrator and from the implied writer, Powers in the end reveals the many different seeds and rhizomes from which his whole forest-like dialogic text has sprouted. Most of the story has grown from a stretch of imagination interweaving Ray and Dorothy's many different readings—those same readings that have inspired Powers himself—their observation of the wildness of trees in their back garden, and the environmental and political crises in the background of their lives—all of which are rooted in recognizable historical events lending the novel its overall realistic anchoring. While disrupting the fictional pact initially established between the omniscient narrative voice and the reader, the metafictional ending nevertheless foregrounds the posthumanist project underlying Powers's resort to liminal realism: "Across the biomes, at all altitudes, the learners come alive at last. [. . .] They will come to think like rivers and forests and mountains. [. . . The] next new species will learn to translate between any human language and the language of green things."[173]

Linking ecopoetics to forms of "reinvention, recreation, or reenchantment," Thomas Pughe explains: "these terms apply to *aesthesis*, that is the capacity—or the incapacity—of a literary text to propose a new outlook on our relationship with the natural world."[174] Powers specifically spurs us to reflect upon how we might come away changed from our enchanting reading experience. Going back to Starhawk's definition of magic as a willful shift of attention affecting our awareness of things, liminal realism can be apprehended as a reenchantment mode that works its magic against the illusions produced by modern thinking. Whether under the pen of Jean Giono, Leslie Marmon Silko, Linda Hogan, Ann Pancake, Starhawk, or Richard Powers, liminal realism in ecopoetic literature thus breaks the spell cast upon our bodyminds by a modern science making us believe in human separation from nature and in the inertness of the latter. Taking my cue from Starhawk, I may conclude as I am nearing the end of my second part that such ecopoetic moments of enchantment work like literary magic, a magic that "teaches us also to break spells, to shatter the ensorcellment that keeps us psychologically locked away from the natural world."[175]

By a careful, heteroglissic interweaving of science, poetry, and myths, ecopoetic novels may answer the widespread call for cross-disciplinary approaches to nature that may help us mend the two-culture split. Reconciling takes found respectively in science, in the arts, and the humanities, liminal realism can help rebraid different ontologies and epistemologies in meaningful and sustainable ways. Ecopoetic fiction thus propels the (re)invention of new stories of multispecies entanglements. As apparent in the mythopoeic fabric of the novels in my corpus, the ecopoet(h)ic role of ecoliterature may lie in part in the creative translation of the findings of scientific investigations as to the nonhuman voices and agencies of the earth,[176] and in the poetic expression of our ecological unconscious.

In addition, liminal experiments can help bring solutions to the current crisis of the imagination at the heart of the ecological crisis. Performing a postmodern reenchantment that is compatible with a secular worldview, ecopoetic literature proves to be that kind of magic that, to take up Starhawk, "might also be called the art of opening our awareness to the consciousnesses that surround us, the art of conversing in the deep language that nature speaks."[177] As a matter of fact, the ecopoets in my corpus draw attention to the anamorphic power of art and science, both fields providing reading grids and myths that can frame or expand our inhabitation of the world. Via the different ways in which these ecopoets rehabilitate the notion of a multispecies 'song of the earth,' they take part in the project of reenchantment delineated by Thomas Berry in his influential *Dream of the Earth*: "The reenchantment with the earth as a living reality is the condition for our rescue of the earth from the impending destruction that we are imposing upon it. To carry this out effectively we must now, in a sense, reinvent the human species within the community of life species. Our sense of reality and of value must consciously shift from anthropocentric to a biocentric norm of reference."[178] As Jonathan Bate insists, "we cannot do without thought-experiments and language-experiments which imagine a return to nature, a reintegration of the human and the Other. [. . . Our] survival as a species may be dependent on our capacity to dream it in the work of our imagination."[179] As I tackle in my next part the different ways in which the texts in my corpus bring to the fore the anamorphic potential of ecopoetics, it should appear that beyond the rehabilitation of the song of the earth as an accurate biocentric topos, it is the very nature of human language itself that is materially reintegrated into the living fabric of the earth.

NOTES

1. Perrin, "Sécheresse," 222; Rumpala, *Décombres,* 100.
2. Starhawk, *Dreaming, Spiral, Earth.*
3. Starhawk, *Fifth,* 124.
4. Ibid, 267.
5. Ibid, 43.
6. Ibid, 47–48.
7. Biologist Rupert Sheldrake has formulated the controversial "hypothesis of formative causation": "This hypothesis postulates that the characteristic forms taken up by molecules, crystals, cells, tissues, organs, and organisms are shaped and maintained by specific fields called *morphogenetic fields* (from the Greek *morphe* meaning *form* and *genesis* meaning *coming-into-being*). The structures of these fields are derived from the morphogenetic fields associated with previous similar systems; the morphogenetic fields of past systems become present to subsequent similar systems by a process called *morphic resonance,*" "Laws," 81–82. In short, while Sheldrake recognizes the role played by the encoding of inherited DNA in the unfolding of biological forms and behavior, he argues that the latter are additionally "given directly by morphic resonance from past organisms of the same species," a phenomenon that is believed to be "not significantly attenuated by spatial or temporal separation," Ibid, 82–83. To illustrate his theory, Sheldrake refers to laboratory rats influencing each other's learning capabilities in non-locality: "If a number of rats, for example, learn to carry out a task that rats have never done before, then other rats everywhere else in the world should be able to learn the same task more easily in the absence of any known type of physical connection or communication," Ibid, 84. See also Sheldrake, *New Science.*
8. Starhawk, *Fifth,* 143.
9. Ibid, 78.
10. Ibid, 171. Considering the many resonances with some of Leslie Marmon Silko's and Paula Gunn Allen's writings, it is very likely that Starhawk's concoction in her syncretic brew of liminal realism was influenced by her indigenous and ecofeminist contemporaries. Like many involved in the counterculture movement of the 1960s and 1970s, Starhawk clearly exhibits awareness of the intersectionality between ecofeminist and indigenous stakes, both in terms of politics and relational ontology.
11. Starhawk, *Fifth,* 130.
12. Ibid, 132.

13. Starhawk, *Dreaming*, 10.
14. Chanady, quoted in Faris, *Ordinary*, 20.
15. Starhawk, *Fifth*, 50.
16. Published in English in 1990, Carlo Ginzburg's insightful historical study of the old pagan goddess cults and later practices of witchcraft in relationship with shamanism may have influenced Starhawk (Ginzburg, *Sabbat*).
17. Starhawk, *Fifth*, 159.
18. Ibid, 301.
19. Ibid, 164.
20. Ibid, 87.
21. Ibid, 88.
22. Ibid, 32.
23. Ibid, 60.
24. Ibid, 70–71.
25. Ibid, 71.
26. Ibid, 61.
27. Ibid, 95.
28. See Donna Fancourt's paper on altered states of consciousness in utopian feminist novels, where she approaches the text as "re-visionary site," "Accessing," 110.
29. Starhawk, *Fifth*, 171.
30. Ibid, 172.
31. Ibid, 202.
32. Ibid, 204.
33. Ibid, 224.
34. Ibid, 226.
35. Turner, *Ritual*, 95.
36. Starhawk, *Fifth*, 224–25.
37. Ibid, 226.
38. Ibid, 227.
39. Ibid, 227.
40. Ibid, 227.
41. Ibid, 258.
42. Descola, *Par-delà*, 245.
43. Such ecopoetic liminal realism does not necessarily depend on magic events in the sense of the supernatural, nor does it necessarily imply such a radical poetization of scientific worldviews as those found in Starhawk's novel or Richard Powers's *The Overstory*. The first chapter in my next part is devoted to the ecopoetic art of anamorphosis.
44. See for instance Bert Hölldobler and E. O. Wilson's consequential study, *The Superorganism: The Beauty, Elegance, and Strangeness of Insect Societies*. Before reading this book, I once attended a conference given by Yves Le Conte, Research Director at INRA, on pheromonal communication and social regulation in Apis mellifera bees and was astounded to realize how precisely Starhawk's ecopoetic thought experiment in fact coalesced with what humans do understand of bee behavior and language.

45. Starhawk, *Dreaming*, 13.
46. Starhawk, *Earth*, 11.
47. Ibid, 14.
48. Shepard, "Ecology," 63.
49. See for instance the documentary *Do Trees Communicate?*, directed by Dan McKinney and starring Suzanne Simard, or *Intelligent Trees*, directed by Julia Dordel and Guido Tölke, featuring both Suzanne Simard and Peter Wohlleben. Simard has also given popular TED talks and written *Finding the Mother Tree: Discovering the Wisdom of the Forest*.
50. On the nature spirituality of these writers, see Taylor, *Dark*.
51. Powers first enrolled into college as a physics major, and then changed to English literature. An avid reader of both the classics and nonfiction, he is also a melomaniac, having learned vocal music as well as how to play the cello, the clarinet, the saxophone, and the guitar. Powers besides followed computer science classes and worked as a computer programmer in his youth. These varied skills have evidently percolated into his dialogic novel.
52. Powers, *Overstory*, 112.
53. Ibid, 112–13.
54. Ibid, 113.
55. Ibid, 119.
56. Ibid, 436.
57. Also called *Oecanthus fultoni*, this North American cricket modulates its rate of stridulation chirp according to the heat of the environment.
58. Powers, *Overstory*, 436.
59. Ibid, 113.
60. Ibid, 114.
61. Kimmerer, *Braiding*, 252.
62. Powers, *Overstory*, 115.
63. Ibid, 114.
64. Ibid, 117.
65. Ibid, 117.
66. Ibid, 139.
67. Ibid, 115.
68. Ibid, 123.
69. Ibid, 122.
70. Ibid, 122.
71. Ibid, 125–26.
72. Ibid, 126.
73. Ibid, 453.
74. Ibid, 446.
75. See Brandon Keim's interview with Suzanne Simard, "Never Underestimate the Intelligence of Trees: Plants Communicate, Nurture their Seedlings and Get Stressed," published online: http://nautil.us/issue/77/underworldsnbsp/never-underestimate-the-intelligence-of-trees?linkId=76593830. See also Mancuso and Viola, *Intelligence*, and Gagliano, *Thus*.

76. Powers, *Overstory*, 424.
77. Ibid, 217.
78. Ibid, 454.
79. Ibid, 453.
80. Ibid, 127.
81. Ibid, 128.
82. Ibid, 3–4.
83. Ibid, 389.
84. Gagliano, *Thus*, 6.
85. Powers, *Overstory*, 129.
86. Ibid, 129.
87. Ibid, 219.
88. Leopold, *Almanach*, 263.
89. Ibid, xix.
90. Powers, *Overstory*, 115.
91. Ibid, 124.
92. Ibid, 124.
93. Ibid, 132.
94. Ibid, 143.
95. I am borrowing Suzanne Simard's metaphor to refer to those ancient trees that sustain many others around them. See her Ted Talk, Ferris Jabr's "The Social Life of Forests," and Simard's *Finding*.
96. Gibson, *Reenchanted*, 2; Taylor, *Dark* 94–95.
97. About the wonders of the Gingko biloba, some of which are covered by Powers, see Nathanael Johnson, *Unseen*, 87–110.
98. Powers, *Overstory*, 136.
99. Ibid, 123.
100. Ibid, 223.
101. Ibid, 141.
102. Ibid, 279–80.
103. Ibid, 141.
104. Ibid, 284.
105. Snyder, *Place*, 188.
106. Powers, *Overstory*, 129.
107. Ibid, 133.
108. Ibid, 140.
109. Ibid, 129.
110. Ibid, 134.
111. Ibid, 134.
112. Cazajous-Augé, "Poetics," 179.
113. Powers, *Overstory*, 124.
114. Ibid, 135.
115. Ibid, 177.
116. Ibid, 177.
117. Ibid, 163.

118. Faris, *Ordinary*, 17.
119. Powers, *Overstory*, 177.
120. Ibid, 146.
121. Ibid, 165.
122. Ibid, https://www.ncbi.nlm.nih.gov/pmc/articles/PMC3836250/#R15. II, §1.
123. See Woods, "Voice-Hearer."
124. Ibid, V, § 1.
125. See Woods on the importance of storytelling for voice-hearers, V, § 1.
126. Powers, *Overstory*, 200.
127. Ibid, 163.
128. Ibid, 150.
129. Ibid, 150.
130. Ibid, 151.
131. Ibid, 157.
132. Ibid, 157.
133. Ibid, 158.
134. Ibid, 114.
135. Ibid, 131.
136. Ibid, 390.
137. Nash, *Wilderness*, 2.
138. Powers, *Overstory*, 458.
139. Ibid, 178.
140. Ibid, 322.
141. Ibid, 493.
142. Hogan and Peterson, *Sightings*, 277.
143. Act 2, scene 2.
144. Vizenor, *Wordarrows*, 93.
145. Ibid, 93.
146. Powers, *Overstory*, 219.
147. Kimmerer, *Braiding*, 18–19.
148. Ibid, 19.
149. Ibid, 19.
150. Powers, *Overstory*, 339.
151. Snyder, *Place*, 190.
152. Abram *Spell*, 13.
153. Powers, *Overstory*, 492.
154. See Frumkin, "Building," Hansen et al. "Shinrin-Yoku," and Li, *Forest*.
155. Powers, *Overstory*, 183.
156. Ibid, 182–83.
157. Ibid, 183.
158. Powers, *Overstory*, 183.
159. Edward O. Wilson defends that owing to the time when we lived in the African savannah, humans have evolved biological responses to more-than-human nature that increased chances of survival in the wild; see *Origins*.
160. Powers, *Overstory*, 183.

161. Jonathan Bate recounts Rilke's experience as follows: "He seemed to become nature itself, to share his being with tree and singing bird as inner and outer were gathered together into a single "uninterrupted space," Bate, *Song*, 263.
162. Powers, *Overstory*, 499.
163. Alaimo, *Bodily*, 17.
164. Davis, "Psychological," quoted in Kahn and Hasbach, "Introduction," 5.
165. Ibid, 5.
166. Ray's readings echo current debates and takes up the argument at the heart of Christopher D. Stone's 1972 essay *Should Trees Have Standing? Law, Morality, and the Environment*. See Powers, *Overstory*, 249–51.
167. Powers, *Overstory*, 250.
168. Ibid, 251.
169. Powers, *Overstory*, 432.
170. Slemon "Magical Realism," 422.
171. Durkheim, *Formes*, 308.
172. Pughe, "Réinventer," 75, translation mine.
173. Powers, *Overstory*, 496.
174. Pughe, "Réinventer," 79, translation mine.
175. Starhawk, *Earth*, 11.
176. While some critics have cringed at the ostentatious didacticism of Powers's novel, it is nevertheless evident that his ecopoetic foray into the world of trees means to educate readers who may otherwise be unwilling to read more "serious" literature.
177. Starhawk, *Earth*, 11.
178. Berry, *Dream*, 21.
179. Bate, *Song*, 37–38.

PART III

Writing and Dwelling Ecopoetically

Chapter Nine

Ecopoets and the Art of Anamorphosis

In my first part, I have dealt with the disenchantment/enchantment dialectics to then emphasize how ecopoetics could contribute to a necessary reenchantment of humans' rapport with the world in a postmodern context. In my second part, I have tackled liminal realism as an initiatory mode allowing for thought-experiments that magically situate us in-between human and other-than-human worlds, reinitiate ecological relationships of interdependence and mutualisms, and ecopoetically rebraid epistemologies and ontologies widely considered irreconcilable. I have underscored how liminal realism makes room for other-than-human ways of knowing and being, endowing ecopoetic texts with the enchanting power to augment the human *Umwelt*. In this third and last part, I shall explore further the connections between writing and dwelling ecopoetically to reveal how ecoliterature has been dedicated to the fixing of a general crisis that hinges on our impoverished imagination (Buell), perception (Sewall), and esthesia (Lacroix, Zhong Mengual and Morizot).[1]

As broached in part 2, Richard Powers has turned to the science on trees to compose an enchanting story entangling humans' and trees' fates, while translating in accessible terms current science on tree behavior and communication. Like Patricia the fictional scientist, Powers negotiates a realm in-between science and myth where he didactically walks "that ultrafine line between numbness and awe,"[2] translating data and numbers into ideas, concepts, and stories that humans can both grasp and feel for. Addressing potential criticism of his ecopoetic stretch of the imagination into liminal realism, Powers delegates justification to his psychologist character. Speaking of storytelling, Adam argues: "'The best arguments in the world won't change a person's mind. The only thing that can do that is a good story.'"[3] This is a point that has been made by several ecocritics, starting with Scott Slovic and Paul Slovic: "Instead of numbing audiences with blizzards of nerveless information, skilled communicators can navigate accurately and vividly between large-scale phenomena and small-scale illustrations, between the remote and the proximate." Literary narrative helps "foster audience *attachment*," while numbers in themselves "can be both empowering and dreadfully desensitizing."[4] All the ecopoets in my corpus exhibit full cognizance that "in order to make data *mean* something (and even perhaps to *count* for something), it seems vital that quantitative discourse be complemented with other modes, such as story and image, which so forcefully inspire human audiences and shape our moral compass."[5]

Dealing with the ecopoetic "reinvention" of nature, Thomas Pughe writes: "Literature does note recreate nature. However, by means of its work on writing, it is ceaselessly reinventing the interactions between humans and nature as well as the representations of nature formed by humankind."[6] Keeping these considerations in mind, I will first examine the notion of ecopoetic anamorphosis which is foregrounded by some of the texts in my corpus. I will explore how this may affect our representations of humans and nature, destabilize our anthropocentric gaze, and reform our apprehension of the world which presently still gives precedence to sight to the detriment of other senses. As I endeavor to show, it could be that via anamorphic processes induced by anthropocentric, and, conversely, biocentric metaphors, comparisons, and shifts in focalization, *ecopoiesis* also achieves a form of liminal realism devoid of any extraordinary phenomena.

My first two chapters show how ecopoets intervene as postmodern shamans, self-consciously using anamorphosis as a magic trick allowing them to mediate between human and other-than-human worlds. Referring to visions of storytelling, dancing, singing, and writing as a celebrating and enacting of our harmonious participation in the earth, I will then consider *ecopoiesis* as a careful reweaving of word to world. Having examined in the third chapter the trope of weaving in relationship to *ecopoiesis*, I will finally focus on the rematerialization of language at work in my corpus. Studying how ecopoetic language provides a medium for humans to articulate—in both senses of the verb "to articulate"—*with* the living world, I will analyze how, concretely, reen-*chant*-ment co-*responds* to poetic translations and echoes of the earth. In this last and longest chapter, I will scrutinize how, further than the constant reframing of our representations and finetuning of our perceptions, it might be that ecopoetic storytelling actively participates in the world's co-becoming, in its mattering, and its "singing." To that purpose, I will be paying close attention to the material embeddedness of *ecopoiesis* in the fabric of the physical world itself—how beyond the recuperation of the ancient topos of a *harmonia mundi*, ecopoetics may reveal a very rational take on the notion of a song of the world. Shedding light on the multispecies "*sympoiesis*" at play in the world both in and outside of literature,[7] I will delve further into the sensuous *poiesis* at work in my corpus. Proposing close readings of key passages, I will analyze the poetic echoes of the polyphonic song of the earth in literary translations of the geophony and biophony arising from the greater-than-human world in the first *place*.

Far from evading the question of the artificiality involved in nature writing, the books in my corpus often stress the crucial role of language and art as mediators between reality and our perception of it. As ecocritic Sharon Cameron has put it, "[to] write about nature is to write about how the mind sees nature, and sometimes about how the mind sees itself."[8] Already, I have broached how liminal realism that involves shapeshifting essentially induces a shifting in perspectives tied to a form of interspecies anamorphosis. From the Greek *ana*, meaning "again," and *morphosis,* for "shaping," the concept of anamorphosis refers to a trans-*formation*—a transfiguration, or a change in the perception of forms occurring through a change in perspectives, and vice versa. In visual art, an anamorphosis is obtained by distorting an image, reverting to an optical trick or via the use of a reflecting surface, forcing observers to shift viewing stances. Only when adopting the position calculated by the artist will viewers recognize a coherent form in what may otherwise seem chaotic or lacking unity. An anamorphosis thus ostensibly problematizes how the human gaze organizes our perception of shapes in space.[9]

Besides, anamorphosis ties in with the metaphoric process at the heart of literary *poiesis*. Anamorphosis is a visual trick inviting us to look at one thing and perceive another, when metaphor spurs us to consider one thing in terms of another with the mind's eye. In *La Métaphore vive*, French philosopher Paul Ricœur says of the figure of speech that it "has been compared to a filter, a screen, a lens, to express how it places things under a certain perspective and teaches us to 'see like . . . '; but metaphor is also a mask that disguises. It has been said that metaphor integrates diversities, but it also leads to categorial confusion."[10] Having said that, and referring to work carried out by Colin Murray Turbayne, Ricœur nonetheless defends the merits of metaphorical language, to be approached using "as if" as a critical pointer. Since truth cannot be described literally, or facts represented as they really are, Ricœur encourages us to settle for expressing not the way reality is, but "what is seems like to us." We can only therefore "replace the masks," but remaining aware that this is precisely what we are doing: "To put it in a nutshell, the critical consciousness of the distinction between use and abuse does not lead us to discard the use of metaphors, but to re-use metaphors, in the endless quest for other metaphors, or possibly for a metaphor that would be the best possible one."[11] This is especially relevant to the Gaian ecopoetics tackled in Chapter Three.

Both anamorphosis and metaphor thus imply some degree of artistic, epistemological, and ethical self-reflexiveness, tied to a double movement of calculated magical distortion and revelation. Ricœur moreover addresses how metaphor induces a "stereoscopic vision."[12] Taking up the concept of stereoscopic vision first coined by Bedell Stanford, Ricœur explores the tension between literal and metaphorical interpretations of a phrase: "the one is literal, bounded by the established value of words, the other is metaphorical, born from the 'torsion' imposed on those words so that they can 'make sense' within the statement as a whole."[13] Ricœur approaches the overlapping of both interpretations as creating a stereoscopic vision that highlights the "being-like" quality of things.

Historically, the notion of anamorphosis is linked to the seminal painting by Hans Holbein "The Ambassadors" (1533), where a bizarre, slanted object appearing in the foreground turns out—once one agrees to look at the painting from an unusual close-up and peripheral perspective—to be a human skull. Rather well-known too is the vase anamorphosis designed by Danish psychologist Edgar Ruben (1915). Depending on what is foregrounded or backgrounded as viewers strive to outline the represented object, Ruben's ambiguous two-dimensional picture forms either a white vase against a dark background, or two human faces in profile, drawn in white against a black background. My contention here is that unsettling acquired habits of perspective, of perceptual outlining, and playing up the ductility of human vision, ecopoetic anamorphoses trouble fixed perceptions and representations of the forms of nature, the better to regenerate them.

To study *ecopoiesis* through the lens of anamorphosis raises several issues addressed in this chapter. First, it provides an operative conceptual framework for the move orchestrated by some of the novels in my corpus from an anthropocentric to a zoocentric or biocentric stance, especially those focalizing a scene through a nonhuman point of view. As we shall see, the literature I am looking at encourages defamiliarizing views of the world where humans appear as peripheral. In so doing, they address that "defiantly self-indulgent anthropocentrism" which E. O. Wilson scourged as the most "[crippling], intellectual vice."[14] In their "attempt to reanimate nature," as Christopher Manes might put it, ecopoets do not hesitate to "[put] at risk the privileged discourse of reason" as they aspire to create "a new language," one that radically displaces "Man"—that "fictional character" that "has occluded the natural world, leaving it voiceless and subjectless."[15]

Second, the concept of an ecopoetic anamorphosis offers a phenomenological/aesthetic take on our cognitive tendency to foreground human matters and, simultaneously, to background nonhuman ones. This tendency depends on an outlook trained to attribute centrality to humans by a culture stipulating a clear separation between humans and their natural "surroundings." Indeed, as Ruben's vase demonstrates, humans make sense of what they see via a dual movement of foregrounding and backgrounding, a mental operation that can only be accomplished by means of delineating the contours of an object perceived sharply against a partly effaced "background," which consequently becomes secondary in perceived significance. As Michel Serres has pointed out, the conceptual conditioning entailed by the very word "environment" is quite problematic. Hence Serres's jettisoning of the word that metaphorically maps out a delusive anthropocentric image schema: "Do forget about the word environment [. . .]. It supposes that we humans sit at the heart of a system of things gravitating around us. It makes us the navel of the universe, the masters and possessors of nature."[16] As I will show, rather than presenting us with images staging clear-cut lines of demarcation between a being and her environment—images that deceptively encourage us to think of humans as beings who can be neatly outlined, individualized as subjects, and potentially extracted from their *oikos*—the texts in my corpus activate images of meshworks, of entanglements, and rhizomatic interlacing between an individual and its milieu.

Last but not least, the *ecopoiesis* in my corpus invites us to experience the world through animal ways granting less precedence to sight than is habitually the case in human perception. While many anamorphoses depend on stereoscopic tricks magically conferring relief to flat images, some ecopoetic anamorphoses go further in that they may restore a fuller perception of the world, the physical depth of which may not appear at first *sight*, without inflecting one's overridingly visual perspective. Compensating for the virtual anosmia, deafness, haptic numbness, and the relative effacing of proprioception which characterizes our predominantly scopic take on nature, ecopoetic anamorphoses sometimes perform a kind of multisensorial stereoscopy. They can indeed superimpose the various complementary textures of the living world, shedding synesthetic light on earthly reliefs that cannot be rightly apprehended through an optical grasp in isolation from the other senses. As a result, ecopoetic anamorphoses tend to reorganize the forms of nature, fleshing them out in a multidimensional, enchanting way, or interlacing all the different senses implicated in our bodily experience of dwelling. In so doing, they invite readers to cross-examine the reliability of an empiricism based first and foremost on what our overbearing human vision may seize of the world. As the texts in my corpus reveal, such faulty empiricism tends to expunge nonvisual dimensions of reality. Mobilizing different senses and different viewpoints at once in our perception of the forms, material fabric, and languages of nature, synesthetic anamorphoses cue us to better apprehend the invisible coherence of the world emerging through its interwoven, sensuous textures.

The question of anamorphic practices surfaces throughout my corpus via references to visual arts that may oppugn, and potentially renovate, our perception of the forms of nature. This is notably the case in Barbara Kingsolver's *Animal Dreams,* when Cosima reflects on her father's pet hobby as a photographer: "he made photographs of things that didn't look like what they actually were. He had hundreds: clouds that looked like animals, landscapes that looked like clouds. [. . .] The photo was my favorite, a hand on a white table. And of course it wasn't a hand, but a clump of five saguaro cacti, oddly curved and bumpy, shot against a clear sky. All turned sideways. Odd as it seemed, this thing he did, there was a great deal of art to it."[17] The book cover for the first edition stages visual equivalents of Homer's photographic art, with clouds looking like a coyote or a hen. Both the cover and the quote at hand evoke the anamorphic practice that most of us have spontaneously played, that is, the imaginative game whereby one looks into the sky for clouds shaped like recognizable forms. This wilful practice forms an exercise in pareidolia—a term designating the human tendency to make out meaningful images in random forms.

In a passage describing the way the sky is "animated" by the wind, Jean Giono gestures to the fundamentally pareidolic, ecopoetic stance of humans:

> The sky from one end to the other is completely inhabited by immense clouds in the shapes of monstrous men, or wild beasts or horses. The wind [. . .] animates them with a great life that is not confined to each cloud, man, beast, horse, but which passes from one to the other without a barrier; so that at any moment the form of a man flows gradually into the spine of a beast, the horse that made a gigantic leap, then he lets his thickened legs trail away; his thighs, his hoofs, join, and he has become a mountain. His name is a forest of trees. Then, again, it flows and slips into all the shapes in the world.[18]

Although the poetic vision is clearly foregrounded as resulting from a visual illusion coproduced by the forms of nature and human perception, it nevertheless reveals the unfurling of a live universe where everything is continuously involved in a cosmic process of co-becoming and shapeshifting. In a typically animistic way, all the forms of nature are blown life into by a unifying principle, here wind. In this sense, Giono's worldview is highly compatible with Linda Hogan's in *Power*, which stages Oni, the wind, as the cosmic breath of life. Both novels, with their insistence on the animating power of wind, may be read in the light of David Abram's cross-cultural and etymological rebraiding of material and spiritual notions of animating psyche, breath, wind, air, soul, mind, spirit, and even speech.[19] All three writers depict a universe ecopoetically entangled with human bodyminds that organize the chaotic dance of energy and of life on earth. As highlighted by Giono, this is one of the essential ways in which humans are spontaneously prone to ecopoetic enchantment.

Getting back to *Animal Dreams*, Cosima's father's ecopoetic art provides a metatextual comment, pinpointing some of the stylistic effects of Kingsolver's prose, which typically makes a pervasive use of both anthropomorphic and biomorphic comparisons. Cosima perceives the world through an anamorphic lens. Focalized through her point of view, plants are described in terms evoking animals or humans, and reversely, humans are frequently described as plants or animals: "The African violets were furred like pets, and the prayer plants had leaves like an old woman's hands, red-veined on the back, that opened wide in the sun and folded primly together in the shade."[20] On top of transgressing conventional boundaries between vegetal, animal, and human lifeforms, the stereoscopic vision here induced underscores the plants' textures, and in the case of the "prayer plants"—the common name of which can in itself transform our vision of them—their physical response to light, to changes in temperature, and their relative motility. Going beyond mere anthropomorphism, the anamorphic vision renders apparent the fact that plants do possess a form of sentience, albeit one that is different from ours. In addition, making great use of comparison in her prose, Kingsolver does not skimp on the anamorphic tendency of the human eye. On the contrary, she systematically uses the comparative "like" ("looked like"; violets "furred like pets"; "leaves like a woman's hands"). Thus, rather than concealing the subjective shaping of reality that is always at play in our apprehension of phenomena, Kingsolver brings to light how humans *make sense* of the world by trying to establish what forms, things, and beings feel or look *like* to us.

In addition, Kingsolver shows, the distorted ways in which we may perceive nature bespeak a tendency to project onto the nonhuman world images belonging to the individual or the collective unconscious. To give a few examples, in the opening chapter of *Animal Dreams*, Cosima stumbles onto a birthday party involving a group of children playfully whacking a peacock pinata, but she envisions instead a sombre scene with children torturing a helpless animal. While it establishes—with a certain dose of comedy—the unreliability of the main character's point of view, it simultaneously betrays her repressed feelings of vulnerability.[21] Her thwarted view appears unconsciously determined by images of death that are latent in her psyche and keep resurfacing throughout her narrative.[22] Cosima's perspective in the beginning thus serves to emphasize some of the complexities potentially involved in anamorphic, metaphorical vision:

> If you've never walked through an old orchard, you have to imagine this: it presents you with an optical illusion. You move through what looks like a hodgepodge thicket of trees, but then at intervals you find yourself at the center

of long, maddeningly straight rows of trees standing like soldiers at attention. There's a graveyard in northern France where all the dead boys from D-Day are buried. [. . .] I remember looking it over and thinking it was a forest of graves. But the rows were like this, dizzying, diagonal, perfectly straight, so after all it wasn't a forest but an orchard of graves. Nothing to do with nature, unless you count human nature.[23]

On top of a dense network of symbolism orienting our interpretation of the story,[24] the anamorphic picture revisions what may be conceptualized as pertaining to either "nature" or "culture." On an epistemological level, the passage underscores the discrepancy between wild nature (a forest) and cultivated nature (an orchard), even though in many people's minds, an orchard may be apprehended as a vision of "nature." The same could easily apply to planted forests versus old-growth forests. Meanwhile, it begs the question of what exactly might be accounted for by the thorny concept of "human nature"—a phrase which, from a modern standpoint, might seem like something of an oxymoron.

In Kingsolver's *Prodigal Summer*, two characters come across so-called "lady-slippers," a type of orchid the common name of which works metaphorically. While underlining the process at hand, Kingsolver induces a second anamorphosis of the flower to debunk the puritanical, patriarchal narrowmindedness of early botanists systematically associating flowers with images tied to romantic ideas of femininity. Transcending conventional gender associations, the ensuing description cues us to shift perspectives onto the orchid, discerning its shape no longer as that of a "lady's slipper" but rather as male genitalia: "dozens of delicately wrinkled oval pouches held erect on stems, all the way up the ridge. She pressed her lips together, inclined to avert her eyes from so many pink scrota."[25] The two characters laugh at "whoever had been the first to pretend this flower looked like a lady's slipper and not a man's testicles. But they both touched the orchid's veined flesh, gingerly, surprised by its cool vegetable texture."[26] Kingsolver thus effectuates an anamorphosis that revisions our visual perception of the orchid, here collapsed moreover with a *feel* for the plant's texture.

In this last example, we may get an inkling of what Lawrence Buell defines as a foreshortening of perception derived from mental/textual practice, which, he claims, is "the negative manifestation" of the "environmental unconscious."[27] "By this," Buell clarifies, he means "to refer in part to the limiting condition of predictable, chronic perceptual underactivation in bringing to awareness, and then to articulation, all of that is to be noticed and expressed."[28] While Buell insists on the negative aspect of the environmental unconscious, i.e., "the impossibility of individual or collective perception coming to full consciousness at whatever level: observation, thought, articulation, and so forth," he nevertheless validates the flip side of environmental unconscious "as potential": "as a residual capacity (of individual humans, authors, texts, readers, communities) to awake to fuller apprehension of physical environment and one's interdependence with it."[29] Along those lines, throughout this chapter, I look at anamorphosis and metaphor as tricks developed by the human imagination to achieve breakthroughs and open up the doors of perception beyond established conceptual or linguistic constraints.

One of Kingsolver's imagined photographic anamorphoses elucidates the human cognitive propensity to identify an object by outlining its contours, the better to distinguish it from its background: "He shrank the frame into focus, shut it off while he slid a rectangle of paper into place and set the timer, and then projected the picture again, burning it into paper. In the center were two old men hunched on a stone wall, backs to the camera. / 'They look like rocks,' I said. 'It's hard to see where the wall ends and the men start.'"[30] Kingsolver relentlessly explores how some of the lines conventionally separating humans from the rest of the world might be erased or reconfigured. In that sense, the ecopoetic organic world she pens can be compared with that crafted by Homer through photography, magically affecting what we see by removing the lines isolating humans from their surroundings: "'Men who look like stones' [. . .] / 'Except for the hat,' I said. 'That's a giveaway.' / 'I'm taking it out.' He held a small spatula into the beam of light and waved little circles over the area of the man's hat, as if he were rubbing it out, which is exactly what he was doing. Photographers' call this 'dodging,' and the spatula was a 'dodging tool,' [. . .] When I was little I called it the Magic Wand."[31] Kingsolver here foregrounds the manipulation of representations characteristic of the way *ecopoiesis* may deliberately distort the vision of what it presents, rendering visible aspects of the world which we tend to overlook.

The anamorphic project at the heart of Powers's *The Overstory* is encapsulated in several interrelated metafictional elements. First, gesturing to the motivation ecopoets may have in anthropomorphizing other-than-human nature, when little Patricia wonders how come humans can be so blind to plant lives, her father explains, "'Adam's curse. We only see things that look like us.'"[32] Second, Nick's artwork hinges on anamorphoses that draw attention to tree existences. As indicated by Nick's "fabulous tromp l'oeil" possessing the faculty to magically transform an old barn wall into an Edenic "glade of broadleaf trees flowering as if it's May,"[33] Powers invites us to probe our bodyminds for how our esthetic sensibility and, therefore, our capacity to perceive the forms of nature, might be impacted by optical and representational illusions. Our warped perception of trees as separate beings for example stems from our above-ground perspective making us blind to underground mycorrhizal networks: "All the ways you imagine us—bewitched mangroves up on stilts, a nutmeg's inverted spade, gnarled baja elephant trunks, the straight-up missile of a sal—are always amputations. Your kind never sees us whole. You miss half of it, and more. There's always as much belowground as above."[34] The opening of the novel thus problematizes how humans inherently en-*vision* and organize the chaotic alien forms of nature by projecting onto them other known forms (a spade, or an elephant trunk onto a tree). Anamorphosis, in that sense, can be seen as a visual equivalent of metaphor, highlighting the *look-alike* of things.[35]

Let us recall Nietzsche's musing on the illusory nature of a language of truth: "What therefore is truth? A mobile army of metaphors, metonymies, anthropomorphisms: in short a sum of human relations which become poetically and rhetorically intensified, metamorphosed, adorned, and after long usage seem to a nation fixed, canonic and binding; truths are illusions of which one has forgotten that they *are* illusions; worn-out metaphors which have become powerless to affect the senses."[36] Essentially, the incipit of the novel where Powers has trees addressing Mimi, and, by proxy, his readers, stresses our overbearing dependance on eyesight and the entailed foreshortening of human perception. Underlying how there is more to the living world than what the human eye can see, the tree voices crafted by Powers entice us to try and *picture* the volatile organic compounds that elude eyesight and yet are part of the conversations going on in the landscape: "Close your eyes and think of willow. The weeping you see will be wrong. Picture an acacia thorn. Nothing in your thought will be sharp enough. What hovers right above you? What floats over your head right now—*now*? [. . .] *That's the trouble with people, their root problem. Life runs alongside them, unseen.*"[37] As announced from the start by "a chorus of living wood," Powers aims to contribute to a greening of the bodymind, making us porous to the invisible sensuous textures of the world which are mostly over*looked* as we *picture* the world with our mind's *eye*, often to the detriment of other senses: "If your mind were only a slightly greener thing, we'd drown you in meaning."[38]

Powers exhibits the anamorphic power of ecopoetic art via Nick's family inheritance—a photographic project consisting in a collection of photos of the same chestnut tree, shot from the same angle every single month of the year over several generations. As a result, the flip-book produces "a magic movie," showing "the black-and-white broccoli [turning . . .] into a sky-probing giant."[39] Replicating the enchanting technique of motion pictures, the art project bridges the rift separating tree time from human time. It allows us to behold the incredible animacy and doing of the chestnut which, when seen through the lens of human activity and relative to human speed, might wrongly appear as inanimate: "Each picture on its own shows nothing but the tree he climbed so often he could do it blind. But flipped through, a Corinthian column of wood swells under his thumb, rousing itself and shaking free."[40] On the one hand, Powers emphasizes the perceptive limits tied to the human experience of time which can affect the way we attribute or deny agency to other lifeforms. If the vegetal world is placed by many on one of the lowest rungs of life, it may well be because of our anthropocentric speed-dependent notions of movement *versus* stillness, of action *versus* inertia.[41] On the other hand, Powers's *mise en abyme* of art highlights how *ecopoeisis* can reveal dimensions of reality that often slip our notice. The amazing complexity of the chestnut tree on Nick's family property only appears to Olivia, thanks to the flipbook of tree drawings that Nick has sketched from the photos, which makes "[the] thing [spiral] up into life" and sharpens her perception of the tree: "Nick leads her across the property, and there the mammoth thing stands, spread out against the porcelain sky. Strange and beautiful math governs the subtending of the hundred branches, thousand twigs, ten thousand twiglets, a beauty that the barn full of art has just primed her to see. [. . .] From the interstate, she failed to notice the thing's thick, tapering grace. The way it flows upward to the first, generous cleave. She wouldn't have noticed, except for the flip-book."[42] Here again, the poetic quality of expression ("the mammoth thing," "the porcelain sky") suggests how metaphors serve as a linguistic relay of anamorphosis. Both processes displace and augment our perception of the world by yoking together the intelligible with the sensitive into a *poetically sensible* vision.

In her exploration of Brazilian old-growth forests, Patricia comes across a spontaneous piece of land art emerging from a gigantic tree:

> No one tells her what to see. A child could make it out. A one-eyed myopic. In knots and whorls, muscles arise from the smooth bole. It's a person, a woman, her torso twisted, her arms lifting from her sides in finger branches. The face, round with alarm, stares so wildly that Patricia looks away. [. . .] What kind of sculptor would pour such skill and effort into a thing so remote it might never be discovered? But it's no carving. [. . .] Just the contours of the tree. [. . .] It's the Virgin, looking on the dying world in horror.[43]

Foregrounding the phenomenological entanglement between the forms of nature and human perception and culture, this passage moreover signals how *ecopoiesis* may strategically exploit the anthropomorphic tendency latent in human perception:

> "Pareidolia," Patricia [. . .] explains: the adaptation that makes people see people in all things. The tendency to turn two knotholes and a gash into a face. [. . .] Patricia looks harder. The figure is *there*. A woman in the coda of life, raising her eyes and lifting her hand in that moment just before fear turns into knowledge. The face may have been formed by the chance efflorescence of a canker, with beetles as cosmetic surgeons. But the arms, the hands, the fingers: family resemblance.[44]

From the Greek *para*, meaning "alongside," or "instead of," and *eidōlon*, for "image," "form," or "shape," pareidolia corresponds to the spontaneous anamorphic propensity of human perception to make meaningful images out of random or ambiguous visual patterns. Contrary to what Patricia states, it is a tendency that is not necessarily anthropomorphic, but more fundamentally anamorphic. Besides, because they offer the possibility of shifting from one vision to another, pareidolia and anarmophosis involve a perceptive liminality. They situate the perceiving subject at an enchanting point in-between two overlapping perceptions. In other words, pareidolia and anamorphosis induce a transfiguration, or shapeshifting, of the world.

Powers thus gives away his own mythopoeic re-*visionist* and postmodernist enterprise. Because we are a poetic and storytelling species, to build an ecology of selves that might halt the blind destruction of our *oikos*, the findings of science ought to be conveyed via ecopoetic myths and metaphors capable of moving us and affecting our worldview, urgently prompting us to re-*vision* the partition between humans and nonhumans:

> The myths come back to her in this tropical upland, stories from her own childhood and the world's. [. . .] She has come across the same stories everywhere she collects seeds—the Philippines, Xinjiang, New Zealand, East Africa, Sri Lanka. People who, in an instant, sink sudden roots and grow bark. Trees that, for a little while, can still speak, lift up their roots, and move.
>
> The word turns odd, foreign in her head. *Myth. Myth.* A mispronunciation. A malaprop. Memories posted forward from people standing on the shores of the great human departure from everything else that lives. [. . .]
>
> Just upriver, the Achuar—people of the palm tree—sing to their gardens and forests, but secretly, in their heads, so only the souls of the plants can hear. Trees are their kin, with hopes, fears, and social codes, and their goal as people has always been to charm and inveigle green things, to win them in symbolic marriage.[45]

Since the dawn of humankind, metaphors and myths have been the way we make *sense* of our inhabitation of the world and of our earthly condition as co-dwellers. As the reference to the Ashuar people suggests, when dealing with interspecies relationships, myths often correspond to cultural forms emerging from nature and which purport to organize multispecies cohabitation. Hence Patricia's husband's suggestion that she deliberately press the buttons of her species' anamorphic and myth-making mindsets the better to provoke awareness: "'Make a poster out of this photo. Put a big caption underneath it: *They're trying to get our attention.*'"[46]

Finally, in a way that rings a bell with Powers' experimental translation of tree language in his own ecopoetic effort to draw attention to trees' agential selves, Nick's environmental activism brings him to act as a tree scribe. He creates anamorphic signs with a lettering composed of tree foliage and branches. The resulting weird, hybrid calligraphy gestures to the possibility for a kind of tree alphabet, with trees forming utterances mediated by human translators. No matter that both characters and readers are perfectly aware of the "pathetic fallacy" at play.[47] The enchanting trick nonetheless captures aspects and dimensions of the world of trees that turn out not so completely fallacious in the end. Ecopoetic metaphors, anamorphoses, and myths, Powers adjudicates, can provide pertinent lenses allowing us to apprehend elusive material phenomena that have been empirically observed, and which it is crucial that we collectively learn to acknowledge in our dealings with the more-than-human world.

This transpires through Patricia's tentative use, when referring to sylvan sentience and multispecies communication—concepts previously unthinkable in a modern context—of those slippery human-related terms such as "awareness," "vocabularies," or "intention."[48] In that sense, Powers reflects eager discussions in the scientific field as to the degree of fallacy in applying up-to-now anthropocentric terms such as "communication," "intelligence," or even "emotions" to refer to trees.[49] As Nick and Olivia synthesize the core of Patricia's book *The Secret Forest*, what prevails in terms of rhetoric is the anamorphic quality of the pictures used to describe other-than-human forms of sentience, intelligence, and purposiveness, conveyed in a lingo acceptable from a scientific standpoint despite its necessarily metaphoric quality:

> They read about how a branch knows when to branch. How a root finds water, even water in a sealed pipe. [. . .] How crowning leaves leave a gap between themselves and their neighbors. How trees see color. They read about the wild stock market trading in handicrafts, aboveground and below. About the complex limited partnerships with other kinds of life. The ingenious designs that loft seeds in the air for hundreds of miles. The tricks of propagation worked upon unsuspecting mobile things tens of millions of years younger than the trees. The bribes for animals who think they're getting lunch for free. [. . .]
>
> They read about trees that migrate. Trees that remember the past and predict the future. [. . .] Trees that summon air forces of insects to come and save them. [. . .] Leaves with fur on the undersides. Thinned petioles that solve the wind.[50]

Mapping onto the world of trees notions, usually referring to human sentience, problem-solving, community building, engineering, economics, advertising, population movement, war, and organized alliances, Powers encourages a cross-species ecopoetics. As a result, we come to picture similarities and connections between human and other-than-human ways of dwelling that transgress the categorial boundaries and hierarchies previously established to posit the alleged superiority of humankind over animal and vegetal realms. For the time of the reading at least, we must think of trees as people, and worthy of at least the same *regard*.

As they personify nonhuman nature, anamorphic tricks strategically call upon our poetic vision to enhance our capacity to think, feel, and dwell with a greater awareness that being really comes down to co-becoming within a multispecies sentient pluriverse. According to Jonathan Bate, "works of art can themselves be imaginary states of nature, imaginary ideal ecosystems, and by reading them, by inhabiting them, we can start to imagine what it might be like to live differently upon the earth."[51] Far from offering a naïve, childish, or pre-scientific view of the world, self-conscious ecopoetic fiction harnesses various means to flaunt the understanding that we urgently need images and stories to reenchant our rapport with the more-than-human world and, accordingly, to intuit the invisible parts of the world as accurately and meaningfully as possible. The poetic vision, ecopoets imply, can and must be reclaimed as complementary with science. As Bate argues, "metaphor is a way of understanding hidden connections, of reunifying the world which scientific understanding has fragmented."[52] Anamorphosis, metaphor, and comparison induce radical changes in perspective, allowing for the chaotic organization of the world to take shape in a way that our bodyminds cannot otherwise wrap themselves around, eventually thus remedying the modern disintegration of the world.

Ecopoiesis deeply affects how we inhabit the earth as it reveals and recasts the pre-existing forms and image-schemas undergirding our perception of the world. In *Philosophy in the Flesh: The Embodied Mind and its Challenge to Western Thought* (1999), George Lakoff and Mark Johnson explore how philosophy, language, and metaphysics all rely on unconscious metaphors that shape our concepts and thereby frame our inhabitation of the world: "philosophical theories are largely the product of the hidden hand of the cognitive unconscious. [. . .] Metaphysics in philosophy is, of course, supposed to characterize what is real—literally real. The irony is that such a conception of the real depends upon unconscious metaphors."[53] By constantly teasing apart the metaphorical mappings made manifest in the language we use, ecopoets rattle and melt the congealed parts of our conceptual systems. Metaphorical language is both "a surface manifestation of conceptual metaphor,"[54] and, when in the hands of ecopoets, a magic wand that can give new shape, features, and texture to the representations and concepts with which we apprehend the world. Once our perception of a thing or being of the world has been modified, adjusted, or vivified, however artificially and momentarily, we can never fully go back to our prior, more circumscribed, or duller, perception of it.

Linda Hogan's poetic use of liminal realism offers a glimpse of what it may mean from an indigenous viewpoint to dwell ecopoetically. Appearing as an "irreducible element of magic," as Wendy Faris would put it, the one aspect of *Power* that may mostly resist a Western reader's adherence lies in the apparition out of nowhere, at the beginning and the end of the story, of "four women from another tribe [coming] down along the road slow as a breeze, shaking their rattles, singing together [. . .], their voices higher than the locusts, [. . .] walking slightly above the ground as if they are gliding and have no feet."[55] The women are "dancing almost, in ancient dresses no one wears anymore, drifting and singing toward [Ama and Omishto]."[56] The narrator comments that "they remind [her] of ghosts," that "they can't be there."[57] Moving somewhere in-between reality and the imagination, the women are said to be part of "a mirage," a kind of optical illusion generated by the heat: "This early in the day there are already heat waves on the road, so it looks like water is on the ground in places. It is a mirage, a pooling of water that isn't really there."[58] Without even looking up, Ama enjoins Omishto, "'See them walking. [. . .] Look at them.'"[59] The seeing at stake takes place mostly through the mind's culturally trained eye: "[Ama] doesn't turn her eyes toward them. It's like the seeing is inside her."[60] Besides, the four women's apparition also involves hearing and proprioception: "it all looks true to me, the women singing slow and then fast, faster, [. . .] and there is the sound of turtle-shell rattles, and I can feel the song in my stomach as they float above the road and seem to have no feet and come toward us."[61]

Significantly, the apparition is juxtaposed with Omishto listening to Ama's voice telling her about her dream of the golden panther "[standing] up like a person" and talking to her. Omishto at that point already has a foot in an enchanting poetic dimension half-way between the world of dreams and concrete reality. Additionally, just before Omishto starts perceiving the women, her auditory focus is on the biophony produced by "the thousand songs of locust all around [them]."[62] Thus, it may be that Omishto's open stance to the world allows for the entanglement between reality and the shimmering world of stories, with their anamorphic power to alter our perception. For the four women seem to materialize out of the heat, the grasshoppers' song, and Ama's stories all at once: "They must be the four women Ama told me about once before, only I didn't believe her then and now I see them for myself, and hear them singing, their voices higher than the locusts."[63] As the heat drops, the vision disperses: "When I look again the women are gone. I want to know where they've gone as much as I want to know where they came from and as I look it is as if I see the space between things like there's a place in between every solid thing where creation takes place."[64] Hogan thus underlines how ontology results from a dynamic creative process, a meshwork of perceptions interweaved with stories and myths that partly determine what we may or may not envision as real.

As it is, a possible interpretation for the function of those "four women from another tribe" lies with the Navajo notion of Hozho, which "is the Navajo word for walking in beauty, walking in a sacred manner, walking with a peaceful heart. Hozho," Paula Gunn Allen explains, "is the heart of the Native American spirit, which is so directly involved with the beauty and living awareness of the land."[65] Evoking the four cardinal directions, the four mythical women form an ecopoetic vision of dwelling which Ama and Omishto welcome because, paradoxically as it may seem, it offers them earthly grounding and guidance. Even readers who do not share an indigenous worldview may embrace the ecopoetic reenchantment at work in Hogan's novel. Indeed, as Jonathan Bate has put it: "Ecopoetics seeks not to enframe literary texts, but to meditate upon them, to thank them, to listen to them, albeit to ask questions of them. [. . .] Ecopoetics renounces the mastery of enframing knowledge and listens instead to the voice of art."[66] Among the common traits between all the books in my corpus, each sheds light on how porous reality can be to the cognitive unconscious underpinning language and myths. Furthermore, each novel calls for an ecopoetics that can potentially reconfigure our conceptual notions and vivify our sensuous perception of the natural world. In that sense, much *ecopoiesis* can be interpreted as a literary, multisensorial practice of anamorphosis.

NOTES

1. Buell, *Environmental*; Sewall, "Skill and "Beauty"; Zhong Mengual and Morizot, "Illisibilité"; and Lacroix, *Devant*.
2. Powers, *Overstory*, 124.
3. Ibid, 336.
4. Slovic and Slovic, *Numbers*, 8.
5. Ibid, 9.
6. Pughe, "Réinventer," 73, translation mine.
7. See Donna Haraway on *sympoiesis*, in *Staying*.
8. Cameron, *Writing*, 44.
9. In biology, the term refers to a gradual change of form through evolution, a limited metamorphosis observed mostly in arthropods, lichen, and fungi.
10. Ricœur, *Métaphore*, 317, translation mine.
11. Ibid, 318.
12. Ibid, 321.
13. Ibid, 375, translation mine.
14. Wilson, *Human*, 17.
15. Manes, "Nature," 24, 26.

16. Serres, *Contrat*, 60.
17. Kingsolver, *Animal*, 70.
18. Giono, *Joy*, 198.
19. Abram, *Spell*, 237–42.
20. Kingsolver, *Animal*, 43.
21. This type of anamorphic process, where the same picture can conjure up different mental images, has inspired the tests conceived by Swiss psychoanalyst Hermann Rorschach in the early 1920s. Banking on pareidolia, patients are presented with random inkblots and are asked what they see in it (a butterfly, a skull). The nature of the perceived form is then read as revealing of what may be on the patients' unconscious minds.
22. For an insightful reading of Cosima's narrative through trauma studies, see Stevenson, "Trauma."
23. Kingsolver, *Animal*, 13.
24. Cosima's mental leap from a picture of an orchard to that of a cemetery moreover betrays her personal obsession with death. This is a repressed aspect of her psyche linked to her mother's death and to a hidden teenage pregnancy ending in still–birth which the novel gradually unravels. Simultaneously, the overlapping between the picture of an orchard and that of a cemetery infuses the landscape with a sense of death that may also symptomize the slow, insidious poisoning of the land which is taking place via the industrial dumping of acid in the local river: "At the first sign of winter the trees began to die. Leaves and aborted fruits fell in thick, brittle handfuls like the hair of a cancer patient" (177). The toxicity of the land is here revealed in the comparison with people undergoing chemotherapy, while the "aborted fruit" image implicitly betrays the return of the repressed inside the protagonist's mind, referring to her willed miscarriage of a six–month–old fetus when she was a teenager.
25. Kingsolver, *Prodigal*, 21.
26. Ibid, 21.
27. Buell, *Endangered*, 22.
28. Ibid, 22.
29. Ibid, 22.
30. Kingsolver, *Animal*, 72.
31. Ibid, 72–73. Cosima's father's tinkering with a photograph to turn men into stones may on a psychological level read as an involuntary self–portrait—a picture of men who, like Doc Homer in the novel, have hardened themselves to appear "like rocks," to embody the men family and community can always depend on. As a result, such men may on the outside seem deprived of sensitivity.
32. Powers, *Overstory*, 114.
33. Ibid, 172.
34. Ibid, 3.
35. I am relying here on Paul Ricœur's theory of the vitality at the heart of metaphor.
36. Nietzsche, in "On Truth," 80, qtd in Lakoff and Turner *More*, 218.
37. Powers, *Overstory*, 3–4.
38. Ibid, 4.

39. Ibid, 19.
40. Ibid, 18.
41. For more on this aspect, see Mancuso and Viola, *Intelligence*.
42. Powers, *Overstory*, 175–76.
43. Ibid, 393.
44. Ibid, 394.
45. Ibid, 394.
46. Ibid, 395.
47. The phrase "pathetic fallacy" was coined by John Ruskin in *Modern Painters* (1856), Vol. III, Part IV. Ruskin used the derogatory term to refer to descriptions of nature where writers ascribed human feelings, or pathos, to what was considered as inanimate. Ruskin disparaged the tendency which, according to him, presented not "the true appearances of things to us," but "extraordinary, or false appearances, when we are under the influence of emotion of contemplative fancy." See Cuddon, *Dictionary*, 692–693.
48. Powers, *Overstory*, 284, 453.
49. Outside of foresters, groundbreaking research is being carried out in the burgeoning field of vegetal neurobiology by Stefano Mancuso, in Italy, who defends plant intelligence as a materially proven fact. While there is now little to argue about plants' attested capacities to perceive their environment and solve vital problems, Mancuso has helped rethink our anthropocentric notion of "intelligence" altogether. See Mancuso and Viola, *Intelligence*, and Mancuso, *Revolutionary Genius*. The merits and pitfalls of anthropomorphism have spawned much debate in France. Whereas French botanist and tree specialist Francis Hallé cautiously refrains from lending plants human feelings, recent contributions have opted for the obverse. This is the case for instance of an enchanting recent study by French novelist Didier Van Cauwelaert who, much influenced by Mancuso's studies, amongst others, deliberately pushes for the anthropomorphizing of the vegetal world, in a book suggestively entitled *Les Emotions cachées des plantes* (which could translate to "The hidden emotions of plants"). Van Cauwelaert claims that plants experience a wide range of emotions and cognitive aptitudes including "fear, humiliation, gratitude, creative imagination, ruse, seduction, jealousy, precaution, compassion, solidarity and anticipation," 14–15. Despite the enticing aspect of the many stories he relates, presented as grounded in scientific evidence, not all of them are reliably referenced, however. This makes it hard at times to distinguish the passages where the writer scrupulously relates scientific studies from those where he might be poetically extrapolating.
50. Powers, *Overstory*, 293–94.
51. Bate, *Song* 250–51.
52. Ibid, 247. Bate is here conjecturing what ecopoet Gary Snyder would say in defense of metaphor, and simultaneously endorsing it. This is part of a chapter titled "What are poets for?" which Bate devotes to *ecopoiesis* as a specifically human way of dwelling within the earth. It is well worth reading as it clearly situates Bate's take on *ecopoiesis* in relationship to some of the greatest thinkers who have wrestled with how poetry contributes to human dwelling, such as Plato, Hölderlin, Heidegger, and Paul Ricœur, to name just those few.

53. Lakoff and Johnson, *Philosophy*, 15. See also Lakoff and Johnson, *Metaphors*, and Lakoff and Turner, *More*.
54. Lakoff, "Contemporary," 36.
55. Hogan, *Power*, 24.
56. Ibid, 25.
57. Ibid, 25.
58. Ibid, 25.
59. Ibid, 25.
60. Ibid, 25.
61. Ibid, 25.
62. Ibid, 24.
63. Ibid, 24.
64. Ibid, 25–26.
65. Allen, "May," xi–xii.
66. Bate, *Song*, 286–69.

Chapter Ten

Postmodern Shamanism
Making Headway Toward Other-than-Human Perspectives

Toying with animal points of view, Jean Giono and Barbara Kingsolver have developed means toward anamorphic liminal experiences that come close to a form of ecopoetic shamanism. This is tied in particular to the way ecoliterature can displace our perspective by adopting other-than-human sentient points of life, while exhibiting a shared tendency to orchestrate interspecies communication.

In *Joy of Man's Desiring,* Giono designs thought experiments whereby his omniscient narrator at times delegates the focalization to an animal agent.[1] The resulting passages immerse readers within a liminal realm in-between human words and animal worlds. In a five-page sequence, the animal sentience imagined by Giono gradually transforms our interpretation of the world, now dominated by olfactory and auditory signs.[2] These few pages are indeed punctuated by an abundance of markers relaying the buck's synesthetic interpretation of the world. Where human focalization mainly emphasizes sight and hearing, the animal focalization here principally turns on smell, with a plethora of expressions such as "he sniffed," or "[the] field smelt strong," as the stag identifies the various "odours" emitted by other animals, humans, and flowers. Thus, we are plunged into the odorscape of a land reeking with wafts and scents which the buck analyzes as he goes along, reckoning from their olfactory signatures which other presences roam the place. To a lesser degree, the passage is also strewn with expressions translating the buck's hearing. Readers must thus *make sense* of the world following the buck's perspective, moving through a differently informed multisensorial world: "Shouts of a man could be heard | . . .]. It was far away. It was like the song of a great bird, modulated and in short spurts."[3] Finally, the animal's vision also comes into play in his deciphering of the animate landscape, highlighted by several uses of the epistemic modal "must" conveying the buck's reasoning: "Far off, in the bright daylight, he saw black dots in a field. They must be three men and a horse."[4] In keeping with his animal *Umwelt*, much of the stag's focus is on the tracks he comes across, mapping out the teeming presences of his co-dwellers: "The stag again found the print of his hooves [. . .]. But he also found the tracks of a hedgehog. [. . .] A little farther along the stag found an otter trail."[5] Giono thus cues us to navigate the landscape the way a stag might.

As Sophie Milcent-Lawson cogently demonstrates in her zoopoetic readings of Giono's prose, such passages testify to the writer's novelistic enterprise aimed at challenging the classic anthropocentric narrative stance by including animal viewpoints.[6] Quoting the writer's lucubration, Milcent-Lawson offers insight into the biocentric project sustaining the "animal sequences" of Giono's prose:

> I have long endeavored to write a novel in which the song of the world could be heard. I find that all current books grant too much room to those mean beings, neglecting to make us perceive the panting of the beautiful dwellers of the universe. [. . .] I am aware that a novel without humans is impossible to conceive, since humans are in the world. What is called for is a positioning of humans in their right place, without making them the center of everything. [. . .] What I would like to do is to place everything where it belongs.[7]

Intermingling human and other-than-human *Umwelten*, Giono augments the flesh of the world available to us. Expanding our intuition of what it might be like, in this very instance, to be a stag,[8] Giono charts a multispecies world that is richer in texture, in meaning, and in presence than that ordinarily marked by most humans.

Furthermore, Giono's rehabilitation of the notion of a song of the world goes beyond the metaphorical. More than just a trope, it concretely refers to the many different voices and sounds produced by animals and elements—what Bernie Krause calls biophony and geophony. The point I would like to drill at this stage has to do with an immersive and synesthetic type of worldly anamorphosis which may percolate into our human ways of dwelling. Fleshing out the olfactory, auditory, and sensuous textures of a world which can never be apprehended at a remove, *ecopoiesis* plunges us into a multidimensional and multispecies world wherein vibrant landscapes are enmeshed with soundscapes, odorscapes, and feelscapes. Foregrounding how perceiving subjects remain constantly interlaced with the world through all their many feelers at once, liminal *ecopoiesis* situating us in-between human and nonhuman worlds heightens our awareness that dwelling can vary in texture and intensity depending on the individual or species at stake. Therefore, ecopoetics operating through liminal realism should investigate how literature stitches together the pluriverse, recreating a dwelling place where human and other-than-human *Umwelten* coexist.

Toward the end, Giono's narrative discreetly shifts again to animal perspectives in a way that might destabilize readers' expectations. In the first nine pages of chapter XXII, the omniscient narrator, it seems, describes the hot, delightful summer where "the bees and the red-, yellow-, and blue-bellied flies" make a droning "throng, blond as a ray of vibrating sunshine," engaged together with the trees in a vibrant "dance of joy" making the larches "purr."[9] "Poetics," Abram argues in a way that fits Giono's prose, "would become the practice of alert, animal attention to the broader conversation that surrounds—to the utterances of sunlight and water and the thrumming reply of the bees or the staccato response of a woodpecker to the hollow creeking of an old trunk—and the attempt to not violate this conversation every time that we speak, but to allow it, to acknowledge it, and sometimes to join in."[10] Turning away from anthropic interests, Giono's focus is now on the lively intertwining of other-than-human existences:

> More and more, the paths of animals interlaced by day and by night, through heath and forest. The foliage of the windless bushes never ceased trembling, brushed by the fox, the weasel, the marten, the rat, or the fussy wings of the owl. [. . .] The bats had come out of their caves [. . . and] scoured the air in awkward and stubborn flight. [. . .] All the varieties of grasshoppers had emerged. All the varieties of butterflies lived their lives, and all the flies and beetles. The lizards all came out of their holes [. . .]. All the snakes had finished their courtship and, uncoiling, went on their solitary ways; the males toward their battles with field mice, rats, birds, and small fishes; the females toward the warm dust of the heath to lay their eggs and to hiss softly while waiting for them to hatch.[11]

In this biocentric and somewhat incantatory passage scurrying with life, which then goes on to mention the courtship and reproduction of foxes, mares, otters, and squirrels, the vibrancy of the natural world is gleaned through a diffuse point of view that, using Gérard Genette's terminology, could correspond to "zero focalization"[12]—that is, a God-like omniscient point of view with access to phenomena taking place anywhere at any time.

Yet, as the buck then enters the picture, the question arises as to whether the previous pages, with their minute attention to the slightest sound and movement of animal lives that are largely invisible to the human eye, might in fact have been focalized all along through an imagined animal sentience. Indeed, relaying what the animal does and feels, the scene where the buck enters the pond is recounted in a long paragraph made of sentences that all begin with the anaphoric repetition of "he saw," reiterated nine times in two short paragraphs, and thereby drilling in the animal perspective.[13] Since the buck can lay claim to the status of a full-blown character, and with an earlier scene already focused through his point of view, such a focalization shift would account for the close attention paid to nonhuman others, breaking away from more traditional omniscient perspectives of classical literature usually adhering to a human form of consciousness.

Further complexifying the focalization at stake, the following pages convey the thoughts, perceptions, and feelings of squirrels, then of ants, and, finally, of a rat.[14] Making us aware of forest and plain ecosystems as multispecies pluriverses, Giono gently brings us to experience other, intersecting ways of dwelling. Complicating the heteroglossia of his liminal realism, Giono lends his squirrels a philosophical questioning that betrays the human consciousness interweaved in the text: "They came out onto the plateau. They stopped to listen. They heard the fox bark. [. . .] They perceived in one direction the trees of the forest where they could go to take shelter among the branches, and there was still time. [. . .] And then they had a formidable burst of courage: what good does it do always to guard one's life carefully like a soft little hazel nut? Can one not, in one good fling, cast it wholly in the direction of one's longing?"[15] To capture the unsettling narrative strategy at work here, we might turn to the "defocalization" concept which Wendy Faris coins to describe "the strangely indeterminate nature of magical realism's generic narrative stance."[16] When under the spell of such disorienting, defocalized passages, "the reader is not sure from what perspective such events are viewed or where such a perspective might originate."[17] This focalizing trick no doubt is part of the enchantment. Creating a kind of hypnotic effect, it seduces us into going along with the ecopoetic illusion of thinking like a stag, a rat, or a squirrel. Adopting an unusual perspective, we are charmed and immersed into an animal existence driven by eating and being eaten, where the joy of life is indissoluble from the fear and violence of death:

> It seemed to [the squirrels] that a great phantom squirrel was saying this to them: a gigantic squirrel, bigger than twenty mountains, and that in listening to its mad voice they would become as big as he. It was only summer speaking to them, and life. They became as big as the phantom squirrel and they went out onto the plateau. The fox of the coombs caught up with them near Mouillure. Toward evening, all that were left took refuge in a willow. That tree had no beech nuts. Here and there they scratched the bark with their claws. Of it came a little flour and a drop of sap. It was magic food. They had never been so happy. [. . .] They remembered the battle with the fox. The sky and the fox were fused in the squirrel's hearts with blood and stars. [. . .] Already the air smelt of sugar and fermentation.[18]

Giono's poetics of multispecies interconnectedness gradually draws our attention first toward smaller and smaller forms of animal lives, and then toward vegetal ones, rhizomatically connected as they are with the earth, the air, and the season: "Moss, oats, firs, cedars, poplars. The tallest tree here rises twenty meters; the deepest root bearing tubers goes to a depth of five meters. Twenty meters above the ground to five meters below, there are more than fifty layers of fruits of all kinds, all nourishing, all gradually enlarged by the patient heat of summer. There are fruits for all teeth and for all tastes."[19] Rehearsing the bounty of the land, Giono's rhapsodic descriptions encompass an abundance of fruits which is desirable from a multispecies perspective:

> The Plain of Roume was covered with round fruit like a blanket over the earth: watermelons, muskmelons, squash. A layer at a metre and a half from the ground of peaches, apricots, green apples, green plums. The orchards groaned under their weight. [. . .] The plateau, higher up in the sky, produced scarcely any round fruits, but spiked grain, beech nuts, clusters of woody seeds. It was also richer in tubers. At certain places the earth cracked, exposing deep in its chasms abscesses of bluish roots toward which hastened a procession of beetles and ants. Sometimes little rodents enlarged the hole with their paws, plunged their heads, and scraped the clotted flour with the tips of their little pointed muzzles. Flocks of birds wheeled whole days above the whitebeam trees, the wild plums, the mulberry bushes. As soon as night came, they alighted.[20]

In addition to offering an enrapturing picture of the land as supporting a biotic communitas, Giono brings us to start thinking and desiring like the fruiting earth itself:

> The sun rolled smoothly from one edge of the empty sky to the other. The husks opened with a crackling sound, the seeds flowed onto the ground, the spike of the oats cast its seeds, the burdocks shed seeds, with a single explosion the mosses expelled little golden grains [that] rode away on the wind]. The datura heads split open, freeing from their shells of white satin the three nuts the colour of night. Cabbages full of moisture and [worked] by the heat gave off a strong odour. Beets, onions, turnips, big carrots rose from the powdered soil, pushed out by their swelling flesh. In all the sun-bathed places in the mountains, wild apricot trees dripped sap and juice. Only some marmot came to lick their syrup.
>
> The wheat was ripe.[21]

Working like a sensuous firework that might turn on many of our carnal appetites at once, this passage entwines anthropic and other-than-human life-drives and production. Even trees and plants, with their fruits and seeds, seem to be calling out to animals and humans, alluring us with their biosemiotic outputs into the flesh of the world. Giono's writing dredges up a sense of rapture—if not "jouissance" to take up Roald Barthes[22]—to be found in restoring one's wildness, reclaiming a way of dwelling that is grounded in our human capacity to *feel with* the earth. Additionally, Giono's harmonious song and dance of the world follow seasonal biorhythms that synchronize life across microcosmic and macrocosmic dimensions, thus stressing organic aspects of the world that are crucial to its material reenchantment.

Another form of liminal *ecopoeisis* spurring us to think like the earth—like a mountain forest, to be more exact—emerges from Kingsolver's post-pastoral *Prodigal Summer*. This is a novel that takes up both Giono's and Rachel Carson's ecopoetics of reenchantment, while ostentatiously reclaiming the complementarity between science and poetry, ecology, and spirituality. Overall, ecofeminist novelist, short story writer, poet, and essayist Barbara Kingsolver writes of the wonders of the world from a position that is greatly informed by her initial training as a biologist and zoologist. The way she puts it, the sciences of nature help give meaning to life: "I think biology is my religion. Understanding the processes of the natural world and how all living things are related is the way that I answer those questions that are the basis of religion. What's the purpose of my being here? What happens when we die?"[23]

Orchestrated as a novel with three intertwined third-person narratives, *Prodigal Summer* follows the lives of three sets of characters who are connected by land and place in Southern Appalachia. Their relationships to nature are filtered through their rural activities, all of which make them tightly dependent on other-than-human lives, as they are engaged in farming, hunting, gardening, animal watching, and nature preservation. Manifesting the political activism Kingsolver infuses in her writing, these focalizing agents carry the voices, diverging points of view, and perceptions that can make readers sensitive to the ecological principles determining the life of a mountain. Dialogism thus takes place, in great part through the voice of a wildlife biologist who, having written a PhD on coyotes, now lives as a recluse working as a National Forest keeper in a game-protection area. Since she is always moving between the world of humans and that of the woods—something inscribed in her name, Deanna *Wolfe*—, her role as mediator once again ties in with that of a postmodern shaman. Throughout, Deanna patiently tracks coyotes, hoping to see them move into the local ecosystem to replace overhunted, bygone red wolves. Yet, as opposed to some of the other books in my corpus that might in a different context be labeled "magical realism," Kingsolver's scientific kind of liminal realism here situates us in-between human and other-than-human worlds by overtly tapping into ecology, biology, and ethology, rather than folklore or non-naturalist ontologies.

Deanna serves as mediator and instills basic notions of ecology, synthesizing readings such as Jonathan Roughgarden's *Theory of Population Genetics and Evolutionary Ecology*, or R. T. Paine's concept of keystone species—how top predators' lives are enmeshed with the entire biotic community of a mountain through trickle down effects of who eats who: "She'd carefully read and reread Pain's famous experiments from the 1960s, in which he'd removed all the starfish from his tidepools and watched the diversity of species drop from many to very few. The starfish preyed on mussels. Without starfish, the mussels boomed and either ate nearly everything else or crowded it out."[24] Reminiscent of Aldo Leopold's famous call to "think like a mountain,"[25] Deanna's project is precisely tied to her capacity to think like a river, and beyond, indeed, like a mountain:

> Plenty of people had watched and recorded the disaster of eliminating a predator from a system. They were watching it here in her own beloved mountains, where North America's richest biological home was losing its richness to one extinction after another, of plants and birds, fish, mammals, moths and stoneflies, and especially the river creatures whose names she collected like beads: sugarspoon, forkshell, acornshell, leafshell. Sixty-five kinds of mussels, twenty now gone for good. [. . .] The main predator of the endangered shellfish was the muskrat, which had overpopulated to pestilence along the riverbanks over the last fifty years. What has kept muskrats in check, historically, was the mink (now mostly coats), the river otter (also nearly gone), and surely, the red wolf.[26]

Hence the synecdoche in what follows, whereby the red wolf, or the coyote, stands for the main organ in the life of the entire mountain, itself compared to a kind of superorganism:

> [Deanna] believed coyotes were succeeding here for a single reason: they were sliding quietly into the niche vacated two hundred years ago by the red wolf. [. . . The coyotes were] insinuating themselves into the ragged hole in this land that needed them to fill it. The ghost of a creature long extinct was coming in on silent footprints, returning to the place it once held in the complex anatomy of this forest like a beating heart returned to its body. This is what she believed she would see, if she watched, at this magical juncture: a restoration.[27]

Deanna's thorough understanding of the complex world of coyotes transmits the kind of knowledge zoologists and ethologists have garnered: "The two predators were hardly distinct: the red wolf may have been a genetic cross between the gray wolf and the coyote. Like the coyote, it was a scent hunter that could track in the dead of the night, unlike the big cats that hunt by sight. It was like a coyote in its reproductive rate, and close in size. In fact, judging from the tracks she'd seen, the coyotes here were nearly red-wolf-sized, and probably getting larger with each generation."[28] Kingsolver thus instructs readers about coyotes—how they eat and sense, the social ways of their packs, how they raise their cubs, communicate, and how, as keystone predators, they play an irreplaceable part in preserving the balance of an ecosystem. To redress the unfavorable image of wolves and coyotes in the popular imagination, Deanna diplomatically tells a different story of them, bringing readers to shift perspectives onto the animals so that they might no longer be regarded as pests to be ruthlessly hunted down, but as dignified and necessary wild creatures humans can learn to cohabit with. In that sense, Kingsolver here comes close to the kind of "diplomacy" between animal and human worlds advocated by French philosopher, wolf-tracker, and ethologist Baptiste Morizot.[29]

Kingsolver moreover embraces a liminal form of realism via a system of echoes stitching together the first and last chapters of her book, respectively focalized through Deanna, and, against all odds, through a female coyote's point of view. Quite audacious in its undertaking to challenge anthropocentrism, the novel's closing chapter gives precedence to a nonhuman point of view, earnestly granting it a kind of personhood: "She paused at the top of the field, inhaling the faint scent of honeysuckle. It seemed odd for someone to be down there, this late at night. [. . .] She was following a trail she couldn't be sure of, and she was used to being sure."[30] In all likelihood, it may take a while for readers to figure out Kingsolver's trick in this very last chapter, surreptitiously shifting to an animal point of view introduced by the third-person feminine pronoun "she." Indeed, we may at first be misled by the duplication of the initial situation from the opening chapter, with Deanna "following tracks in the mud she couldn't identify," when "[she] was used to being sure."[31] In addition to the unanticipated pronominal shifting here at work, the repetition of entire sentences between the two chapters orchestrates salient echoes making the change in focalizing agents even more jolting. Deanna and the coyote are indeed attributed similar situations, thoughts, questions, perceptions, motions, and even emotions. In the first chapter, Deanna's relish for the wild and her tracking style evoke animal ways:

> She loved the air after a hard rain [. . .] Her body was free to follow its own rules: a long-legged gait too fast for companionship, unself-conscious squats in the path where she needed to touch the broken foliage [. . .]. Her limbs rejoiced to be outdoors again. All morning the animal trail had led her uphill, ascending the mountain, skirting the rhododendron slick, and now climbing into an old-growth forest whose steepness had spared it from ever being logged. [. . .] She found a spot where it had circled a chestnut stump, probably for scent marking. [. . .] She squatted, [. . .] and pressed her face to the musky old wood. Inhaled.
>
> "Cat," she said softly, to nobody.[32]

Evidently, her tracking competence she owes partly to her acquired capacity to think like a coyote. Hence her keen sniffing, allowing her to identify which other predators might have marked the shared territory with their olfactory signatures.

Furthermore, Deanna's close observation of the greater-than-human world together with her scientific training have taught her to read the invisible signs of interspecies connections: "She studied the stump: an old giant, raggedly rotting its way backward into the ground since its death by ax or blight. Toadstools dotted the humus at its base, tiny ones, brilliant orange, with delicately ridged caps like open parasols. The downpour would have obliterated such fragile things; these must have popped up in the few hours since the rain stopped—after the animal was here, then. Inspired by its ammonia."[33] And yet, far from feeding into the long-held illusion that humans may ever fully grasp the essence of nature and capture others' ways of being, Deanna recognizes the elusive quality of an animal's mind from her human standpoint: "Male or female, [a coyote] had paused by this stump to notice the bobcat's mark, which might have intrigued or offended or maybe meant nothing at all to it. Hard for a human ever to know that mind."[34] Note the hypothetical modality ("might," "or maybe") underscoring how plagued with difficulty humans feel as they try to conjecture the perceptions, emotions, and thoughts of a nonhuman animal. The gap barring humans direct access to an animal mind is moreover suggested by the deictic "that" in this passage ("Hard for a human ever to know *that* mind"), implying respectful distance with regard to the animal's specific world.

The unforeseen focalization switch is first suggested by the shift from daytime to moonlight tracking, then by the bodily posture of the anonymous "she," and finally by the movement verbs evoking animal ways: "She lowered her nose and picked up speed, skirting the top of the long field that lined this whole valley, ducking easily through the barbed wire fences."[35] However, the narrator translates emotions recalling the human thoughts and feelings of the *incipit*: "She loved the air after a hard rain, and a solo expedition on which her body was free to run in a gait too fast for companionship. She could stop in the path wherever she needed to take time with a tempting cluster of blackberries or the fascinating news contained in a scent that hadn't been here yesterday."[36] On top of the references to the animal's omnivorous diet, the synesthetic narrative validates the transition to an animal *Umwelt* by homing in on the various textures of the landscape allowing the coyote to decipher the alien inscriptions of anthropic activities. Those include the anthropophony and odorscape generated by humans:

> She had never been able to reconcile herself to the cacophony of sensations that hung in the air around these farms; the restless bickering of hounds penned

behind houses, howling across one valley to another, and the whine of the perilous freeway in the distance, above the sharp, outlandish scents of human enterprise. Now here, [. . .] there was gasoline wafting up from the road, and something else, a crop of dust of some kind that burned her nose, drowning out even the memorable pungency of pregnant livestock in the field below.[37]

Consolidating the parallel situations of the coyote and of the woman in the beginning, the animal's attention is drawn to some of the same biosemiotic signs, sparking a kind of interspecies communication. Like Deanna in the first chapter, the coyote interprets the delineating of territory through the language of scent-marking common to territorial predators: "She crossed back into the woods and then stopped again to put her nose against a giant, ragged old stump that had a garden of acid-scented fungus sprouting permanently from its base. Usually, this stump smelled of cat. But she found he had not been there lately."[38] Empathizing with the point of view of the animal driven toward a mate, the reader is vicariously brought to ponder the huge stakes of the unfolding story ending:

> She paused [. . .] picking up the scent she'd followed for a while earlier tonight [. . .]. It was a male, and particularly interesting because he wasn't part of her clan [. . .]. She paused again, sniffing, but that trail wasn't going to reveal itself to her now, no matter how hard she tried to find it. And on this sweet, damp night at the beginning of the world, that was fine with her. She could be a patient tracker. By the time cold weather came on hard, and then began to soften into mating season, they would all know each other's whereabouts.[39]

The epistemic uses of the modal "would" translate the wolf's capacity to project herself into the future, forming plans a couple of seasons ahead. If this may betray an anthropomorphic projection,[40] it nonetheless has the advantage of cueing readers to step inside the animal's skin. In addition to making us care from a coyote's standpoint, it coaxes us to compute the implications tied to the coyote's fate for the biotic community at large. Eventually, this scene could adumbrate the fulfillment of Deanna's dream in the long run—a dream, which, lending our own guts to the organless character, we, as readers, have come to share—to see the whole mountain restored.

Backgrounding the partially unresolved human drama at the heart of the plot—with many of the potential outcomes previously brushed out and left suspended—the ending invites readers to consider the coyote as one of the main characters in the story. Rather than whether or not Bondo will own up to the child he has unwittingly fathered, for example, or whether Deanna will end up raising their child with the help of nearby single women, what seems more importantly at stake appears as we must think beyond solely human matters, from the greater perspective of the mountain.

The closing paragraphs echo the *incipit verbatim*, framing the 444 pages of the book with a biocentric view of the pluriverse. Debunking the Cartesian notion of human exceptionalism, the opening and closing lines indeed create a loop foregrounding both the interconnectedness of all lifeforms and the biocentric stance fostered ever since the *in medias res* beginning: "Her body moved with the frankness that comes from solitary habits. But solitude is only a human presumption. Every quiet step is thunder to beetle life underfoot; every choice is a world made new for the chosen."[41] Through dramatic irony, the parallel between Deanna and the coyote's situations turns out quite ominous. Indeed, we realize at the very end that rifle-toting, coyote-hunter Eddie Bondo is stalking the female coyote, just as he had stealthily been watching Deanna in the very first chapter:

> He might have watched her for a long time, until he believed himself and this other restless life in his sight to be the only two creatures left here in this forest of dripping leaves, breathing in some separate atmosphere that was somehow more rarefied and important than the world of air silently exhaled by the leaves all around them. But he would have been wrong. Solitude is a human presumption. Every quiet step is thunder to beetle underfoot, a tug of impalpable thread on the web pulling mate to mate and predator to prey, a beginning or an end. Every choice is a world made new for the chosen.[42]

The prominent system of echoes and the embedded tracking situation involving both animals and humans create tension: whether Deanna has managed during the time of their relationship to win Eddie Bondo over to thinking like a mountain remains for the reader to speculate. With the female coyote's life hanging on Eddie's unpredictable decision to pull the trigger or not, there comes a crucial sense of ecological responsibility. Whether Bondo acts upon hubris or humility could indeed catapult an entirely different outcome for the future of the multispecies community at the heart of the book which, by that point, readers must have come to feel attached to.

A constant throughout Kingsolver's work lies in her upbraiding of the disenchanting narratives produced by Modernity. Whether the world she pens pivots on scientific, historiographic, or mythopoeic approaches—and it often interweaves all three of those threads at once—Kingsolver's writing relentlessly interrogates our naturcultural identities, teasing out how the stories we tell impact the way humans fit within the world. Presumably because her initial training is in biology and zoology, by no means does her takeaway from science betray any form of disenchantment. In Kingsolver's literature, science leads onto the obverse road—one where the grounding of human nature *and* culture in the greater-than-human world is an inexhaustible source of wonder and meaning. If champions of human exceptionalism may chide her ecofeminist stance as a form of regression, one of the ways Kingsolver reenchants the world, paradoxically, is to punctuate all the unconscious ways in which we are biologically attuned to the complex rhythms and textures of the more-than-human world.

One of the lines of demarcation which Kingsolver constantly attacks is that meant to delineate the world of humans from that of animals. In *Animal Dreams*—mark the ambiguous, oxymoronic dimension of the title—for instance, Cosima reflects on her involuntary, animal response to the danger alert picked up from the biophony around the house where she is baby-sitting for a friend:

> A pack of coyotes set up a racket near the house, yipping and howling, so close by they sounded like they had us surrounded. When a hunting pack corners a rabbit they go into a blood frenzy, making human-sounding screams. The baby sighed and stirred in his crib. At seven months, he was just the size of a big jack-rabbit–the same amount of meat. The back of my scalp and neck prickled. It's an involuntary muscle contraction that causes that, setting the hair follicle on edge; if we had manes they would bristle exactly like a growling dog's. We're animals. We're born like every other mammal and we live our whole lives around disguised animal thoughts.[43]

Far from being reductive, the eye-opening vision Kingsolver encourages is one where consciousness of our own animality breeds more compassion and respect for other-than-human forms of sentience,[44] while it also helps humans wholesomely reintegrate the more-than-human world. Kingsolver upholds that it is precisely by acknowledging our earthly natures that we might make the best of our own potential as humans, whose specificity may be, as others have suggested before her, to dwell ecopoet(h)ically.[45] With reference to David Abram's eponymous book, Kingsolver's characters may be approached as engaged in a process of "becoming-animal."[46] As Abram puts it, "the phrase speaks first and foremost to the matter of becoming more deeply human by acknowledging, affirming, and growing into our animality."[47]

In line with ecofeminism's asseveration that we must urgently reconnect with our bodily natures, Kingsolver reclaims our corporeal connections with the more-than-human world. While they may seem either execrable or fabulous to those attached to the notion of full human autonomy and control, most of these connections simply express our animal natures. In *Prodigal Summer*, for example, Deanna muses about "the obvious animal facts people [refuse] to know about their kind."[48] Bondo's curiosity is piqued: how might a woman living in the woods be aware of ovulating? Meanwhile, Deanna thinks to herself that any woman would know "if she was paying attention."[49] Deanna unabashedly instructs Bondo that one of the unconscious motives for women's strong sexual appeal to men during their ovulation period may be accounted for by the bodily production of pheromones. Thus, Deanna insists, humans unknowingly respond to forms of communication taking place through scent.[50] Furthermore, she apprises him of the fact that women's fertility cycles naturally synchronize with lunar cycles: "Any woman will ovulate with the full moon if she's exposed to enough moonlight. It's the pituitary gland does it."[51] Albeit elucidated scientifically, these hardly perceptible corporeal connections may bolster our sense of wonder at the material complexity and the welter of agentic entanglements at play in our earthly existences.

While human affairs mostly occupy the center stage in Kingsolver's novel, the overall vision systematically interlaces those with the fates of other-than-human lifeforms, all caught together in the same, "prodigal summer." This is hinted at also in the titles respectively heading the three interweaved narratives, that is, "Predators," "Moth Love," and "Old Chestnuts," drawing attention to the lives of animals, insects, and trees. As a matter of fact, nonhuman matters also appear in the limelight, as for instance those of "the bright golden Io moth" that Deanna observes in the first chapter, "hanging torpid on the window screen."[52] As it is, Kingsolver frequently works on what Lawrence Buell calls "the dignification of the overlooked."[53] She humbly suggests that human matters may compare in importance with those of tiny critters, each responding differently to circadian rhythms:

> The creature had finished its night of moth foraging or moth love and now, moved by the first warmth of morning, would look for a place to fold its wings and wait out the useless daylight hours. She watched it crawl slowly up the screen on furry yellow legs. It suddenly twitched, opening its wings to reveal the dark eyes on its underwings meant to startle predators, and then it flew off to some safer hideout.[54]

A similar attention is paid to tiny, vulnerable lacewings, whose life-drive is not without mirroring the attractions taking place on the human stage: "[. . .] a fresh hatch of lacewings seemed to be filling up the air between branches. [. . .] They were everywhere suddenly, dancing on sunbeams in the upper story, trembling with the brief, grave duty of their adulthood: to live for a day on sunlight and coitus. [. . . Their] new, winged silhouettes rose up like carnal fairies to the urgent search for mates, egg laying, and eternal life."[55]

In a way that might recall Giono's writing, the same vital fluxes urge all lifeforms forward throughout Kingsolver's novel. Perceptible in the initial bonding leading to intercourse between Deanna and Eddie Bondo, the forceful attraction between them acquires a form of agency beyond their control: "[There was] this new thing between them, their clasped hands, alive with nerve endings like some fresh animal born with its own volition, pulling them forward."[56] Whether human love or reproduction instinct, the life-drive Kingsolver evokes gets nearly fleshed out. Endowed with an agency of its own that will influence the course of both characters' lives, the vibrant magnetism between them simultaneously foreshadows Deanna's unexpected late pregnancy at the end of the novel. As if following a choreography worked out by nature and transcribed by the ecopoet, Deanna and Bondo soon surrender to the irresistible pull and tug of sexual attraction, "falling together like a pair of hawks."[57] Working against speciesism, courting rituals between humans and between animals are likened, engaging both in dances that are choreographed by a ubiquitous reproduction drive: "[Bondo and she]'d had their peculiar courtship: the display, the withdrawal, the dance of a three-day obsession."[58] Inviting us to reconsider human dwelling from a more humble stance, the behavior of Kingsolver's human animals is observed through the same lens as that used for other species, all of whom are pictured falling or rising in love, "yielding to earthly gravity," and responding to the same seasonal pulse.

Lusa's character is another scientific shaman. An entomologist specialized in moths, she turns to farming after marrying into a rural family. Claiming "moth love" as the object of her study, Lusa's insight into the world of moths vicariously cues her to be more sharply aware of the odorscapes in which she moves: "Lusa was alone, curled in an armchair and reading furtively [. . .] when the power of a fragrance stopped all her thoughts. [. . . She] was lifted out of her life. / She closed her eyes, turning her face to the open window and breathing deeply. Honeysuckle."[59] Even though a quarter of a mile away from where she is sitting, the plant manifests its existence across the distance, impressing itself on Lusa's bodymind via the olfactory signature it wafts in the air. The mere evocation of the familiar pungent smell might also suffice to intoxicate readers by proxy. The word can indeed act as a linguistic prompt mentally activating our corporeal memory, which we may experience again as a ghost smell. Thus, albeit purely through the magic power of language, we readers may experience the same, uplifting transport as Lusa, making us feel in *touch* with the plant through our noses. Let us note in passing that the very name "honeysuckle" speaks volumes of the sweet olfactory spell cast onto humans by the plant—a spell that must have played a great part in enticing humans along their travels to help the exotic plant colonize Appalachia and all those other parts of the world honeysuckle is not endemic to.[60] As with Powers' tree people, fragrance accounts for an enchanting form of interspecies communication, magically changing one's consciousness of our tentacular interconnections with an animate world extending far beyond the visible.

The sense of smell, Kingsolver drills, is one of the ways humans and critters respond to similar forms of stimuli: "[Lusa] told [Cole] about the scent cues animals use to find and identify their mates. Pheromones. [. . .] He was interested in moth love. More interested still when she explained to him that even humans seem to rely on certain pheromonal cues, though most have little inclination to know the details. Cole would, she thought. Cole, the man who buried his face in every fold of her skin to inhale her scent."[61] Reclaiming our animal natures, Kingsolver questions some of the culturally constructed notions of femininity that may seem aberrant in the light of biology: "'How come a woman will do everything humanly possible to cover up what she really smells like?'" asks Cole. Underlining the socially imposed rejection of our animality, Lusa's thoughts on the matter impel us to reconsider the notion that hair on a woman might be visually repellent, whereas in terms of smell, it happens to play in favor of attraction: "Even shaving armpits defeats the purpose. The whole point of pubic hair is to increase the surface area for scent molecules."[62]

Part of the glue cementing Lusa and Cole's couple despite their wildly different backgrounds lies in their shared appreciation of their earthly selves: "Cole made love like a farmer, which is not to say he was coarse. On the contrary, he had a fine intelligence for the physical that drove him toward earthly scents, seeking out with his furred mouth her soft, damp places [. . .]. Her body [. . .] became something new in the embrace of a man who judged breeding animals with his hands. He gave her to know what she'd never before understood: she was voluptuous."[63] Honoring Cole the farmer as "a scholar of his own kind,"[64] with a fine knowledge of earthly matters, Kingsolver in many ways writes an apology for our bodily natures. Deanna's proficiency as a tracker of large mammals makes her heed the part played by smell in interspecies communication. She diplomatically eschews soap altogether "because it [assaults] the noses of deer and other animals with the only human smell they [know], that of hunters—the scent of a predator."[65] Much of the physical attraction between Deanna and Bondo is also explained by the enthralling power of scent: "She was surprised [. . .] by her own acute physical response to his body held so offhandedly close to hers. She could smell the washed-wool scent of his damp hair and the skin above his collar. This dry ache she felt was deeper than hunger, more like thirst."[66] If Deanna is ostensibly sensitive to how beautiful Bondo *looks*, much of the overwhelming charm he casts upon her works through her nostrils: "That wool intoxication made her think once again of thirst, if she could name it something, but a thirst of eons that no one living could keep from reaching to slake, once water was at hand."[67]

In addition to heightening our perception of the multisensorial world we have evolved with, Kingsolver's fiction further points to the ecological applications that may be derived from unravelling the spell of the sensuous world. This is exemplified for instance by the biomimetic traps ingenuously developed by Lusa's father, who studies "the pheromones of codling moths, notorious pests of apple trees." Rather than using pesticides that destroy much more than the targeted pest, scientists, we are told, have "learned to fool the males into mating with scent-baited traps so their virgin brides might vainly cover the world's apples with empty, harmless eggs."[68] Following that direction, scientists are indeed learning from the way nature spontaneously banks on biomimesis, which has inspired a diplomatic kind of interspecies communication designed to lure one species into cooperating with another via the use of biofences.[69]

Biomimetic tricks and anamorphoses first emerged in the natural world. Hence the symbiosis between the pollinator bee and the orchid which Deanna ponders: "'The bee smells something sweet and goes inside and then he's trapped in there unless he can find the one door out. So he'll spread the pollen over the place where the flower wants it.'"[70] It might be remarked at this point that the symbiogenetic relationships of orchids and bees, or wasps, can depend in some cases on a biomimetic anamorphosis playing on olfactory *and* visual illusions at once.[71] Such orchids offer a case in point for a type of "becoming-animal" based on an anamorphic trick evolved by the flower, whereby it imitates both the pheromonal emissions and the shape of a female wasp.[72] The orchid cunningly baits male insects hoodwinked by the biosemiotic anamorphosis: seeking to fertilize their female fellows, the charmed critters pollinize the plant.

The art of interspecies anamorphosis then cannot be circumscribed to the realm of human *poiesis*. Thus, may we ponder with Cosima, as she watches her father reveal his photographic anamorphoses in his dark room, to what extent human *poiesis* may really differ from the spontaneous inventions of the greater-than-human world:

> We stood without talking and watched a gray image grow on the paper like some fungus with a mind of its own. I thought about the complex chemistry of vision, remembering from medical school the textbook diagrams of an image projected through the eyeball temporarily inscribed on the retina.
>
> "I never thought about how printing a photograph duplicates eyesight," I said. "It's the same exact process in slow motion." [. . .]
>
> "Probably there's no invention in the modern world," he said. "Just a good deal of elaboration on nature."[73]

Further than that, if we rely on the adaptive, symbiotic purpose of anamorphosis in nature, might it not be logically ventured that humans might have developed an anamorphic propensity in response to the forms of nature the better to evolve multispecies partnerships that increase the chances of survival of each? In any case, what is underscored throughout by Giono, Kingsolver, and the other writers in my corpus, is the fact that human lives are enmeshed with many of the same forces, rhythms, expressions, and equilibria also regulating our symbiotic planet, the intricacy and aliveness of which can be apprehended through its various feelscapes.

NOTES

1. My reading of literary aspects pertaining to the field of zoopoetics has no doubt been oriented in part by my ongoing discussions over the past few years with French Scholar Anne Simon, who has founded the field of zoopoetics in France. See for instance her most recent monograph, *Bête*, where she takes up some of the lectures held at the University of Perpignan.
2. Giono, *Joy*, 101–05.
3. Ibid, 101.
4. Ibid, 101.
5. Ibid, 103.
6. Milcent-Lawson, "Point de vue" and "Tournant animal." In her luminous studies of the zoopoetics at play in Giono's oeuvre, Sophie Milcent-Lawson also analyzes such shifts to animal points of view in other works by Giono.
7. Quoted in Milcent-Lawson, "Tournant animal," translation mine.
8. I am here implicitly referring to Thomas Nagel's seminal study, "What Is It Like to Be a Bat?" The philosopher reflects on the subjective character of experience which is tied to the specific bodymind of the subject wherein being and consciousness are located.
9. Giono, *Joy*, 404.
10. Abram, *Becoming*, 291.
11. Giono, *Joy*, 404.
12. Genette, *Figures III*.
13. Giono, *Joy*, 407–08.
14. Ibid, 409–10.
15. Ibid 409.
16. Faris, *Ordinary*, 43.
17. Ibid, 46–47.
18. Giono, *Joy*, 409.
19. Ibid, 410.
20. Ibid, 410–11.
21. Ibid, 411–12.
22. In *Le plaisir du texte*, Barthes defines jouissance as that extreme and intense feeling of faltering, of being totally shaken by a text, the reading of which sways readers from their usual sense of identity. Linked with the uncanny, texts that trigger jouissance momentarily dissolve the reader's sociocultural, constructed sense of self as they abide by a language, a code, and values that destabilize entrenched norms.

23. Perry, "Interview," 147. Part of what follows stems from an earlier paper, Meillon, "Measured."
24. Kingsolver, *Prodigal*, 62.
25. Leopold, *Almanach*, 140.
26. Kingsolver, *Prodigal*, 63.
27. Ibid, 63–64.
28. Ibid, 63.
29. Morizot, *Diplomates*.
30. Kingsolver, *Prodigal*, 441.
31. Ibid, 1.
32. Ibid, 1–3.
33. Ibid, 3.
34. Ibid, 7.
35. Ibid, 441.
36. Ibid, 441–42.
37. Ibid, 442.
38. Ibid, 443.
39. Ibid, 443.
40. I have considered elsewhere how Kingsolver's deliberate tendency to anthropomorphize other-than-human nature is carefully counterbalanced by an ecofeminist reclaiming of the animal nature of humans, both male and female; Meillon, "Measured."
41. Kingsolver, *Prodigal*, 1. As in much of Kingsolver's writing, one may detect the influence of Henry David Thoreau, namely here of his essay "Solitude," from *Walden*.
42. Kingsolver, *Animal*, 444.
43. Ibid, 117–18.
44. In *Animal Dreams*, the similarities between animal and human sentience induce a shifting of perspectives onto cockfighting. The practice is common in certain milieux where it is valued as a sport and an occasion for gambling. Cocks involved in those "games" are handled by their trainers until they die in agony of hemorrhage. In a way that may extend to other culturally admitted practices such as bullfighting, or certain types of slaughtering where animals are not spared undue suffering, Kingsolver's novel exposes how such practices can go on only as long as humans have no regard whatsoever for the plight of sentient animals.
45. Not to rehash what has been widely discussed by my predecessors, I am referring readers interested in the ties between Hölderlin's poetry, Heidegger's reflections on the latter's assertion that humans dwell poetically, and ecopoetics, to Jonathan Bate's pertinent chapter "What are Poets for?" in *Song*, 243–83.
46. In an extensive footnote, Abram refers to Deleuze and Guattari's elaboration on the notion of "becoming-animal." He compares his approach specifically with Deleuze's in the following way: "We share several aims, including a wish to undermine an array of unnoticed, other-worldly assumptions that structure a great deal of contemporary thought, and a consequent commitment to a kind of radical immanence—even to *materialism* (or what I call 'matter-realism') in a dramatically

reconceived sense of the term. My work also shares with him a keen resistance to whatever unnecessarily impedes the erotic creativity of matter. / Despite the commonality of some aims however, our strategies are drastically different. [. . .] As a phenomenologist, I am far too taken with lived experience—with the felt encounter between our sensate body and the animate earth—to suit his philosophical taste. As a metaphysician, Deleuze is far too given to the production of abstract concepts to suit mine." *Becoming*, 10.

47. Abram, *Becoming* 10.
48. Kingsolver, *Prodigal*, 93.
49. Ibid, 93.
50. "[She] knew some truth about human scent. She'd walked down city streets in Knoxville and turned men's heads, one after another, on the middle day of her cycle. They didn't know why, knew only that they wanted her. That was how pheromones seemed to work, in humans at least—nobody liked to talk about it." Ibid, 92.
51. Ibid, 93.
52. Ibid, 28.
53. Buell, *Environmental*, 184. In Kingsolver's climate change novel *Flight Behavior* much wonder is focused on the monarch butterfly. See Flys-Junquera, "Conversations."
54. Kingsolver, *Prodigal*, 28.
55. Ibid, 16.
56. Ibid, 20.
57. Ibid, 22.
58. Ibid, 26.
59. Ibid, 30.
60. Stefano Mancuso hypothesizes in that direction. Recalling the basic principle that nothing emerges in nature and lasts that does not somehow serve the purpose of evolution, Mancuso and Viola go so far as venturing the possibility that willful plants have developed flowers, fruits, fragrance, colors, and shapes meant to mesmerize and seduce humans into interspecies partnerships, whereby humans help them disseminate, thrive, and reproduce.
61. Kingsolver, *Prodigal*, 37.
62. Ibid, 38.
63. Ibid, 37.
64. Ibid, 36.
65. Ibid, 6.
66. Ibid, 21.
67. Ibid, 23.
68. Ibid, 37.
69. On biofences, see also Morizot, *Diplomates*, 117–22.
70. Kingsolver, *Prodigal*, 22.
71. This is the case of *Drakea elastica*, an orchid endemic of Western Australia.
72. It may be noted in passing that it is by way of this example that French philosophers Deleuze and Guattari introduce their famous concepts of the rhizome, of

deterritorialization and reterritorialization, and of becoming-animal in *Mille Plateaux*, 17–20.

73. Kingsolver, *Animal*, 73.

Chapter Eleven

Reweaving Word to World

Ecopoets as Instruments of the Sympoietic Song of the Earth

Having dealt at length with the various modes, poetic processes, and devices employed throughout my corpus to reenchant our rapport with the world on a phenomenological level, in this last chapter I would like to unravel aspects of the *ecopoiesis* under study which might corroborate Jonathan Bate's contention that "[poetry] is the song of the earth."[1] Taking up Heidegger's distinction between "world" and "earth," Bate argues:

> Ricœur's "world"—the abstract, disembodied zone of possibility—is a building inside the head. It is not synonymous with any actual dwelling-place upon the earth. [. . .] If "world" is, as Ricœur has it, a panoply of possible experiences and imaginings projected through the infinite potentiality of writing, then our world, our home, is not earth but language. And if writing is the archetypal place of severance—of alienation—from immediate situatedness, then how can it speak of the condition of ecological belonging? Heidegger replies with the other half of the paradox: there is a special kind of writing, called poetry, which has the peculiar power to speak 'earth.' Poetry is the song of the earth.[2]

Going back to both Hölderlin and Heidegger, Bate emphasizes the role of *ecopoiesis* as an antidote to the era of "enframing" and disenchantment ushered in by the advent of modern technology. "Poetry," he claims, "is our way of stepping outside the frame of the technological, of reawakening the momentary wonder of unconcealment. [. . .] What is distinctive about the way in which humankind inhabits the earth [. . .] is that we dwell poetically."[3]

Rebounding on Heidegger and Bate's shared notion of the ecopoet's vocation to "speak earth," and by doing so, to ground us within it, this chapter delves into the concrete ways in which, beyond the anamorphic reframing of our perception and conceptualization of the world, ecopoetic prose fiction can furthermore perform "a presencing, not [just] a representation, a form of being, not [just] of mapping."[4] To that purpose, I will first sift my corpus looking for confirmation of the notion of ecopoets as instruments playing the song of the earth. Exposing further the transdisciplinary theoretical framework subtending my take on an ecopoetics of reenchantment, I will approach *ecopoiesis* as a stringing together of physical and discursive matter, a reweaving of world to word. In his illuminating study, *Ecopoetics: The Language of Nature, the Nature of Language*, Scott Knickerbocker starts from the recollection of his thirteen-month-old's tentative speech as the little one strived to name the trees around him. This, Knickerbocker reflects, was his child's first engagement with "the old human habit of weaving world to word."[5] In the wake of Knickerbocker's deep and subtle exploration of the workings of *ecopoiesis*, I would say that liminal realism and *ecopoiesis* of reenchantment pave the way toward an answer to the scholar's starting question: "Is nature on the other side of language, or can language, despite its mediating function between the human and nonhuman, weave us to nature?"[6]

Native American thinker, activist, and poet Jack D. Forbes spells out an ecopoetic stance consonant with the texts I have called upon so far: "The Universe is our Holy Book/ The Earth our Genesis/ The Sky our sacred scroll/ The Animals our teachers/ The Mountains our prophets/ [. . .] The Waters our testaments/ The World our study."[7] Forbes calls onto the Mother Earth trope as a way to emphasize, on the one hand, how human life emerged from the biosphere and, on the other hand, how humans can never be weaned from the *oikos* sustaining them.[8] From an indigenous perspective, there can be no separation between humans and nature: "It is empirically obvious," Forbes writes, "that we are not only children, sucking at our earth-mother's breast all of our lives, but that we are also mixed with, and part of, that which Europeans choose to call the environment. *For us, truly, there are no 'surroundings.'*"[9] In his long poem, Forbes propounds that all earth beings together form an orchestra: "We and all the animals and/ living things/ We complete the world/ We are its skin/ its membranes [. . .]/ We are its flutes/ its drumheads/ We are its maracas/ its voices/ We are not alone,/ not separate. / If the world be a drum/ we are its taut skin/ vibrating/ with its messages."[10] Taking this proposition seriously, ecopoeticians can seek to read texts as musical scores calling the reader in to sound the humming, beats, and songs which are co-composed by the biophony, the geophony, and the anthropophony that together partake in producing the soundscapes of the world.

Starhawk's musician and witch character, Bird, precisely embodies the conception of humans as instruments of the land's wild music. As they can feel the pulse of the living world and magically ground themselves into the various elements and energies of the physical world, Starhawk's characters exhibit a sharp sensitivity to the many outputs coming from the earth—a sensitivity that accounts for the way her protagonists are constantly incorporated into their environment and, simultaneously, incorporating it. Let us look for instance at when Bird escapes into the open after years of forced seclusion and labor:

> Bird sniffed the air, all his senses suddenly so alive he ached all over, like the aching of salivary glands tasting food after long hunger. [. . .] Whatever might happen next, he would have had this moment to stand once again on the living earth, to feel her like a vibrating body under his feet, to breathe air and feel a wind on his face that had blown free and unobstructed over the Pacific, to smell the compound incense of leaves, dust and ocean salt, the tang of bay laurel and sage, to see living things in their soft colors, blue and green, and umber earth below. He wanted to cry but he didn't dare. Instead, he took a deep breath, drinking in the life that flooded back into his body.[11]

As Bird synesthetically revels in his renewed corporeal interlacing with the land, sight is here the last sense to be called upon. Presumably, it is because Bird has not succumbed to the "ocular hypertrophy" that characterizes us moderns that he has turned out particularly gifted for music. Consequently, his "pattern" for grounding is a musical one: "His pattern, he remembered, was nothing visual: it was a riff of music, four bars of an Irish reel he'd been picking out on the guitar that morning [. . .]. He hummed the tune to himself, and yes, he was there, at the level where he could *see* the lines of energy running through the metal on his wrist."[12] Bird is literally attuned to the earth, whose rhythms provide the source of his musical inspiration. He can feel the "vibrating body" of the earth, with "her life humming beneath him," and his capacity to "draw her strength into his body" depends on subtle sound waves, which he channels through his body as if through a gigantic ear, and which he can pick up and translate into music.

Throughout my corpus, the notion of a song of the world resurfaces with many allusions to the world's "humming"—a term that is particularly interesting in its polysemy, potentially referring to a form of wordless singing or to a droning, rumbling, or vibrating sound or sensation. As Murray Schafer underscores, while "touch is the most personal of all our senses [. . . , hearing] is a way of touching at a distance," specifically in the case of "low frequencies where sounds become vibration."[13] Whether produced by the vibration of vocal cords, in the case of human humming for instance, or by the circulation of energy, via the air or the earth for instance, "humming" onomatopoeically captures a physical vibration turning into sound waves characterized by a certain regularity. These sound waves are picked up by our ears—and potentially by our whole bodies—and deciphered by our brains as rhythmic sound patterns or melodies. What we perceive as the song of crickets or cicadas for instance is but the result of stridulation produced by vibrating bodies.

When tied with the human perception of biophony or geophony, the notion of "humming" may recall the language of trees relayed in Powers's novel. It may ring a bell too with the "humming" sensed by Mogey, in Pancake's novel, through which he can detect the presence of animals, and even of smaller and more static earthlings: "It's like a higher hum than the still things, trees and ground and rock, although I only call it hum because I don't got no other word for it. It's not something caught by ear. As I got older, I'd catch it off small creatures, too, and after I got to be a man, [. . .] I could catch it, just quieter, even off trees and dirt and stone."[14] If the narrator claims that the subtle "humming" he feels coming from the buck and other earthlings is "not something caught by ear," it might be wondered to what extent the fine human inner ear—which plays a part in our sense of hearing as well as in proprioception, enabling us to situate ourselves in space and keep our balance—could subliminally pick up vibrations in the air triggered by minute movement or breathing, or potentially magnetic fields, yet occurring on frequencies too low for full human awareness. At the same time, we may be reminded of people who are technically deaf, and who can yet feel music pouring through their whole bodies—something proving that there is more than just the ear involved in our physical response to sound.

The song of the earth, in other words, here corresponds to the vibrancy of matter as it is translated into sound waves by our sensing bodies. The land's music thus emerges from the entangling of forces, bodies, and elements engaged in a multispecies co-becoming. As Australian ecocritic and ecopoet Mark Tredinnick reminds us, taking his cue from David James Duncan, "'[the] physical universe as we now understand it' [. . .] is a symphony of forces woven through galaxies, unseen fields, synapses, subatomic particles and cells.' And each of us (writers) is 'steeped in and assailed' by that symphony of generative and defining forces, the hum of things here, 'even as we study it.'"[15]

Tapping into Julia Kristeva's notion of the semiotic and giving it an ecofeminist twist that is coherent with David Abram's theory of language, I have hypothesized the innate constitution, via our earliest experiences of dwelling, of an ecopoetic semiotic order that is responsive to and regulated by the *chora* of the earth. Relying on Kristeva's seminal contribution to our understanding of poetic language, I am interested in an ecopoetic semiotic order that precedes and slips through the symbolic, and which is regulated through our sensitive bodies rather than through our logical, representational experiences of language, communication, and dwelling. As Abram puts it, "[the] invisible shapes of smells, rhythms of cricket songs, and the movement of shadows all, in a sense, provide the subtle body of our thoughts."[16] Kristeva posits a semiotic modality of signification tied to the pregrammatical language infants respond to and speak before they can consciously separate from the mother's body. According to Kristeva, this pregrammatical language can still be traced through the symbolic language that we are then taught to use as a structured mode of communication. A musical, dancing part of language, the semiotic *chora* percolates in the textured, rhythmic qualities of speech, where it keeps on transcribing somatic and affective responses to movement and voice which are not first and foremost mediated by the logos.[17]

Building on Kristeva, the ecopoetic qualities of the texts in my corpus bring me to ponder how, resisting the thetic phase through which the break between subject and object, between signified and signifier, and between conscious individual and live *oikos* is achieved, language might be invested as an umbilical cord keeping us tied to the subtle rhythms and many voices of our earthly matrix. "Human language," Abram insists, "arose not only as a means of attunement between persons, but also between ourselves and the animate landscape."[18] Expressive of the fact that the earth has never ceased to cradle and rock the human experience of dwelling, ecopoetic language, in its material reweaving of word to world, may to a certain extent resist the disarticulation between humans and nature which modern patriarchal systems have initiated. From this standpoint, it could be that *ecopoiesis* helps humans partly withstand alienation by keeping them connected to the *chora* of Mother Earth, following the rhythms and movements of her polyphonic being and multispecies body.

The semiotic dimension of language may subsequently account for the co-composing of a resistant song of the earth, forever in the making. In *Refuge*, Utah nature writer Terry Tempest Williams pens such a passage corroborating the ecopoet's capacity to plug into the *chora* of Mother Earth and let it flow through her own writing:

> The heartbeats that I felt in the womb—two heartbeats, at once, my mother's and my own—are heartbeats of the land. All of life drums and beats, at once, sustaining a rhythm audible only to the spirit. I can drum my heartbeat back into the Earth, beating, hearts beating, my hands on the Earth—like a ruffled grouse on a log, beating, hearts beating—like a bittern in the marsh, beating, hearts beating. My hands on the Earth beating, hearts beating. I drum back my return.[19]

At that point in her essay, Williams is wondering how to "correspond with the land when paper and ink won't do," and how "to empathize with the Earth when so much is ravaging her."[20] Abiding by the somatic logic of the musical body, and connecting the beats of her own prose to the sustaining beating of land, Williams here engages in a rhythmic translation that sparks a conversation with the earth.

According to the hypothesized capacity for the ecopoet to "speak earth," *ecopoiesis* emerges not simply from our cognitive responses to what is, but from the somatic locus of the body. Speaking earth is a way of feeling earth—a way of being in the world where meaning comes from sensuous co-becoming. The melodic and phonetic texture of language, its motile body, thus orchestrates a grammar of the sensitive that is heard and felt through language more than it is conceptualized. Because, as Abram puts it, "each terrain, each ecology, seems to have its own particular intelligence, its unique vernacular of soil and leaf and sky," human language, specifically in its most vernacular forms, must then translate the specific songs of the earth which can be heard in different places. In *Refuge* for instance, Terry Tempest Williams composes an *ecopoiesis* that is directly inspired by the song of Great Salt Lake City, with "[wind] and waves [. . .] like African drums driving the rhythm home." The remarkably repetitive, alliterative quality of the prose (in [w], [dr] and [r], and [ð]) here and elsewhere takes up the beating and drumming of wind, lake, and land: "Wind and waves. A sigh and a surge."[21]

Reclaiming this unbreakable bond with the *chora* of Mother Earth, the ecopoets in my corpus defy and partly undo the alienation from the natural world that has been imposed by a patriarchal symbolic order, language, ideology, and system.[22] The ecofeminist stance inherent to such a reviving of one's corporeal connections with the earth is made clear by Williams, as she reinscribes an Earth Goddess into her Mormon version of the Holy Trinity: "I believe the Holy Ghost is female, although she has remained hidden, invisible, deprived of a body, she is the spirit that seeps into our hearts and directs us to the well. The 'still, small voice' I was taught to listen to as a child was 'The gift of the Holy Ghost.' Today I choose to recognize this presence as holy intuition, the gift of the Mother." Tempest suggests that through "the Motherbody," one may hear the voice of the earth, and thus find "a spiritual counterpart to the Godhead." As a result, she conjectures, "perhaps our inspiration and devotion would no longer be directed at the stars, but our worship could return to the Earth."[23]

As a matter of fact, Williams's enthralling going along with the pounding of the wind and waves of Great Salt Lake, by dint of which she invites the drumming of the land into her own prose, is described as an uplifting, mystical experience: "I am spun, supported, and possessed by the spirit who dwells here. Great Salt Lake is a spiritual magnet that will not let me go. Dogma doesn't hold me. Wildness does. A spiral of emotion. It is ecstasy without adrenaline."[24] The spiritual dimension is redoubled with a physical sense of communion, sensuously enjoyed and described as a form of lovemaking with the land:

> My hair is tossed, curls are blown across my face and eyes, much like the whitecaps cresting over waves.
>
> Wind and waves. Wind and waves. The smell of brine is burning in my lungs. I can taste it on my lips. I want more brine, more salt. Wet hands. I lick my fingers until I am sucking them dry. I close my eyes. The smell and taste combined reminds me of making love in the Basin; flesh slippery with sweat in the heat of the desert. Wind and waves. A sigh and a surge.[25]

Language, with its sensuous textures, its exquisite alliterations and assonances in [b], [w], [k], [l], [ɜːr], [eɪ], and [aɪ], here offers an ecoerotic way to relive and revel in an ecopoetic jouissance the cresting of which is experienced in a trance-like state of enchantment, combining physical and spiritual senses of continuum and communion with the earth.

Rehabilitating the poetry inherent in Appalachian vernacular, the lyrical ending of Ann Pancake's novel provides solid evidence for how ecofeminist poetics can resist the hegemony of patriarchal systems and their alienating structures of thought and language. As touched upon before, to overcome the dislocation of their lives brought about by MTR, Bant, and her feisty mother, Lace, choose to continue to inhabit their ravaged mountains. Despite the various forms of discouragement, pressure, and intimidation which they have borne the brunt of, the two women simply will not be removed, both literally and affectively speaking, from the mountains where they were born and raised. At the end, Bant's synesthetic, mellifluous, and polyphonic narrative becomes literally enchanting as it turns into an incantation. Keeping close to the ground, Bant invokes many endemic plants, mostly the yellowroot and ginseng, which, like herself—like the language that she speaks and like the generations that have come before her—are literally rooted in the mountain:

> Then I was moving the way I used to in the woods, before the distance came between me and it, the way I moved in woods and woods only. All the clumsy I felt around people, and buildings and pavement and flat, it used to fall away from me in the woods, and it fell away now. I could feel what was nearby, its closeness, its give, beech, poplar, oak, holly hickory hemlock laurel, touching nothing, tripping nowhere, what Mogey always said about the hum. October smell in my head, and me and Grandma, sassafras and pawpaw and beechnut, like sunflower seeds under your feet, I pushed myself harder, *don't care*, and yellowroot, too, after the sap went down *Some folks, they use yellowroot for just about anything*. They brought me up here before I was born, and I do remember, smell of November wetter-leaved than October smell *don't care* [. . .].[26]

Whether the people's lifestyle, their vernacular Appalachian English, or the food ensuring their subsistence, all of it has grown out of one and the same homeplace, nested in the mountain forests. The asyndeton in the list of trees called upon by Bant ("beech, poplar, oak, holly hickory hemlock laurel") frees the signifiers from the constraints of conventional linguistics, drawing attention to the material presence on the page of the unruly, resistant words. As the required commas between tree names gradually disappear, the undisciplined punctuation and syntax create a sustained staccato rhythm evoking the throbbing heart of the wild forest ("beech, poplar, oak, holly hickory hemlock laurel, touching nothing, tripping nowhere"). Simultaneously, the alliterations in [h], [k], and [l] ("poplar, oak, holly hickory hemlock laurel") goad readers to engage their phonatory organs into a form of panting by dint of which we may experience the suffocation of the mountain. The brisk rhythm of the sentence translates a sense of urgency as the mountain's breathing organs—the trees—keep getting decimated. The reading thus takes us through the gasping of the agonizing mountain forest. And yet, through this struggling respiration, one might also get a feel of the stamina and resilience of a whole ecosystem fighting to regain a more measured pace and balance—as expressed by the parallel construction at the end, "touching nothing, tripping nowhere." With this ecofeminist tour de force, Pancake reclaims the poetic power of vernacular language.

Bant's rebellion against modern, urban power and worldviews moreover shows in her emancipation from the rules of prescriptive grammar—whether those dictating the use of punctuation and conjunctions for the coordination of nouns, as above, or those regulating the formation of compound adjectives and past participles ("smell of November wetter-leaved than October smell"). Pancake here resorts to the stream of consciousness technique allowing her to translate the many different thoughts, memories, voices, and affects surging all at once in Bant's frantic bodymind. Beyond the demise of her grandmother, and of some the native plants, the invocation of "sassafras and pawpaw and beechnut" draws attention to the delectable quality of the indigenous plant names, which evoke the tasty substance of the actual things, somewhat preserved here via sensuous *poiesis*. Simultaneously, the accumulative polysyndeton ("and . . . and") conveys a sense of abundance—there used to be plenty—and resistance. The grandmother and plants might be gone, yet their songs and voices shall be remembered and persist.

Through the semiotic and sensuous qualities of her ecopoetic language, Pancake here strikes some of the affective and musical chords deep within us, chords relating us to an endangered world. The song of Gaia is heard at the end through the song of ginseng: "Grandma showing me to chew for the good taste in it *Now that was our chewing gum,* me carrying my little sack, head weed-high, *Close to the ground like you are, Bant, you can see stuff bettern me. Red berries is what you're looking for.* Senging. Sanging. Sing, sang, sung. *You can live off these mountains. Put you in a little garden, and you can live off these here.*"²⁷ While yellowroot gave its evocative name to one of the nearby mountains in the novel, the name ginseng comes from Chinese, where it means "human-root" ("rénshēn"; "rén," meaning "man"; and "shēn," meaning "root")—an anamorphic name the plant has originally earned due to the anthropomorphic shape the root often takes. In the characters' earlier lives, both plants used to be valued, gathered with care, used in herbal medicine for their properties as immune system boosters, and sold to make a humble living, thus adding to the manifold entanglements between humans and ginseng. Functioning as synecdoches for the mountains they have emerged from, these native plants both literally and metaphorically embody the vitality of an entire bioregional community. Their tragic disappearance from the mountains in the wake of MTR consequently symbolizes the potential extinction of the life of a whole biocentric community, the health of which is depicted throughout the novel as being grounded in humility, in relationships of respect and interdependence with humus—in other words, in human response-ability leading to ecological responsibility.

Relying on the diffractive potential inherent in poetic echoes of the earth, Pancake uses Bant's voice to will the roots of a whole naturculture into a song of survivance.²⁸ Toward the end of Bant's recalcitrant narrative, ginseng gets truncated from its human component and is referred to simply as "sang," thereby taking up the colloquial idiom for the plant. But then, the natural root ("sang") gets chanted into the conjugation of the irregular verb "to sing": "Sing, sang, sung." Gaining poetic ground and agency, the noun ginseng turns into a shapeshifting gerund ("Senging. Sanging"), giving birth to paronomasia and poetic neologisms sounding the song of ginseng. The ecopoetic neologisms evoke all at once the activities of gathering ginseng, of re-rooting oneself, of singing, and of ecopoetic regeneration. Just as Bant's moving through the mountains, earlier in the novel, seems an echolocated response to the woods, co-choreographed by her and the land, with "the curve and dip of the ground echo-shaping the curve and shape of [her] body,"²⁹ Pancake's *ecopoiesis* here arises from the ground up. It is co-produced by the resisting vitality of the many lifeforms and voices of the earth, here more specifically those of Appalachia in West Virginia. The plants are called upon to bring new vitality to the land and its dwellers, to stand up for and reenchant mountains that have been silenced, now implicated in Pancake's song of resistance. Pancake's writing thus exhibits awareness of the part played by *ecopoiesis* in preserving the life and healthy balance of a place. Taking up British explorer Bruce Chatwin's observations of the colonial dislocation processes at work in Australia, it might be said that Pancake shows awareness that "an unsung land is a dead land: since, if the songs are forgotten, the land itself will die. To allow that to happen [is] the worst possible of crimes."³⁰

In Starhawk's novel, it is precisely Bird's ingrained connection to the earth's music that redeems him, awaking him from a prolonged period of amnesia during which he has been tortured and drugged:

> His hands on the broom handle seemed stiff and clumsy, the fingers somehow misshapen, as if they had been broken and not set right. That disturbed him in some way, almost more than anything else, as if it represented the loss of something so basic that he had to protect himself of the memory. It teased at the back of his mind, though, like liquid notes of music; like rippling melodies flowing off the strings of his guitar. And then it hit him, with a force almost physical that left him sweating and clutching the broom handle for balance. He could remember his fingers, deft and fluid, not so much making music as matching what already existed and poured through him, his hands one with his instrument and the great singing voice inside him.[31]

Like ecopoets, Bird makes himself an instrument of the song of the earth. To him, music is the expressive texture of an invisible yet tangible aspect of reality ("like liquid notes of music; like rippling melodies flowing off the strings of his guitar")—a material energetic flow which he gives himself over to, becoming a mediator for it, just like his guitar. The passage is marked by alliteration in [l] ("like liquid," like rippling," "melodies flowing"), combined with the anaphoric repetition of "like," with internal rhymes in <ing>, assonance throughout in [ɪ], and eye-rhymes in <i>. Together with the onomatopoeic quality of dynamic process verbs such as "rippling" or "flowing off," these poetic qualities confer a material thickness to the prose which fleshes out the musical texture of language. To a certain extent, readers themselves are cued to respond physically to the synesthetic call of the earth thus mediated by the ecopoet. Like Bird, we are likely to find ourselves yielding to the musicality of language, "hit [. . .] with a force almost physical," with its own "rippling" and "flowing off," and rocked by the tonic stress patterns and the alliterative, assonantal, and incantatory aspects of Starhawk's prose.

The implied ecopoet emerges from such lyrical passages, striking the physical chords of language. In his luminous study of contemporary French writers such as Yves Bonnefoy, Philippe Jaccottet, and Jacques Réda, essayist Jean-Claude Pinson draws from Heiddeger, Merleau-Ponty, Ricœur, and Rilke to shed light on *ecopoiesis* as an embodied response to dwelling which intimates that "the world, ordered like a cosmos, abides by a measure rendering it apt to change into a song."[32] Just like Bate's in the Anglophone world, Pinson's work is seminal in France. Although Pinson did not yet claim the term "ecopoetics," his 1995 monograph on *Dwelling as a Poet* has nonetheless established how part of French poetry and prose aims at "reattuning language" to the world.[33] Already Pinson underscored the materiality of words, tied as it is to our own corporeality. "By calling upon the stuff of words, by playing with rhythms and assonances, one reawakens via poetry that part of words which they owe to sedimented gestures and breaths. Language is thus driven back to its corporeal dimension."[34] The lyrical dimension of language, Pinson insists, plays an essential role in returning us, via language, to the "sensitive thickness" that is characteristic of our fundamentally embodied experience of dwelling—a dimension that concepts, and what Pinson decries as "logolatry," can deprive us of. This provides an inkling of that double dimension of *ecopoiesis* which may help us *both think and sing*, or, to take up Yves-Charles Grandjeat's felicitous phrase, "whose thinking is done by singing."[35] Foregrounding language's concrete, musical participation into the world's *physis*, these matters tie in with the concept of "sensuous poesis" which Knickerbocker has expounded on, that is "the process of rematerializing language specifically as a response to nonhuman nature."[36]

Relying on excerpts where the poetic prose rematerializes language as part of the earth's soundscape, I argue that *ecopoiesis* effectively reen-*chants* the world as it calls us into the presence of the vibrant world it articulates with. Taking up Charles Bernstein's notion of sound as being "language's flesh," it can be argued that, "[in] sounding language we ground ourselves as sentient, material beings, obtruding into the world with the same obdurate thingness as rocks or soil or flesh."[37] As Knickerbocker underlines, poetic arrangements of sound effects entangle the ecopoet, the poem, the reader, and even the ecopoetician in the experiencing of the flesh of language, itself a continuation of the flesh of the world: "When a poet skillfully experiments with various sound effects in a poem, when a reader revels in the sensuous pleasure a poem provides, and when a literary critic deepens our understanding or appreciation of the way a poem's form—its body—shapes meaning, these writers and readers experience both their own and the poem's embodiment, even when silently sounding the poem in one's inner ear."[38] Insisting that reading affects us beyond our cognitive capacity to form representations and activate our conceptual minds, that in fact, reading speaks directly to our bodies thereby triggering immediate, somatic responses, Knickerbocker brings in the reception theory advanced by his fellow scholar Garret Stewart. Shedding light on the implication of the "reading body," Stewart writes: "This somatic locus of soundless reception includes of course the brain but must be said to encompass as well the organs of vocal production, from the diaphragm up through throat to tongue and palate [. . . The] body is the site of silent reading, subtending all conception."[39]

In a similar vein, during a ceremony at the beginning of Starhawk's novel, Maya the oratress betrays the implied writer's skillful attention to the magic spell of language, making hearers and readers physically aware of some of the invisible, yet tangible aspects of the world: "The drums began to beat, a trance rhythm, steady but slightly syncopated, to lead the mind and then shift it in unexpected directions. Maya spoke, her voice rhythmic, musical, crooning an incantation."[40] Once again, readers are called to respond to the semiotic order giving the sentences their measure. The opening iambic rhythm ("The drums began to beat") works the magic conjointly with alliterations in [b] and [t] to sound the beating of drums, while the rest of the unfolding sentences pound out tonic stresses and alliterations performing the "steady but slightly syncopated" rhythm of Maya's enchanting voice, "crooning an incantation." This may recall Bate's definition of *ecopoiesis*, as "language's most direct path of return to the oikos," with rhythmic qualities answering those of nature and thus echoing the song of the earth.[41]

At the end of Starhawk's novel, as Bird decides he would rather get executed rather than become an instrument of war, the internal focalization draws a parallel between his musical art and *ecopoiesis*:

> He thought he'd lost the music that was in him, but now it worked his lips and pried his mouth open and forced its way out of him in croaks and gasps. [. . .] That was all he had to do, to sing—to his grandmother and his lover and his enemies and his executioners. He had found his ground to stand on, and, yes, there was a bottom place, a place where who he was and what he could not do was stronger even than hope. He understood now that he could never lose the music. It grew in him as the silence grew around him. They had broken his hands, but they had not broken his voice, they had broken his will, but they had not broken his ears, and if they took his ears they could never take the inner ear, the inner voice. And even when his voice was silenced, some voice would still continue to sing. For he realized now that he was wrong in thinking the music was in him.
>
> He was in the music, and it would always find an instrument.[42]

Translating the land's wild music into *ecopoiesis*, Starhawk suggests, ecopoets act as mediators for the polyphonic song of a resilient symbiotic earth which we are always immersed in. Besides, it might be worth pondering that our inner ear determines both our sense of hearing and our sense of balance. Considered by some as our sixth sense, proprioception depends in great part on the inner ear which affects our psychomotor capacity to locate ourselves, to find verticality and equilibrium when moving through space while negotiating gravity, and to locate the sources of the various stimuli we are constantly assailed with. Quite astonishingly, research into the proprioceptive capacities of blind people has evidenced that, albeit less developed than in bats, humans can in fact practice a certain amount of echolocation in their negotiation of space and movement. This is particularly the case for sightless individuals, some of whom have developed the extraordinary ability to echolocate via palate clicking.[43] From there, it could be hypothesized that *ecopoiesis* might be a poeming of the world which is produced by subjects whose auditory and proprioceptive hypersensitivity has allowed them to develop a form of echolocation, which they then turn into a form of poetic echo-vocalization.

According to Joachim-Ernst Berendt, as goes the title of his book, "the world is sound." Studying "the formation of chaotic rhythms"[44] within "the harmonic structure of the universe,"[45] Berendt argues that "the tendency toward harmony, immanent in music, in a way is nothing else but a reflection of the same tendency outside of music, in almost all fields."[46] It is a tendency "of everything that vibrates [. . .], a tendency of the universe to share rhythms, that is, to vibrate in harmony."[47] As Berendt demonstrates, the human ear, body, and mind are physically predisposed to follow the harmonies of the cosmos, through the laws of resonance.[48] Relying on musicology and physics, Berendt seeks support for his intuition that the ancient notion of a *harmonia mundi* matters far more importantly than as a mere trope or myth. Hence the endorsement of his work by Austrian-American physicist Fritjof Capra, who, in the preface, lends credit to a non-mechanistic, holistic vision of the world as an ongoing choreography, or, for that matter, as music: "Matter, at a subatomic level, consists of energy patterns continually changing into one another—a continuous dance of energy. To unify recent insights in physics and in the life sciences into a coherent description of reality," he argues, "a conceptual shift from structure to rhythm seems to be very useful. Rhythmic patterns appear throughout the universe, from the very small to the very large."[49] Backing up Berendt's adamant call to reawaken to our auditory intelligence, Capra insists:

> It has been said since ancient times that the nature of reality is much closer to music than to a machine, and this is confirmed by many discoveries in modern science. The essence of melody does not lie in its notes; it lies in the relationship between the notes, in the intervals, frequencies and rhythms. When a string is set vibrating we hear not only a single tone but also its overtones–an entire scale is sounded. Thus each note involves all the others, according to current ideas in particle physics.[50]

It is tempting to draw the tentative conclusion that, arising from contact with everything around, *ecopoiesis*—whether in the form of music, prose, or poetry—reverberates and diffracts the sound waves emitted by vibrant matter both within and without. Giving it musical or linguistic shape, *ecopoiesis* makes the song of the world more accessible to the less attuned ear, the less alert body, or the less contemplative mind. Bearing in mind the way birds nestle their acoustic signatures, or dwelling songs, into the niches available to them in the surrounding soundscape, *ecopoiesis* might be envisioned as a form of singing of one's way into the ongoing symphony of the world, arising from a practice that conjugates forms of echolocation and echo-vocalization. From that perspective then, the notion of a *harmonia mundi* could be envisioned as referring to the emergence of the polyphonic song of the earth formed by the various harmonies and voices occupying different, complementary acoustic niches. As demonstrated by Bernie Krause, the earthly orchestra indeed includes human and other than-human instruments, altogether playing a score potentially involving anthropophony, geophony, and biophony all at once.[51]

Also referring to recent insights into particle physics the better to defend that "the world is music," Mark Tredinnick's eponymous study on "the land's wild music" sheds light on how the latter resonates through nature writing. Tredinnick elaborates on the lyrical dimension of nature writing. He contends that "[these writers] are taking part in the land, listening and being changed. [. . .] The rhythm we readers hear is an articulated listening—the reverberation of an intelligence to the life of a place. Their work," Tredinnick goes on, "is not dislocated from the land, but a kind of participation and a continuation of it."[52] However, Tredinnick warns, "[the] song of the text will never *be* the song of the earth. Yet perhaps it may echo it, ring true to it, always in a human voice, always shaped by letters on a page, while letting the fragmented music of the real world sound on, out of the poet's ultimate hearing."[53]

Now the task that lies ahead for ecopoeticians is to reveal the technical ways in which, if we are to follow Tredinnick's hunch, "[the] work of art [. . . .] may [. . .] sing in tune with the place from which it arises, to which it looks, whose soul it shares, whose chant it is a response to—call and response, call and response, making one song."[54] As Yves-Charles Grandjeat cogently propounds in his paper on the "poetic-co-operation" between writer and earth at work in nature writing, ecopoeticians may approach *"the poetic statement as a ritualistic offering of material and psychic energy* [. . .] taking 'place' in the ecosystem at large." Mobilizing a particularly evocative simile, Grandjeat draws the conclusion that "the nature writers [. . .], calling on powerful images to release their songs from the clutches of the self-centered, anthropocentric ego, then striking them like so many meditation bowls, and letting them fill up with the dense substance and energetic flow of sound, can turn them into ritualistic channels for poetic co-operation between [humans] and the world."[55] While grasping the musical and affective resonance at play in *ecopoiesis*, Grandjeat's compelling image of the striking of a meditation bowl moreover speaks volumes in terms of the possibility for a secular spiritual reenchantment that adheres to a new materialistic onto-epistemology. Resisting the complex alienation process resulting from Modernity and Postmodernity, *ecopoiesis* foregrounds language as one of the humble media that can reinitiate humans into a deep sense of belonging to, and living-dancing-singing in harmony with, the wider earth.

Already, in its most Dionysian scene, Giono's *Joy* staged the enrollment of humans in the great orchestra of the earth. Loosened up with the wine flowing freely at the table and coursing through their veins, the group assembled for a banquet start vibrating collectively from within "as if from the distant woods and forests came the dull, monotonous thumping of a dance drum at whose rhythm they would soon be forced to turn and dance."[56] Tuning in the pulse of the earth—potentially an echo of the sap tides pulsing through the trees all around them—Marthe "[hears] the rumbling of the drum that [filters] louder from the woods, the forests, the trees, the grass, and one might even say from the earth itself." The polysyndeton here expresses the rhythmic coordination running through the forest, enfolding even humans within the cosmic, seasonal cadence of "the earth itself," and thus abolishing the distance between humans and nature: "It seemed as though one could feel its throbs in the ground, here beneath one's feet, here under the table, like the violent throbbing in the arteries of men inflamed."[57] Soon, this shared cadence of a great terrestrial dance and song ripples into Giono's writing, rewilding his prose and freeing it from the habitual syntactical norms meant to reign in the composition of classically measured sentences. Relaying the tipsiness felt by the characters, Giono's drunken prose starts jamming, composing an assonantal and alliterative polysyndeton, the tonic stress pattern of which translates the "throbbing" and "pulsating [. . .] sound of [their blood]: "[They] heard the rumbling and the drum beats and the cadence and the wild rhythm of the dance drum."[58] As each of the characters sensitively flock to one and the same contagious and quickening tempo, the musicality of the world meanwhile percolates into the prose, filling it with poetic echoes of the land's wild music. Giono's prose itself sounds the "*tum, tum*" of the "savage drum"[59]: "And the blood beat dully, boom, boom, boom, like the savage dance of all the earth."[60]

Finely attuned to the earth as they are in that moment, the characters come close to fully embracing their own wildness, which is catalyzed here through their blood pulsing in synchrony with the earth as if rolling along with a call-and-response song. As if it were a metronome or a drum set, the pace is set by the earth, driving the tempo of all human and other-than-human activity:

> But, above all, there was that drumming of the blood, the rumbling of the blood. It thumped in both men and women upon a deep-sounding drum. [. . .] Each felt bound to this cadence. It was like the blades of a threshing machine. It was like the flail that strikes the grain, flies back. It was like the travail of the man leaping in the wine vat. It was like the steady gallop of a horse. [. . .] Always, always, without ceasing, because blood does not cease to beat, to explore, to gallop, and to demand with its black blood to join the dance.[61]

Marked by repetition—with many alliterations in [bl], [d] ("drumming," "blood," "rumbling," "blood," "deep-sounding drum," "blades," "blood," "black blood") and [l] throughout, the recurring [ʌm] syllable which translates the "rumbling" of the "drum" rippling through Giono's prose, the anaphoric repetition of "It was like" at the beginning of four consecutive sentences, and with a few other instances of internal rhymes, assonants, and paronomasia ("flail/flies," "flail/travail," "like/strikes/flies/like/wine/like," "back/vat")—this passage offers both a response and a call to wildness. Whereas the characters hold back, incapable as they are of letting go of their usual restraint and composed social postures, Giono's writing drums a beat calling for "the true dance"—that which turns "the calls of the body" into "calls of joy."[62] This harmonious dance is one whereby reason is measured through blood: "Blood knows what's right, after all."[63]

Despite the call they can all feel resonating through their bodies, the characters remain at that stage inhibited by their learned sense of propriety. They are torn between rational thinking, and embodied knowledge—what they have been taught in a modern context, and what they know *by nature*, from within. At the end of the day, despite their momentary access of euphoria and collective effervescence, Giono's characters stick to their trained allegiance to a sociocultural superego that clips their natural inclinations. Their ecopoetic flight along with the earth's wild music is impeded, as they remain unable to yield to the full to the momentary enchantment:

> It calls and one does not dare to respond. And it calls and one does not know if one must . . . and one's body is filled with desire and one suffers. One does not know and one knows. Yes, vaguely, one realizes that it would be good, that the world would be beautiful, that it would be paradise, happiness for everyone and joy. To be guided by one's blood, let one's self be beaten, explored, let one's self be carried away by the galloping of one's blood [. . .]. And one would hear galloping, galloping, beating, beating, exploring, exploring, and the thundering drum beneath the great black plain of the pulsating blood.[64]

The characters' ambivalence betrays their fear of letting go and experiencing a dissolution of the boundaries of the self into the wider vibrancy of the world. This uplifting passage is not without evoking a kind of mystical, yet earthly—because grounded in the earth—trance. Taking place still early in the book, it ends with a dysphoric note turning it into an aborted moment of joy. Nevertheless, the scene offers a glimpse of the enchanting epiphany that will trickle down later into the plot. In addition, if readers may feel let down by the characters' silencing of their trues selves at that point, the frustration caused by their resistance to their ecological unconscious is nonetheless overridden by the ecopoetic jouissance triggered by Giono's lyrical prose.

Besides, an earlier passage points to the transcending power of art, helping one process even the negative feelings that dwelling will inevitably bring. Before getting his humans to learn such wisdom, Giono first casts the wild dance of the buck, made doleful by the human world: "Then the stag danced for himself. He was on the barren heath. He felt sad as he thought of the horse. He lifted his feet one after another. He lowered his head, he raised it again. He sneezed. He was sad. The naked land, the spring-time, the absence of females, the horse, the old man, the young man who was irrigating."[65] Through dancing, a great transformation occurs:

> He danced the dance of the old man, he danced the dance of the young man with gentle eyes. He danced the dance of the unhappy horse and the sad deer. He danced the moor. He danced his springtime desire. He danced the mist and the sky. He danced all the odours and all he saw, and all that touched his eyes, his ears, his nostrils, and his skin. He danced the world that thus entered his being. He danced what he would have danced if he had been happy. And he became happy.[66]

While this animal sequence certainly involves anthropomorphizing, Giono thus encourages a practice of ecopoetic co-becoming. Through the anaphoric repetition of "He danced the world," and "He danced," working into the prose a type of meditative scansion, Giono gestures to a kind of ecopoetic trance, or ceremony, with the enchanting power to transmute frustration into desire, alienation into participation, and melancholia into joy.

Furthermore, there are three key passages later in the novel where characters finally do embrace the light glimpsed together in that first intense moment of revelation. The characters eventually give in to their animal natures as they modulate human forms of *poiesis* in cadence with the summer pulse of the earth. They act like the birds sitting at earth level, in nests that are "hidden in the grass," rocked by the "savage" music playing inside the trees: "[The birds] listened to the sound of the big roots as they awoke, and the flow of the sap in the stems, and the hissing of the sap as it spread in its leaves, and the lapping of the sap which came out in bubbles from all the cracks in the bud."[67] Meanwhile, also responding to the pulse of life coursing through all, with the wind blowing through a "fragment of cloud [. . .], successively filling the barn with shadow and sunlight, then more shadow and sunlight," Bobi yields to his summer-driven desire and leaps into a series of somersaults, "the barn floor [. . .] rumbling under him like the skin of a drum."[68] Thus taking part in the dance and song of the earth, Bobi feels "intoxicated" and "happy," standing under "the gentle blue sky sweeping silently over the earth like the wing of a bird."[69] No need for wine here to feel uplifted by a euphoric feeling of belonging and enchantment. Poetically engaging the rhythms of the earth, its summer saults, suffices to work the magic.

Likewise exerting themselves at the cedar loom the men have built for Marthe, the women seem to become one with their instrument, moving rhythmically, their weaving composing its own dance and song: "The handles of the warp rods clicked. The shuttle hissed as it left the right hand, struck the left hand; hissed upon leaving it, struck the right hand; hissed, struck; with a very noble, very slow rhythm of force and care."[70] Giono glorifies the old craft of weaving as an ecopoetic practice. Marthe's focused, measured, and hand-powered art is elevated half-way between a dance and a song, or a becoming-bird: "She advanced and withdrew in cadence. She was like a great bird dancing on one spot and fanning its breast."[71] When Barbe in turn stands at the loom, she plays it differently, marking her own, personal style: "The sound was swift and clear. Barbe made scarcely any gestures. [. . .] The hand did not shut and the shuttle flew all alone toward the left palm, like a bird that alights and then darts away."[72] While onlookers stare at her, mesmerized, Barbe starts singing to the music of her own weaving: "'Love joy, love joy,' then the clacking sound of the rods of the shuttle, of the bar, the dull trembling of the uprights, then: / 'Love joy, love joy!'"[73]

Lastly, there is one crucial scene where anthropic activity is elevated into a form of *ecopoeisis*, when Randoulet's skill at harvesting induces a collective sense of grace:

> If you have ever seen a swallow, in a swift plunge, brush his breast over the water, and fly up and swoop and rise, without ever wetting the tip of a wing . . . [. . .] It was really something beautiful. [. . .] It seemed as if Randoulet sensed [the stones] in advance. His scythe wavered, never twice the same rhythm. It was a work of taking one stalk at a time, slow and precise. Randoulet's muscles were all constantly in full play with waiting or action. He swung the scythe, held it back, made it pass flat over the stones, plunged the tip of the point, raised it, swung it once more. Each time the necessary gestures came exact and perfectly timed for the scythe to be safeguarded and for the wheat to be cut off at the ground. It was a joy to watch. They had all gathered, men and women. They took a step when Randoulet took a step. There was a single mower.[74]

Just like writing, weaving, or singing, working the land becomes a means for humans to humbly partake in the joyful song and dance of the earth. When carried out ecopoet(h)ically, Giono suggests, human activity can prove elating, performing similar functions as those ascribed to religious rituals, in that they celebrate and reinstate relationships of harmony, reciprocity, and care between humans and the greater-than-human world.

Ecopoiesis, whatever its medium, thus renews the Sacred Hoop, as Paula Gunn Allen or Dhyani Ywahoo might say: "The quality of our laughter and joy, the knowledge of our voices, thoughts, and actions are weaving beauty around the land. There is a harmony; there is a song. [. . .] Our thoughts make sound waves on the planet, wind currents upon the stream. As our thoughts become clarified in the wind of personal experience so, too, they become clarified around the planet through bioresonance."[75] While the concept of "bioresonance" might sound esoteric, it yet provides an interesting take on the diffraction of sounds, images, movement, thoughts, feelings, and ethics which occurs as our feelings and thoughts—themselves the result of interplay between our bodyminds and the larger, biophysical world—get translated into action and *poiesis*. Hence Ywanoo's claim that "we can shape the world around us with our thoughts and feelings," that, by imagining ourselves as a "great lake sending forth endless ripples of compassion and care,"[76] by translating our psychic, vital energies into action and *ecopoiesis*, we may help reweave the sacred hoop of life the signature of which are harmonious landscapes, odorscapes, feelscapes, and soundscapes.

The last passage of Giono's novel is focalized through the buck. When the animal bellows, sending out his own, recognizable acoustic signature into the multispecies, biophonic soundscape, the cacophony that erupts in response is filtered through the animal's perception:

> The foxes stopped on every path, listened, replied, and began to run. The badgers *growled*. The otters began to *miaow* and sharpen their claws against the willow trunks. The tawny *owls called* to one another. The squirrels screamed in the branches, the rats ran like mad into the swamp, *jostling* the reeds; the ducks flew up again and then settled back. The *weasels*, the field mice, the *grasshoppers,* the lizards, *the beetles,* and the great moths were aroused. The *silent snakes* darted their lighted tongues above their eggs, and the doe *called a long, long [, long] call* with a great warm voice full of love.[77]

Resting in great part on onomatopoeia, Giono's prose rings with many such poetic echoes of the voices of the earth. Here, the musicality more specifically relays the biophony at play, translating interspecies communication within the animal world.[78] As a result, Giono's multispecies writing becomes densely liminal, situating readers somewhere in-between animal and human points of life, and crafting a middle voice halfway between human and animal realms. Giono thus casts light on *ecopoiesis* as an incantatory practice of reen-*chant*ment blending human and other-than-human echo-vocalizations.

By creatively focalizing scenes through an animal sentience, Giono makes room for the myriad voices of the earth picked up by the buck, the detail of which most human ears could not so easily disentangle and identify. In reading such passages, we may feel spurred to emulate the buck's superior capacity to interpret biophony as he navigates the world's soundscapes. Moreover, as the buck stands at attention, ears pricked, and engages into a practice of multispecies call-and-response, it can be argued that Giono is here discretely theorizing his own ecopoetic practice as a human—his own joyful attempt to acknowledge and respond to other-than-human voices: "Then the stag came ashore on the opposite side, the world grew calm, and gradually silence reigned. He listened. [. . .] More than twenty times he crossed and recrossed the pond [. . . and at] each crossing he bellowed, and each time the vast world replied. Finally, overcome by fatigue and so burning with joy that he smoked like a brazier, he lay down in the grass."[79] Giono's experimentation with animal viewpoints thus makes headway into the ecopoetic joy of singing oneself into the world, thereby partaking in the great, multispecies symphony of the earth.

With hindsight, it may be noted that Giono anticipated Donna Haraway's *sympoiesis*. In her recent, provocative essay where she argues that a better name for the Anthropocene might be the "Chthulucene," Haraway pushes for the concept of *sympoiesis*: "It is a word for worlding-with, in company. Sympoiesis enfolds autopoeiesis and generatively unfurls and extends it."[80] Enjoining us to "stay with the trouble," Haraway proposes a vision of our climate change era that remains open to a multispecies reknitting and becoming-with:

> To renew the biodiverse powers of terra is the sympoietic work and play of the Chthulucene. Specifically, unlike either the Anthropocene or the Capitalocene, the Chthulucene is made up of ongoing multispecies stories and practices of becoming-with in times that remain at stake [. . .] Unlike the dominant dramas of Anthropocene and Capitalocene discourse, human beings are not the only important actors in the Chthulucene, with all other beings able simply to react. The order is reknitted: human beings are with and of the earth, and the biotic and abiotic powers of this earth are the main story.[81]

NOTES

1. Bate, *Song*, 251.
2. Ibid, 251.
3. Ibid, 258. Bate is here relying on Heidegger's notion of modern technology as a form of "enframing" and concealment, while *poiesis* consists in revealing—a form of unconcealment, or of "bringing-forth into presence." Bate, *Song*, 253–62.
4. Ibid, 262.
5. Knickerbocker, *Ecopoetics*, 1.
6. Ibid, 2.
7. Forbes, *Columbus*, 193.

8. "And our Mother and Grandmother is the Earth/ upon which we graze/ upon whose breast, / it is said, / we suckle all our lives / never being weaned." Ibid, 195.
9. Ibid, 181.
10. Ibid, 202. The imagery Forbes activates ties in with bioacoutician Bernie Krause's work, where he explores the biophony and geophony at play in the world. See *Orchestre* and *Chansons*.
11. Starhawk, *Fifth*, 58.
12. Ibid, 64.
13. Schafer, *Paysage*, 34, translation mine.
14. Pancake, *Strange,* 169. Mogey's claim that the world's humming is "not something caught by ear," points to the subliminal dimension of some of our perceptions, such as proprioception, and, simultaneously, to the fact that proprioception in fact involves the inner ear as well as a myriad captors located all over the skin and throughout the body. Although we are usually unaware of it, our inner ears are constantly registering the world around us, helping us locate the sources of the various stimuli we receive.
15. Tredinnick, *Wild*, 288–89.
16. Abram, *Spell*, 262.
17. See Julia Kristeva's essay where she elaborates her semiotic approach to language, and more specifically, to the poetic dimension of language. In her distinction between "geno-text" and "pheno-text," Kristeva is already quite attentive to the ecological and embodied continua that are articulated via the "geno-text"—the non-symbolic signifying "process" that subtends the "pheno-text"—a term which Kristeva uses to designate communicational language as it is socially regulated and structured. According to Kristeva, significance results from the entanglements between pheno-text and genotext in linguistic expression (Kristeva, *Révolution*, 83–100).
18. Abram, *Spell*, 263.
19. Williams, *Refuge*, 85.
20. Ibid, 84–85.
21. Ibid, 240.
22. Readers interested in exploring Kristeva's contribution from an ecopoetic standpoint may also refer to Anne Elvey's essay, which draws on Kristeva's theory of an articulation between the semiotic and the symbolic within the signifying process. Albeit not explicitly connected with ecofeminist theoretical frameworks, both Kristeva's and Elvey's work nevertheless highlight the resistant and transgressive potential of ecopoetic language against patriarchal systems:

Kristeva's work is pertinent, therefore, for a critical affirmation of human interrelationship [. . .] where her work announces as central to the process of signification (1) the continuity and difference between self and other; and (2) the gap between desire and gratification (and an attendant unsettling of an imaginary of human agency as control of the other, particularly the other-than-human). While the focus on castration suggests a founding violence in signification, problematic from both feminist and ecological perspectives, Kristeva's analysis of signification describes a patriarchal social economy and its accompanying imaginary (or worldview). She offers one understanding of the imaginary's simultaneous resilience and instability.

This interplay of resilience and instability creates echoes across the boundary between the symbolic and the semiotic. By an invocation of language open to the semiotic and tending toward an imitation of its object, *mimesis* can transgress the thetic break from the *chora*. [. . .] While affirming that the unifying and structuring character of the symbolic is necessary for a text to be a signifying practice, Kristeva signals the transgression of the thetic through poetic language, especially when it departs from grammatical construction. (Elvey, "Matter," 185)

23. Williams, *Refuge*, 240–41.
24. Ibid, 240.
25. Ibid, 240.
26. Pancake, *Strange*, 355.
27. Ibid, 355.
28. I am taking up Gerald Vizenor's portmanteau word, which he originally coined to express how survival and resistance are both inextricable and essential for Native Americans to move beyond tragedy and victimry; see Vizenor, *Manifest*, vii.
29. Pancake, *Strange*, 37.
30. Chatwin, *Songlines*, 52. In the particular context of Aboriginal worldview, Chatwin explains, places are mapped by "Dreaming-tracks" or "Songlines," which indicate the "Way of the Law," mapping out the "footprints of the Ancestors": "Aboriginal Creation myths tell of the legendary totemic beings who had wandered over the continent in Dreamtime, singing out the name of everything that crossed their path—birds, animals, plants, rocks, waterholes—and so singing the world into existence." Ibid, 2.
31. Starhawk, *Fifth*, 27–28.
32. Pinson, *Habiter*, 182, translation mine.
33. Ibid, 160, translation mine.
34. Ibid, 161, translation mine.
35. Grandjeat, "Volcano," 7.
36. Knickerbocker, *Ecopoetics*, 2.
37. Bernstein, *Close*, 21.
38. Knickerbocker, *Ecopoetics*, 7.
39. Stewart, *Reading*, 17. Again, I am relying on Knickerbocker's seminal study of sensuous *poiesis*.
40. Starhawk, *Fifth*, 16.
41. Bate, *Song*, 75–76.
42. Starhawk, *Fifth*, 473–74.
43. In blind humans, the inner ear and its connections with the visual brain can indeed be used for echolocation. See groundbreaking work by "Bat Man" Daniel Kish and researcher Johanna Hook, *Echolocation*. See also Thaler and Goodale's paper "Echolocation in Humans," or, in France, Olivier Després's PhD thesis on the interaction between visual and auditory systems in the construction of spatial representation in humans.
44. Berendt, *Sound,* 121.
45. Ibid, 127.
46. Ibid, 116.

47. Ibid, 116–17.
48. Ibid, 51–71.
49. Capra, "Foreword," xi–xii. Not surprisingly, Capra's transdisciplinary approach has its foundations in particle physics and systems theory and extends into ecology, metaphysics, spirituality, and ecoliteracy. Capra has written at great length about the cross-over between Eastern mysticism and physics. As an upholder of holistic paradigms, he may be viewed as a defender of reenchantment. Capra is a co-founder of the Berkeley Center for Ecoliteracy, and he has coauthored a book with ecofeminist scholar Charlene Spretnak, *Green Politics*.
50. Capra, "Foreword," xii.
51. Krause, *Orchestre*.
52. Tredinnick, *Wild*, 293.
53. Ibid, 240.
54. Ibid, 240.
55. Grandjeat, "Poetic."
56. Giono, *Joy*, 163.
57. Ibid, 163.
58. Ibid, 164.
59. Ibid, 165.
60. Ibid, 167.
61. Ibid, 171. The translator has evidently put much effort in translating the lyricism of Giono's prose, overall doing justice to the original: "Mais, par-dessus tout, il y avait le tambour du sang, le grondement du sang. Il tapait sur un sombre tambour dans les hommes et dans les femmes. [. . .] On se sentait lié à cette cadence. C'était comme le volant des batteuses qui battent le blé. C'était comme le fléau qui bat le blé, vole, bat le blé, vole. [. . .] Toujours, toujours, sans arrêt, parce que le sang ne s'arrête pas de battre, et de fouler, et de galoper, et de demander avec son tambour noir d'entrer dans la danse." Giono, *Joie*, 156.
62. Giono, *Joy*, 172.
63. I have here proposed my own translation, which I believe to be truer to the original ("Le sang a raison, somme toute." Giono, *Joie*, 157) than the English translation ("In short, the blood has won." 172), which, to my mind, considering how the characters refrain from surrendering to the wild music of their blood and therefore, instead of experiencing euphoria, are stuck with sadness at the end of the day, runs counter to the original.
64. Giono, *Joy*, 171–72.
65. Ibid, 103.
66. Ibid, 104.
67. Ibid, 342. In their book on plant sentience and communication, Mancuso and Viola refer to research proving that plant roots actually do emit sounds via a kind of "clicking" while they also perceive and identify the vibrations emitted at various frequencies through the soil. See *Intelligence*, 106–7.
68. Giono, *Joy*, 343–44.
69. Ibid, 344.
70. Ibid, 373.

71. Ibid, 373.
72. Ibid, 375.
73. Ibid 375.
74. Ibid, 414–15.
75. Ywanoo, "Renewing," 274.
76. Ibid, 274, 279.
77. Giono, *Joy*, 408, emphasis added. In the original, the adverb "longuement" is repeated thrice, whereas the English translation has dropped the third iteration here reintroduced. The process creates imitative harmony, the syntax stretching out with the same repeated, lingering adverb, as if to translate the insistent calling of the longing doe.
78. The zoopoetic aspects of Giono's work have been brilliantly tackled by Sophie Milcent-Lawson, who ventures fine close readings of the prosopopoeia and stylistic effects at work in Giono's prose. See Lawson, "Bruits," "Point de vue," "Tournant animal," and "Chant du monde."
79. Giono, *Joy*, 408.
80. Haraway, *Staying*, 58.
81. Ibid, 55.

Chapter Twelve

Restor(y)ing and Rewor(l)ding
Writing in a Grounded Middle Voice

Written from an indigenous viewpoint, Susan Power's novel *The Grass Dancer* (1994) stages *ecopoiesis* as achieved through sympoietic ritual dancing.[1] Like Linda Hogan or Robin Wall Kimmerer, Power is concerned with reweaving indigenous stories, practices, and worldviews into the present world of her readers via storytelling. She stages the performance of a young female grass dancer at a powwow, accomplished in honor of sweetgrass. In the Sioux tradition, the sacred plant holds a central ceremonial role.[2] Pumpkin's modernity is visible in her regeneration of traditional roles—grass dancing conventionally being a man's function. Via a choreographed anamorphosis, Pumpkin dances herself into sweetgrass, performing ecopoetic enchantment: "She was the best grass dancer on the field; she became a flexible stem, twisting toward the sky, dipping to the ground, bending with the wind. She was dry and brittle, shattered by drought, and then she was heavy with rain."[3] In her becoming-grass, Pumpkin is mesmerizing to onlookers. Her practice is a translation of the dance taking place between the sacred grass, earth, water, and wind. Here again, Pumpkin's practice is a liminal one: "'As a grass dancer,'" she tells her young friend, 'I'm trying to become something else. I step outside of myself.'"[4] Indeed, people watching her seem to fall under the spell of sweetgrass: "[Herod] was watching not a girl, he thought, but the spirit of grass weaving its way through a mortal dancer. Pumpkin was the color of blazing grass: grass that is offered to the sky in prayer."[5]

In Barbara Kingsolver's *Animal Dreams,* as Cosima reflects over the power emanating from the deer dancers at a powwow, what is highlighted is also the anamorphic and liminal potential of such ecopoetic practices:

> There were boys with black shirts and leggings, white kilts, and deer antlers. Their human features disappeared behind a horizontal band of black paint across the eyes. They moved like deer. They held long sticks in front of them,

imitating the deer's cautious, long-legged grace, and they moved their heads anxiously to the side: listening, listening. Sniffing the wind. The woman in black stepped forward shaking her gourd rattle, and they followed her. They *became* deer. They looked exactly as deer would look if you surprised them in a secret rite in the forest, moving in unison, following the irresistible hiss of a maiden's gourd rattle.[6]

In both novels, the sacred dances are animated from within by the interpreters' capacity to embody the deer's or the grass's ways of dwelling—their natural grace. The dancers translate what French literary theorist Marielle Macé would call the "styles" of deer or sweetgrass.[7] The dancing, like the music earlier, offers poetic echoes of other-than-human dwelling styles.

Constructing Pumpkin as a liminal character,[8] Power has her constantly incorporating the world through her acute, interlaced senses. Pumpkin exhibits both the skills of a fine grass dancer and an ecopoet's hypersentivity. Her perception of the wind's every movement is a subtle elemental one: "The wind was like water, rippling through the car's open windows, and Pumpkin imagined she tasted the ocean. It was actually the smell of rain invading the car."[9] Pumpkin's synesthesia—the "rippling" of the wind generates haptic, gustatory, and olfactory sensations—predisposes her to become an ecopoetic voice hearer. She can hear for instance the enticing voice of Lake Michigan: "Pumpkin looked into the water. It lapped gently against the breakwater of piled rocks, pulsing in and out, regular as a metronome. It had its own way of speaking, and that particular day Pumpkin heard its voice. *Cool water*, it bubbled. *Lovely cool water.*"[10] As a result of stylistic echoes, the lapping of the water achieves material presence through alliteration in [l], engaging the reader's tongue into a lapping of her teeth and palate in a way that fleshes out the voice of the water, thus articulated through one's mouth. In addition, the bubbling of the water is translated in part through alliterations in [p] and [b], two bilabial consonants the pronunciation of which comes close to bubble-making with one's lips.

What is evidenced here by this relaying of the voice of water is how *ecopoiesis* concretely possesses the power to reen-*chant* the world. This is suggested in a dream scene of a spelling contest, when Pumpkin opens her mouth to spell "grass," and "a cluster of tiny black birds the size of thimbles [flutter] from her mouth."[11] Following the surrealistic logic of dreams, the birds spilling out of Pumpkin's mouth embody the incantatory magic of language, inherent in the live, singing quality of the letters, phonemes, and syllables with which we utter and spell, and, simultaneously, under the spell of which we find ourselves when listening to a song, a story, or poetry. Moreover, birds providing a synecdoche for the song of the earth, this liminal realist passage symbolically gestures to the role of ecopoets as mediators between human and other-than-human voices, whose writing and utterances constitute a bringing-forth of other lifeforms.

As it is, Power's ecopoetic liminal realism substantiates Mark Tredinnick's theory of nature writing as the emergence of a "middle voice" arising from the intra-actions between the ecopoet and the more-than-human world. As recapitulated by Yves-Charles Grandjeat, "Paul Carter's *The Lie of the Land* offers [Tredinnick] the notion of the 'middle voice,' a voice which, Carter writes, 'dissolves the subject-object relation,' so that the two are seen 'grounding each in the other, continuously redefining both in terms of each other.' Then, Tredinnick writes, 'the voice we hear belongs to them both [. . .]. The middle voice is the voice of reciprocity, of intersubjectivity.'" In a lapidary formula, Grandjeat concludes that "[The middle voice] is, clearly, the voice of shared expression."[12] Drawing connections with material ecocriticism, Grandjeat moreover makes a crucial point as to the material "grounding" of *ecopoiesis* in *physis*: "This middle voice arises from a shared reality, a 'trans-corporeality' which, Stacy Alaimo reminds us, stems from 'a new materialistic and posthumanist sense of the human as substantially and perpetually interconnected with the flows of substances and the agencies of environment.'"[13] Emerging from *sympoiesis*, or co-authorship, the middle voice is recognizable in a "mixed writing style" through which nature writes itself.[14]

As with Bobi's death in Giono's *Joy*, there are aspects of Pumpkin's characterization that signal, on the one hand, her exceptional being implicated in totemic relationships, and, on the other hand, the ritualistic function of ecopoetic storytelling. As tragic as it may seem, Pumpkin's precocious death in a car accident is transcended in a two-fold manner—first, as a totemic sacrifice returning her to the spirit world, and second, and relatedly, as an offering renewing reciprocal relationships between her people and sweetgrass. Through a dialogue, readers learn early on that "there's two kinds of grass dancing [. . .]. There's the grass dancer who prepares the field for a powwow the old-time way, turning the grass over with his feet to flatten it down. Then, there's the spiritual dancer, who wants to learn grass secrets by imitating it, moving his body with the wind."[15] It is suggested that the young woman had actually learned to braid the two kinds of grass dancing—translating its dance *and* paving the way for the next dancers, or ecopoets. In the car crash where Pumpkin and her three grass-dancing fellows die, the protagonist ends up "flying across the Badlands, [. . .] shedding fears and insecurities like old skins, until she [is] distilled to a cool, creamy vapor." Eventually, we are told: "Pumpkin melted into the sky, and so she never came down."[16] On top of the imagery evoking at once images of rebirth and dissolution into a greater whole, this death scene is interpreted via some of the remaining characters, including the elder (Herold) and younger generations (Frank and Harley). As with the anamorphosis enfolding the four singing and dancing women in

Linda Hogan's *Power*, the final description of the field where the accident has taken place speaks worlds in terms of the onto-epistemological, reenchanting power of *ecopoeisis*—providing us with meaning, grounding, direction, and vision:

> Herold seemed relieved to find the scorched earth surrounded by a wide swath of flattened prairie grass. [. . .] "This reminds me of the powwow grounds from when I was little. Those old-time grass dancers did a good job of pounding the field flat. They churned through waist-high stalks like they were wading into a river, and it went down like that. Just like that." Herod pointed again to the pressed plain extending beyond the point of impact. [. . .] "Those kids. Those four Menominees. Now they're the true kind of grass dancers. Now they really know how to prepare the way."
>
> Frank looked skeptical, but Harley believed the old man, because the last time he glanced over his shoulder before climbing into the truck, he thought he saw the four figures, graceful as waves, dancing the grass into a carpet.[17]

It is suggested that the immortal "spirit of grass"—conceptualized within a totemic worldview—weaves itself through mortal dancers, storytellers, and poets dedicated to the dynamic entangling of discursive and poetic practices. In this respect, my analyses side with Grandjeat's definition of ecopoetics as a pondering "through a metafictional inquiry," "of what is at stake in the *work* and *energy* of the *representing*," while "[seeking] to respond to the energy of poetic discourse." Whether in the form of a poem, a dance, or a song, what transpires throughout my corpus is indeed that ecopoetic art, as Grandjeat has it, makes "a tiny statement in the broad cultural conversations through which the planet constantly reorganizes itself."[18]

The onto-epistemological, material, *and* spiritual ties knotted together in storytelling are specifically shed light upon in Hogan's *Power*. Hence the stance adopted by Annie Hide, one of the traditional elders in the book, both a healer and storyteller. These two functions are presented as complementary facets of the same activity since words, like medicine, possess their own kind of physical bodies and agency:

> She knows that words are part of this strength of hers. They too, are a person; they come from the birth throes of dream and thought. When spoken, words stand up straight as a stick before her, standing like thin gods, and if she stays by tradition, as if it too, is a person, then something newly born and alive will remain in air, in water, in this world. That something will be here in times of need. Its movements will brush against a person as surely as if it had wings and was flying past, or as if something invisible was breathing on human skin. It will

walk toward us and put its arm about us in that old embrace of something Annie Hide felt to be, knew to be, love. It is something that will sustain.[19]

The comparisons ("as if it too, is a person," "as surely as if it had wings and was flying past," "as if something invisible was breathing on human skin") evoke the aliveness of storytelling, with its power to bring forth, to inspire, comfort, and nurture us with a sense of connection which, in turn, lends meaning and direction to our actions. From this perspective, words, like incantations, do possess actual agency—an enchanting "power" encapsulated in the eponymous book title. Indeed, and as formulated from a new materialistic stance by Serenella Iovino and Serpil Oppermann, "discursive practices 'intra-act' and are co-extensive with material processes in the many ways the world 'articulates' itself."[20] In a similar spirit, but from an indigenous stance, Paula Gunn Allen expounds on the interlacing between reality and storytelling as part of the Sacred Hoop of being: "The words articulate reality—not 'psychological' or imagined reality, not emotive reality captured metaphorically in an attempt to fuse thought and feeling, but that reality where thought and feeling are one, where objective and subjective are one, where speaker and listener are one, where sound and sense are one."[21]

Fulfilling the function of myth, words and stories of interdependence and intra-actions can indeed change our vision of the world, and, at the end of the day, they have the power to affect the world itself. Words and stories impact the way we relate to our co-dwellers, including all forms of nonhuman earth others. They shape the worldviews and ethics that one passes on to the next generation, thus determining how future human societies will treat the rest of the living community they belong to—either as something to be subdued, mastered, exploited, and possessed, or as a thriving *oikos*, vibrant with life, to be honored, treated with humility and respect, and via relationships of mutualisms and reciprocity. This is what transpires throughout Hogan's novel, as the homodiegetic narrator gradually steps into the shoes of the next generation's animal tracker *and* storyteller. Combining the two roles, Omishto concretely reweaves word to world in an ongoing process of rewor(l)ding. She provides spiritual and moral guidance, reentangling within her ecopoetic narrative the fates of humans with that of the Florida panther, and with the entire naturcultural ecosystem of the Everglades, past, present, and future: "[Annie Hide] holds to the thought that if she lives long enough and can tell what she knows to a younger person, there will always be this shining in the world, an unbroken thread of light from a past where we were beautiful. It will curve around and into the present. And in another embrace, it will encircle the future and bring it all whole and together as one."[22] The act of reweaving via storytelling, as made clear in this last quote, is also concerned with the responsibility of harmoniously interweaving the past, the present, and the future.

The path that Omishto has been treading with Ama leads toward healing—a theme that is central to the books in my corpus. In the novel, the task is taken on by Annie: "She is the healer, the one who takes care of human wounds and broken things. Annie is the peacemaker, the mender of fights [. . .]."[23] Clearly, by the end of the book, the female figures moving between different human worlds and between different natural realms provide Omishto, and readers in her wake, with role models laboring toward a restoration of the world through reenchantment: "[All] I can tell for sure is that there is a fracture in the world; a gap between Ama and those for which the world is silent and dead. I reach for words or thoughts that will fill this gap, stitch it together like thread sewing two unmatched pieces of cloth into one."[24] In this conclusion lies a metatextual comment as to the function of ecopoetic literature, with the potential to stitch world and words back together into living tissue, into a harmonious, organic, and multispecies earth fabric. Like Starhawk and her witch-characters' magic rituals, as they draw protective circles and practice "grounding," *ecopoiesis* delineates a safe dwelling place in-between various ontologies and epistemologies, allowing for an enchanted world filled with human and other-than-human songs and voices, which can thereby keep on co-becoming. Thus, ecopoetic reenchantment takes part in a restoration process that, as ethnobotanist Gary Nhaban and Robin Wall Kimmerer defend, is inextricable, from "re-story-ation": "our relationship with land cannot heal until we hear its stories. But who will tell them?"[25] asks Kimmerer. Ecopoets will, the voices in my corpus reply in unison.

Also shedding light on ecopoetic processes, Kimmerer explores the agency of ceremony, whereby "the active force of love of land is made visible," and whereby "attention becomes intention."[26] Stressing the practical reverence at the heart of ceremonies that, beyond the renewal of a spiritual connection between humans and land, often help create or maintain biodiversity on a practical level, Kimmerer studies how ceremonies can "magnify life" in "powerfully pragmatic" ways.[27] What is true about ceremony holds true for an ecopoetics of reenchantment—something that many of us ecocritics will no doubt have felt within the communities of scholars and artists which we have built over the years around our shared interest in earth poet(h)ics. Kimmerer highlights the true power of ceremony: "The ceremony reminds [us] of where [we] come from and [our] responsibilities to the community that has supported [us. . . . It] inspires [us]."[28] Adhering to a similar spirit, as *Power* comes to a close, the last vision is that of Omishto, the storyteller, who, having followed her vision of the four women guiding her toward the place where her elders live according to tradition, joins into a dance and song meant to restore the world by taking part in its unfurling, creative energy: "soon there is a drum, the younger men drumming, [. . .] and I dance and as the wind stirs in the trees, someone sings the song that says the world will go on living."²⁹ As forms of *ecopoeisis*, singing and dancing actively take part in the world's co-becoming.

Dancing, singing, storytelling, and ecopoetic writing all offer a kind of participative ceremony, reweaving our human selves and languages within the earth. The title of Hogan's *Power* gradually unfolds as a metapoietic comment on the power of words, of storytelling, and ritual to translate and keep alive the breath and song of the earth. Hogan pointedly foregrounds the materiality of language itself in relationship to air: "[through] air, words and voices are carried."³⁰ Via Omishto's rejuvenating and retelling of the old stories, we are constantly reminded of the material vibrancy of speech and of chanting: how the cries and words emitted by humans and animals, even the noises produced by the wind or the rain both "tell a story" of what is going on in the phenomenal world, and simultaneously retain a form of agency:

> What is spoken travels by air and the old people say even thoughts travel and are carried with it [. . .]. Lying here in the boat, the world reeling around me, I look into the thick white fog and think of all the people at the place of old law, and of my mother and sister. I think of words and songs, the power of every breath to keep life, and their thoughts move toward me with the wind, in just the same way older people say spirits are summoned. [. . .] It seems that wind blows their thoughts toward me as I float.³¹

Storytelling produces sound waves that travel through the air so they can be heard by our sensing ears and made sense of by our bodyminds. The animated words and stories heard through our sensing bodies and stored in our memories stay alive within us. They are thought forms that both contain and shape thought and feelings, while they direct our affects and can remap cognitive pathways through language. In other words, stories can reboot reality. As *Power* demonstrates, even when heard in the past, stories retain an agential kind of power as they can influence the shape of the present and the future. Suspended in her boat between water and air, Omishto has entered the liminal world of initiands, where all categories can potentially be reshuffled.³² Negotiating different ways of inhabiting the earth, Omishto considers how, according to Janie Soto, her Panther Clan leader, "our every act, word, and thought is of great significance in the round shape of this world and there are consequences for each."³³ As Hogan's novel demonstrates, the stories we choose to silence or tell determine our capacity to perceive and organize the living world, thereby defining our response-abilities to its many textures and harmonies, and, eventually, our responsibilities in our dealings with the more-than-human world.

Reaching deeper than the levels of story, theme, and characterization, Hogan's prose composes at times with "the drum of rain" and the "the roaring voice of the storm."³⁴ Throughout the storm episode setting the events in motion at the beginning of the novel, geophony is relayed in great part through an accumulation of onomatopoeic sound and movement verbs, with a series of gerunds creating a tense rhythm, itself conveying a sense of urgency and fighting ("pushing," "slashing," "roaring and screaming," "hitting," "howling," "gusting"). Additionally, the musicality of certain passages is dominated by alliterations in plosive and fricative consonants, producing imitative harmony: "It forces the air out of my lungs and I see the dead and drowning insects being blown right off the earth in front of my faces. There isn't time to seek shelter in the dark light and slashing rain of the storm. It's roaring and screaming at me. Somewhere there is what sounds like the clatter of chains hitting against each other."³⁵

More impressive in its sympoietic ode to the wind is Michel Serres's short story titled "Vent" ("Wind," in French), which articulates the middle voice arising from contact between ecopoet and elements. Written in an epistolary mode, the narrator claims that "[his] message is being birthed by the tornado."³⁶ Substantiating his insistence that "[his] sentence emerges from the howling of the squall,"³⁷ as he writes his lover, the protagonist explores the onomatopoeic quality of the lexical field of sounds:

> Le vacarme du vent croît; il craque, claque, clame, crépite, détone, explose, gronde, vrombit, tourbillonne, varie et module . . . quel tintamarre de tous les diables! Qui parle dans ce brouhaha formidable? [. . .] Depuis quatre jours, l'ouragan saute et cingle, ronfle et siffle, crie et gueule, encourage et domine, mais jamais ne caresse . . . étendu, immobilisé, aveuglé, insensibilisé par le froid, mon corps devient une gigantesque oreille. De ces clameurs d'enfer, je traduis ce que j'entends.³⁸

If I have first provided the French version to allow my readers to appreciate its lyrical dimension, I might attempt the following translation of Serres's sensuous language, the semantics of which pushes through the phonetics:

> The din made by the wind keeps getting louder and louder–it cracks, claps, clammers, crackles, explodes, booms, rumbles, roars, swirls, varies and modulates . . . What a devilish hullabaloo! Who's talking through this formidable brouhaha? [. . .] For the past four days, the hurricane has been leaping and lashing, snoring and hissing, screaming and bellowing, nudging and pushing, but not once has it caressed . . . Lying still, paralyzed, blinded, and numb with cold, my body is turning into a gigantic ear. I am translating what I can hear from this clamorous hell.

The man's letter turns into a "musical score" as its writer, deprived as he is of sight and movement, magically metamorphoses into an attentive "gigantic ear"—in great part because he is lying in a tent at high altitude and knows that his own survival hinges on the evolution of the hurricane.

In one of her essays, Robin Wall Kimmerer experiments with "the voices of the rain."[39] In her meditative observation of the entanglements between, rain water, trees, lichen, river, and the human bodymind trying to appreciate the differences in formation, size, and sound of individual drops of water, "depending on the relationship between the water and the plant," Kimmerer pays heed to and relays the geophonic and visual diffractions caused by rain falling through trees into a river pool: "[The backwater pool] is a mirror for the falling rain and is textured all over by the fine and steady fall. I strain to hear only rain whisper among the many sounds, and find that I can. It arrives with a high sprickley sound, a *shurrr* so light that it only blurs the glassy surface but does not disrupt the reflection."[40] The repetition of fricatives works together with the neologism "sprickley" coined by the ecopoet to evoke the sound of raindrops falling in a gentle sprinkle, as well as with the onomatopoeia "*shurr,*" the long vowel of which is further lengthened by the redoubled <r>. Moreover, the onomatopoeia that is called upon to give the rain a poetic voice is immediately echoed via paronomasia ("a *shurrr* so light that it only blurs the surface"), with a concatenation of [ʃ], [s], [z] and [ɜːr] sounds interlacing the ecopoet and the rain's voices within the thickly textured dialogic prose.

Kimmerer moreover transcribes the musical parts played by the different instruments in her orchestra, as water falls from maple, hemlock, and alder, into a pool of water, making sound waves that ripple through the air, through the listener's bodymind, and, finally, that are diffracted again into her poetic language. She distinguishes for instance the "rapid pulse" made by hemlock and water, "[releasing] in a steady *pit, pit, pit, pit, pit*, drawing a dotted line in the water below,"[41] from the "[plummeting]" from maple stems of "drips" that are "big and heavy": "They hit with such a force that the drop makes a deep and hollow sound. *Bloink.*"[42] In addition to the onomatopoeia ("*Bloink*") that, forming a whole sentence, sounds the solid statement made as the water drops encounter other bodies, the preceding sentence, composed solely of monosyllabic words but for the word "hollow," spells out the geophony at play, as the words themselves form rhythmic drops of sound. By and by, nouns are substituted for onomatopoeia expressing the acoustic signatures of each instrument: "*Schhhhh* from rain, *pitpitpit* from hemlock, *bloink* from maple, and lastly, *popp* of falling alder water. Alder drops make a slow music. [. . .] The drops aren't as big as maple drops, not enough to splash, but the *popp* ripples the surface and sends out concentric rings." Relying on synesthesia, the ec(h)opoetic and impressionistic diffraction at work involves at once visual, haptic, and acoustic effects: "The reflecting surface of the pool is textured with their signatures, each one different in pace and resonance."[43]

NOTES

1. Born in 1961 in Chicago, Susan Power is an enrolled member of the Standing Rock Sioux Tribe.
2. See Kimmerer *Braiding*, on the sacred plant in Potawatomi tradition.
3. Power, *Grass*, 38.
4. Ibid, 38.
5. Ibid, 48.
6. Kingsolver, *Animal*, 237.
7. See Macé, "Styles."
8. Pumpkin's totemic relationship with sweetgrass and Power's novel could cogently be analyzed through the liminal realism lens polished in Part Two.
9. Power, *Grass*, 50.
10. Ibid, 32.
11. Ibid, 53.
12. Grandjeat, "Poetic."
13. Ibid.
14. Gagliano, *Thus*, 6.
15. Power, *Grass*, 30.
16. Ibid, 51.
17. Ibid, 54.
18. Grandjeat, "Poetic."
19. Hogan, *Power*, 181.
20. Iovino & Opperman, "Theorizing," 454.
21. Allen, *Sacred Hoop*, 71.
22. Hogan, *Power*, 181.
23. Ibid, 225.
24. Ibid, 198.
25. Kimmerer, *Braiding*, 9.
26. Ibid, 248–49.
27. Ibid, 249.
28. Ibid, 250.
29. Hogan, *Power*, 235.
30. Ibid, 178.
31. Ibid, 180.
32. She moreover stands in between the old world of tradition—encapsulated in the village where the Taiga people of her clan live—and the newer, modern world brought to them via the schools, television, and Christian churches—a white world the codes, language, and values of which Omishto's mother and sister have adopted.
33. Ibid, 183.
34. Ibid, 32 and 37.
35. Ibid, 34.
36. Serres, "Vents," 59, translation mine.
37. Ibid, 60, translation mine.
38. Ibid, 60–61.

39. Kimmerer, *Braiding*, 299.
40. Ibid, 298.
41. Ibid, 298–99.
42. Ibid, 299.
43. Ibid, 299.

Chapter Thirteen

Translating the Song of the Earth
Reen-chanting *Earthly Harmonies*

In her essay *The Sense of Wonder*, Rachel Carson's plea for the cultivation of our capacity for enchantment partly hinges on our capacity to hear "the voices of the earth and what they mean."[1] Carson explores how being steeped in nature at an early age helps train our sensitivity to "the voices of living things." "No child should grow up unaware of the dawn chorus of the birds in spring,"[2] she writes. While nothing could replace the actual experience, Carson's prose nonetheless works as an intermediary between that "dawn chorus" and readers who may or may not have heard it firsthand:

> The voices are heard before daybreak. It is easy to pick out these first, solitary singers. Perhaps a few cardinals are uttering their clear, rising whistles, like someone calling a dog. Then the song of a whitethroat, pure and ethereal, with the dreamy quality of remembered joy. Off in some distant patch of woods a whippoorwill continues his monotonous chant, rhythmic and insistent, *sound that is felt almost more than heard*. Robins, thrushes, song sparrow, jays, vireos add their voices. The chorus picks up volume as more and more robins join in, contributing a fierce rhythm of their own that soon becomes dominant in the wild medley of voices. In that dawn chorus one hears the throb of life.[3]

As with Ann Pancake, insisting that the "humming" of the earth is something that is "felt almost more than heard," Carson stresses the haptic, rhythmic quality that characterizes the somatic experience of responding to the earth's song.

In *Prodigal Summer*, Barbara Kingsolver takes up Carson's endeavor to translate the "dawn chorus of the birds in spring." The attunement of her narrative to biophony and geophony is manifest from the start. Deanna, we are told "[loves] the air after a hard rain, and the way a forest of dripping leaves fills itself with a sibilant percussion that empties your head of words."[4]

Kingsolver's prose strikes the physical chords of language. While activating the referential function of language, her *ecopoiesis* resounds with a semiotic quality relaying the somatic experience of being interlaced with the world's flesh. Articulated by the poet, the "sibilant percussion" that "fills" the "forest of dripping leaves" is heard through the assonants in sibilants ([s]), fricatives ([f]), and plosives ([d], [b], and [p]).

Let us note in passing that Kingsolver's husband, Steven Hopp, happens to be both an ornithologist and a guitarist, while she herself is also a trained zoologist and musician—mostly a piano and clarinet player. Their musical sensitivity and animal expertise have no doubt trickled into Kingsolver's prose, specifically through Deanna's character, who takes after the writer in many ways. Attentive to bird calls, Deanna listens "to the opening chorus of the day." As if carrying on the conversation started years before by Carson, Deanna observes: "In the high season of courtship and mating, this music was like the earth itself opening its mouth to sing."[5] Speaking to the human ear and entire body, Kingsolver's cadenced and sensuous *ecopoiesis* gives shape to a musical prose orchestrating the many instruments of the world:

> Its crescendo crept forward slowly as the daylight roused one bird and then another: the black-capped and Carolina chickadees came next, first cousins who whistled their notes on separate pitches, close together, distinguishable to any chickadee but to very few humans, especially among the choir of other voices. Deanna smiled to hear the first veery, whose song sounded like a thumb run down the tines of a comb. [. . .]
>
> The dawn chorus was a whistling roar by now, the sound of a thousand males calling out love to a thousand silent females ready to choose and make the world new. It was nothing but heady cacophony unless you paid attention to the individual entries: a rose-breasted grosbeak with his sweet, complicated little sonnet; a vireo with his repetitious bursts of eighteen notes and triplets. And then came the wood thrush, with his tone poem of birdsong.[6]

This passage shows how *ecopoiesis* can produce a translation of the polyphonic song composed by the greater-than-human world. Kingsolver's prose indeed performs echoes of a rose-breasted grosbeak's "complicated little sonnet," or a wood thrush's "tone poem of a birdsong." In addition to the lexical field of musicality giving distinct tones and textures to all the different voices at play, the language itself produces a musical score. The veery's song is voiced via monosyllabic words creating imitative harmony, reproducing the sound of "a thumb run down the tines of a comb." Consonance moreover arises from the marked alliteration and assonance in [k] and [æ] ("the black-capped and Carolina chickadees came next"), or from the repetition of the diphthong [əʊ] and the long [i:] vowel, as in "a rose-breasted grosbeak with his sweet." The latter nominal clause moreover contains a parallelism, with the compound adjective "rose breasted" visually mirroring the name "grosbeak" via eye rhymes ("rose-breasted grosbeak"). These poetic elements add up so that the prose diffracts the musical texture of the bird's "complicated little sonnet."

The high lyricism at play euphonizes the wild "cacophony," thus harmonizing the dissonance which a bewildered untrained ear might at first perceive in the forest ("nothing but heady cacophony unless you paid attention to the individual entries"). Deciphering and rearranging the various pitches, melodies, and rhythms involved in sylvan polyphony, the ecopoet gently brings readers to perceive not chaos, but the measured songs of the earth. Taking up the pun formulated by French scholar Aline Bergé about the sylvan poetry of Robert Marteau, I would argue that Kingsolver's writing forms an "echo-system" to the ecosystem of the forest[7]. Uplifting the reader from her ordinary stance, the enchanting poetic devices enfolded in Kingsolver's prose engage readers' aural and phonatory organs, our entire sensing bodies and brains, attuning us to biophony. As a result, our own sensations and representations merge with the images, cadences, and the overall sensuous *poiesis* at play, gently prodding readers to experience a kind of becoming-bird, becoming-forest, and becoming-musical.

Turning her attention to elemental sensuousness, Kingsolver elsewhere reveals the companionship which humans may derive from inhabiting mountains. Kingsolver taps into and rehabilitates the inhabitational wisdom and ecological intuition poetically entwined in Appalachian vernacular: "People in Appalachia insisted that the mountains breathed, and it was true; the steep hollow behind the farmhouse took one long slow inhalation every morning and let it back down through their open windows and across the fields throughout evening—just one full, deep breath every day."[8] Not only is the local saying accurate, with mountain vegetation breathing oxygen into our lungs during the day and rejecting CO_2 at night, it furthermore highlights how attuned to its natural environment vernacular parlance can be.

The characters' fleshy involvement with their *oikos* makes for a sensuous grounding that infuses their idioms: "When Lusa first visited Cole here she'd listened to talk of mountains breathing with a tolerant smile. She had some respect for the poetry of country people's language, if not the veracity of their perceptions [. . .]. But when she married Cole and moved her life to this house, the inhalations of Zebulon Mountain touched her face all morning, and finally she understood."[9] Kingsolver here toys with the notion of the earth as an other-than-human sensuous partner: "She learned to tell time with her skin, as morning turned to afternoon and the mountain's breath began to bear gently on the back of her neck. By early evening it was insistent as a lover's sigh, sweetened by the damp woods, cooling her nape and shoulders whenever she paused her work in the kitchen to lift her sweat-damp curls off her neck. She had come to think of Zebulon Mountain as another man in her life, steadier than any other companion she had known."[10] Readers may find themselves physically enraptured by the voluptuousness at stake in this ecoerotic evocation of earthly companionship. Assonants in bilabial [b] phonemes may indeed feel like engaging one's lips into giving light pecks, like kisses, and the many sibilants and fricatives in [ʃ]; [Θ], [ð], [s] and [z] reproduce the airflows inherent in breathing. Mobilizing the mouth and vocal chords, the poetic prose prompts an alternation between voiced and unvoiced phonemes in a process evoking the sighing described. Consequently, readers may feel aroused by the suspiration of the mountain seeping through the *poeisis* and reaching us like a breathy caress.[11]

In *Above the Waterfall* (2015), Appalachian writer Ron Rash has crafted a homodiegetic narrator who is besides an ecopoet. As such, her point of view imbues many a scene with a sense of enchantment. An avid reader of Emily Dickinson and Gerard Manley Hopkins, the character named Becky serves as the ideal focalizing agent to redeem a view of a world filled with wonder. As it is, the concept of sensuous *poiesis* first put forward by Scott Knickerbocker originally stems from the latter's thorough analysis of both Dickinson's and Hopkins' poetry, of which he writes "rather than mirror the world, their poems enact through formal devices such as sound effects the speaker's experience of the complexity, mystery, and beauty of nature." Influenced by wildness, "their language takes on its *own* wildness and materiality distinct from but still a response to nature."[12] Could it be that, similarly, Rash's ecopoetic language arises as a translation of the world's wild music, simultaneously expressing one's enchanting transactions with nature, and the subsequent rewilding of language itself?

In addition to the thrilling story subtending Rash's novel, the terse beauty of his writing makes for an exhilarating experience of en-*chant*-ment. Taking part as a guest writer in the Perpignan 2016 international ecopoetics conference that was convened around the theme "Dwellings of Enchantment: Writing and Reenchanting the Earth," before reading from his novel, Ron Rash performed the reading of an excerpt from Jean Giono's *Chant du monde*—a choice that speaks volumes in terms of ecopoetic filiation.[13] Definitely going beyond simple *mimesis*, Rash's writing fiddles with language so as to turn it into a musical continuation of the land.[14] For example, the contemplative second chapter of the novel displays a purple prose and an incantatory mode invoking the musicality of earth itself: "Though sunlight tinges the mountains, black leather-winged bodies swing low. First fireflies blink languidly. Beyond this meadow, cicadas rev and slow like sewing machines. All else ready for night except night itself. I watch last light lift off level land."[15] The effect produced on the reader is that we cannot simply *see* the picture, as in front of a painting or a photograph. Instead, we find ourselves entirely immersed, through all our senses, in the animate landscape. On top of the painterly qualities of Rash's landscape description,[16] the latter is endowed with a vibrant, material thickness achieved through sonority and spelling. Combining internal and visual rhymes (in <ing> for instance in the first sentence: "tinges," "winged," "swing"; and between "low," "meadow" and "slow "), chiming ("wing/swing," "night/light," "low/slow/sewing"), alliterations (in [l]: "sunlight," "black leather," "low," "fireflies blink languidly"; in [b]: "black, 'bodies,' beyond"; in [f]: "first fireflies," "lift off"; and in [t]: "last light," "lift"), and assonances (specifically in [aɪ], [iː], and [ɪ]), and relying on a plethora of occurrences of the vowel <i> throughout, this passage drills its way into our sensing bodies, both through our mouths, our eyes, and our sense of rhythm.

The synesthesia worked into the text awakes all our senses. The description of the night landscape calls on sight via the dabbing of color and touches of light ("sunlight tinges the mountains," "black leather-winged bodies," "fireflies blink"), on haptic evocations (via the "leather-winged" bats), and hearing (with "cicadas" that "rev and slow like sewing machines"). The rhythmic stridulation of the cicadas—critters owing their very name to onomatopoeia—is taken up by the balanced stress pattern and phonemic repetition in "rev and slow like sewing machines," punctuating the sentence at regular intervals in imitative harmony. Taking the felt scansion at play to a climax, the exquisite language must be savored to the full as it explodes in our mouths in an alliterative apotheosis (in [l], but also in [t] and [f]) that brings a taste of the landscape onto one's tongue: "I watch last light lift off level land." Reen-*chant*-ment, it is implied, banks on the tra la la potential of language. This fountain of chiming monosyllabic words attunes the reader to the musicality springing from Rash's prose, here heralding its own, dense lyricism. The light, elements, and insects exhibit a sonorous quality enfolding us into the vibrant flesh of the world, while human language echo-vocalizes the earth's soundscape. As Becky emphasizes, the practice of *ecopoeisis* spells a form of enchantment by means of which the landscape turns into a compenetrating feelscape: "You don't have to understand the words," Becky tells the school children she takes on outdoor ecopoetic excursions. "Just let the sounds enter you, the same as everything else you see and smell and touch today."[17]

Gesturing toward the reenchanting power of *ecopoiesis*, Becky's complex character development is rather telling. Having suffered from a traumatic confrontation with human violence—an experience making her mute as a child after she has witnessed a terrorist attack in which her schoolteacher was shot—she then heals through an ecopoetic epiphany. The dumbfounding effect of the school shooting on the child is expressed in terms obliquely recalling the petrification of Lot's wife in the Old Testament. While the teacher is nervously trying to usher all the children to a place of shelter, making them promise not to utter "*a single word*," Becky the child cannot hold her tears back, nor can she refrain from looking at the scene of horror—something which she seems to hold responsible for her teacher's death:

> Almost there when her shhhh stills us. Footsteps come halfway down the stairs and pause. Both my hands clutch Ms Abernathy's. Another footstep and a shoe and pants cuff appear. The pipe drips loud and my first tears well. I try to squeeze the tears back inside me but the first one falls and I know he has heard it. . . . Ms. Abernathy stand in the basement door, blocking the exist as I run. Close your eyes, a policeman says as he grabs me. But I look back and when I do my tongue turns to salt.[18]

Astonished by guilt and terror, the child then stays mute for a prolonged period. As if to indicate the ecopsychological capacity for resilience inherent in humans—and the ecopoetic dimension of this resilience—Becky's cure takes place as she is uplifted by the magic power connecting word to world: "I had not spoken since the morning of the shooting. Then one day in July my grandparents' neighbor nodded at the ridge gap and said *watershed*. I'd followed the creek upstream, thinking wood and tin over a spring, found instead a granite rock face shedding water. I'd touched the wet slow slide, touched the word itself, like the girl named Helen that Ms. Abernathy told us about, whose first word gushed from a well pump. I'd closed my eyes and felt the stone tears."[19] Banking on polysemic condensation, the word "watershed" refers at once to the geological basin, to the waterfall, and to those tears shed during the devastating attack leaving the girl dumbstruck.

Rash's narrative intertextually refers to the famous novel by Hellen Keller, *The Story of my Life*, where the blind and deaf homodiegetic narrator recounts how she was initiated to the magic of language by a teacher who came up with the brilliant idea to connect spelling inside the child's palm with her proprioceptive engagement with the world:

> Someone was drawing water and my teacher placed my hand under the spout. As the cool stream gushed over one hand she spelled into the other the word water, first slowly, then rapidly. I stood still, my whole attention fixed upon the motions of her fingers. Suddenly I felt a misty consciousness as of something forgotten—a thrill of returning thought; and somehow the mystery of language was revealed to me. I knew then that "w-a-t-e-r" meant the wonderful cool something that was flowing over my hand. That living word awakened my soul, gave it light, hope, joy, set it free! There were barriers still, it is true, but barriers that could in time be swept away.

I left the well-house eager to learn. Everything had a name, and each name gave birth to a new thought. As we returned to the house every object which I touched seemed to quiver with life. That was because I saw everything with the strange, new sight that had come to me.[20]

In both cases, the flow of water courses through the children's senses and cognitive circuits. Connecting the linguistic value of signs with the corporeal experience of the world, words acquire the magic power to entangle signifier and signified, not via the abstract representation of a concept, but, rather, via immediate embodied cognition and contact with a physical reality that cannot be unraveled from the word itself. Moreover, felt language suddenly provides a watershed allowing inner and outer worlds to trickle into one another. In both instances too, the children experience an intense moment of joyful wonder that hinges on the power of words to reveal the vibrancy of life itself. The transporting experience moves them to embrace to the full their participation in the world. In Becky's case, this crucial moment marks a turning point in her life. Alleviating the horror that had plunged her into an alienating silence, she embarks on a healing process that pivots on the reenchanting power of *ecopoiesis*. She herself later turns into a poet and a guide for other children, encouraging them to learn how to partake in the world by restoring the broken links between world and words, and to thus access the interrelated animacy of both.

The ecopoetic formulae crafted by Dickinson, Manley Hopkins, or Ron Rash call upon an entirely different kind of common *sense*. They co-compose with the more-than-human animate landscape an inner landscape of words— an "*[inscape] of words*,"[21] which is no escape from the real world, but, on the contrary, an enchanted landscape consistently fusing the inner self with the flesh of the world and that of language. Rash's novel thus attains a metapoietic dimension, revealing how an ecopoetics of reenchantment springs from the diffractive power of sensuous *poiesis*, reverberating the entanglements between self and *oikos*. In a way that chimes with Jack Forbes's aforementioned vision of humans as the world's drums, or the taut skins of the world, "vibrating with its messages," Gerald Manley Hopkins for his part has eloquently located "Earth's eye, tongue, or heart" in humans.[22]

As if to corroborate such a stance, Rash incorporates Becky into an instrumental landscape. Via a defamiliarizing anamorphosis resulting from an anthimeria—turning a noun into a verb—the rolling mountains morph into an accordion: "Mountains accordion into Tennessee. Beyond the second ripple, a meadow where I'd camped in June. Just a sleeping bag, no tent. Above me that night tiny lights brightened and dimmed, brightened and dimmed. *Photinus carolinus*. Fireflies synchronized to make a single meadow-wide flash, then all dark between. Like being inside the earth's pulsing heart. I'd slowed my bloodbeat to that rhythm. So much *in* the world that night.[23] On top of recounting the experience of synchronizing one's pulse with the tempo of the earth, the wonderful syncopated flashing of the fireflies takes part in the land's wild accordion music and is relayed in a language beating an imitative melodic rhythm.[24] The music of the earth audibly ripples into the alliterative, consonantic, and rhyming patterns of Rash's prose, which is replete with sensuous effects by dint of which the visual, the aural, and felt textures of the landscape merge with the poetic language that is played. Seething with eye rhymes and ear rhymes that exhilarate the bodymind, Rash's writing makes us move to the rhythm of the land. Once again, it can be concluded that the *chora* of the land percolates into the semiotic dimension of Rash's poetic echoes of the earth.

Subsuming many of the different aspects of the ecopoetics of enchantment I have been elaborating throughout this book, Rash elsewhere connects poetic reen-*chant*-ment with the enchantment induced at once by the world's animacy, by the anamorphoses that spontaneously arise for humans who are naturally inclined toward an ecopoetic stance of wonder, and by the song and dance of the world inspiring our own choreo-*graphies*:

> A mown hay field appears, its blond stubble blackened by a flock of starlings. As I pass, the field seems to lift, peek to see what's under itself, then resettle. A pickup passes from the other direction. The flock lifts again and this time keeps rising, a narrowing swirl as if sucked through a pipe and then an unfurl of rhythm suddenly sprung, becoming one entity as it wrinkles, smooths out, drifts down like a snapped bedsheet. Then swerves and shifts, gathers and twists. *Murmuration*: ornithology's word-poem for what I see. Two hundred starlings at most, but in Europe sometimes ten thousand, enough to punctuate a sky. What might a child see? A magic carpet made suddenly real? Ocean fish-schools swimming air? The flock turns west and disappears.[25]

First, this passage foregrounds the anamorphic propensity ecopoets may have retained from childhood ("the field seems to lift, peek to see what's under itself, then resettle," "a narrowing swirl as if sucked through a pipe," "What might a child see? A magic carpet made suddenly real? Ocean fish-schools swimming air?"). Second, while triggering flights of fancy in the wake of the birds' real flight, the mesmerizing aerial ballet speaks worlds about the collective intelligence arising from each individual's fine-tuned attention to its immediate surroundings. As scientists have shown, the harmonious, collective choreographies initiated by starling murmurations and schools of fish emerge from a kind of deep proprioceptive listening allowing each individual to perceive and respond to the movements of the seven nearest specimens.[26] Consequently, via a propagation effect, the movement of one entails the movement of all the others. As with the hybrid swarm intelligence that Starhawk has imagined in her cli-fi novel casting human-bee alliances, Rash here offers a vision that metonymically implies humans' capacity to stay sensitively attuned to our immediate surroundings, to move as a flock by means of empathy and emotional contagion, and to thus take part in the harmonious choreographies and murmurations of the earth.

Drawing attention to the ongoing composition of an ever-unfurling world, Rash stresses other-than-human agency, coordination, and writing. The synchronized flock of starlings preforms an aerial choreography, a picture appearing through the piling up of verb movements: "rising, a narrowing swirl"; "then an unfurl of rhythm suddenly sprung, becoming one entity as it wrinkles, smooths out, drifts down"; "swerves and shifts, gathers and twists." It also forms an enchanting kind of writing ("enough to punctuate a sky")—as when augurs in ancient Rome deciphered omens written in the sky by bird movement. And, finally, it plays a kind of music, or "murmuration." As emphasized by the italics and the glossing ("ornithology's word-poem for what I see"), even the scientific term 'murmuration' is derived from a poetic echo of the startling phenomenon. With its onomatopoeic quality, the term "murmuration" indeed resonates with the polyphonic song of the flock. The word thereby embodies the muttering of matter which may have mothered many a myth and "word-poem" into existence, captured by humans standing at attention to the wonders of the world. Ecopoets, this passage suggests, strive to relay the murmur of the world through their own human means, creating languages that dance and sing along with the earth.

It may be inferred that the shared etymology at the root of the words "matter," "mutter," and "mother," might also be connected to the word "myth." According to Paula Gunn Allen, *"mythos"* in Greek means "a mystery," a "secret (thing muttered)," or "one who is initiated."[27] Allen explains that the Greek terms from which the world "myth" is descended go back to the Indo-Germanic root, MU. She draws the connection with the Greek words for "a sound of muttering"—from whence came the Latin words *muttum and mutum,* and later the French "mot" for "word"—meaning "a slight sound," and which, quite intriguingly, can signify at once "muttering" and "muteness." As with Rash's Becky, it could be that without ecopoets to decipher and prolong its muttering and song, the world turns mute from the standpoint of humans, no longer capable of interlacing themselves within the audible part of the world's flesh. Reversely, the etymological kinship of these words could suggest that, as David Abram contends, the human mutterings that have gradually been structured into languages and myths have originally sprung forth from the world's wild music.[28] In Rash's novel, Becky's pondering over the complex wor(l)ding at work in the word "murmuration" is concordant with Abram's essay. Drawing from Italian philosopher Giambattista Vico,

Abram approaches language as "arising from expressive gestures." Relying on Vico, Abram posits that "the earliest and most basic words [have] taken shape from expletives uttered in startled response to powerful natural events, or from frightened, stuttering mimesis of such events."[29] Abram moreover relies on Jean-Jacques Rousseau's take on "gestures and spontaneous expression of feeling as the earliest forms of language," and on German philosopher and poet Johann Gottfried Herder's thesis that "language originates in our sensuous receptivity to the sounds and shapes of the natural environment."[30] Finally, Rash's ecopoetic prose substantiates Merleau-Ponty's view of linguistic meaning as "rooted in the felt experience induced by the specific sounds and sound-shapes as they echo and contrast with one another, each language a kind of song, a particular way of 'singing the world.'"[31]

The starlings' murmuration counts amongst the treasures Becky has collected and which help her overcome her mute phases. In her adult life, Becky's ecoterrorist lover dies a sudden death, again leaving her at a loss to make any sense of a deranged world. Trying to reweave world to word, to reinstill a sense of coherence into her life, Becky turns her attention to "spiders spinning webbed words, whip-poor-wills and white owls, wooly worms and snake skins, the sink of a star."[32] The onomatopoeic quality of the bird names "whip-poor-wills and white owls" ostensibly translates the cry of these endangered birds. Moreover, the dense lyricism worked into the language ecopoetically diffracts Becky's observation that "all [has] resonance, meaning."[33] Rash's poetic prose here rings with interweaved alliterations in [sp], [w], and [s], stressing the materiality of language, which is moreover thickened by the proliferating repetitions inside the words of [d], [p], [l], and [k] phonemes. Even on the page, the remarkable reiteration of the same letters spins together these "webbed words" into a sentence. As a result, the sentence reads like a song threaded together by these consonants and vowels running through it all and lending it co-*hear*-ance.

Meanwhile, the syntax is ambiguous in its coordination, making room for ecopoetic polysemy. The commas tying together the nominal clauses after "spiders spinning webbed words," may indeed read as either listing various wonders—"spiders spinning," "whip-poor-wills," and "white owls," and so on—or as part and parcel of the spiders' spinning together of webbed words and worldly wonders. For the many readers familiar with the children's tale *Charlotte's Web*, where the spider heroine saves her pig-friend from slaughter by spinning into her web human words designating the pig's precious qualities, thereby signaling to humans the value of her fellow's existence, it is tempting to read this passage intertextually—that is, as revealing Rash's spelling out of the wonderful qualities of the endangered birds and other creatures he would like to call our attention to.

In a way that might bring back to mind the archetype of the creation Goddess Spider Woman, weaving the world into being through her thought-poems or vibrant chanting, Rash's spinning critters may read as avatars of the ecopoet weaving together world and word. Hence the end of this passage, with Becky musing that *"word* and *wonder* and *world* [can] be one."[34] Again here, the voiced alliteration in the semi-vowel [w] and the consonant in [d] strike the sonorous chords of language. The chiming conveyed by the polysyndeton moreover involves both paranomasia ("word/ world") and alphabetical diffraction (of <wo>, <r>, and <d>). The rippling of the words "word," "wonder," and "world" into one another thus aurally and visually performs the bringing into one at the core of the matter ("could be one"). Moreover, the place occupied by the word "wonder," at the heart of the polysyndeton, intimates the centrality of wonder in ecopoetic wor(l)ding. Finally, it must be observed that the phoneme [w] breathed through Rash's poetic formula is derived from the quizzical, Indo-European [kw], conveying puzzlement, or questioning.[35] The breathy whispered quality of Rash's murmuring prose thus translates the utter wonder at the world's many cryptic and polyphonic utterances, the mystery and muttering of which palpitates at the heart of *ecopoiesis.*

NOTES

1. Carson, *Wonder*, 84.
2. Ibid, 85.
3. Ibid, 85, emphasis added.
4. Kingsolver, *Prodigal*, 1–2.
5. Ibid, 51–52.
6. Ibid, 52.
7. Bergé, "Geste," 127.
8. Kingsolver, *Prodigal*, 31.
9. Ibid, 31.
10. Ibid, 31–32.
11. This passage may bring to mind the ecoeroticism that imbues much of Terry Tempest Williams's writing, as she often describes corporeal engagement with land, water, and air as a form of lovemaking.
12. Knickerbocker, *Ecopoetics*, 13.
13. Following the same conference, French Scholar Frédérique Spill published a brilliant monograph on the poetics of wonder at the heart of Rash's work (*Radiance*). Her chapter on *Above the Waterfall* tackles aspects of the novel that are closely connected to and complementary with mine. I am originally indebted to her in my discovery of Ron Rash's writing. Moreover, I have found great pleasure in our conversations

around her beautiful chapter "Ron Rash's *Above the Waterfall*, or the Square Root of Wonderful," included in *Dwellings*, a collection of essays I have edited.

14. The following pages expand upon analyses first developed for a paper published in French: Meillon "Chant."

15. Rash, *Waterfall*, 3.

16. For more on this aspect, see Spill, "Square," and *Radiance*.

17. Rash, *Waterfall*, 178.

18. Ibid, 119.

19. Ibid, 13.

20. Keller, *Story*, 11.

21. Rash, *Waterfall*, 103.

22. See Hopkins' poem, "Ribblesdale": "And what is Earth's eye, tongue, or heart else, where/ Else, but in dear and dogged man?"

23. Rash, *Waterfall*, 13.

24. This scene is not without similarities with the enchanting moment David Abram recounts, when he contemplates the flashing of fireflies reflected over rice paddies at night, triggering an anamorphosis that shifts his center of gravity and gaze onto the world, and inducing a trance-like state, *Spell*, 3–4.

25. Rash, *Waterfall*, 104.

26. See Andrea Cavagna et al., "Starling," or Eric Bonabeau et al., *Swarm*.

27. Allen, *Sacred*, 103.

28. While David Abram's work has been in the limelight of recent studies, there are many earlier seminal studies subtending my take on an ecopoetics of reenchantment, reviving the primal entanglements between the song of the earth and human speech. During my Postmaster's degree at the University of Toulouse, France, I was exposed to fascinating studies in diachronic linguistics going back to Indo-European and raising new questions on the emergence of language. I am particularly indebted to Professor Dennis Philps for encouraging me to read Merlin Donald's comprehensive study, *Origins of the Modern Mind: Three Stages in the Evolution of culture and Cognition*. Donald's robust chapter on pre-Darwinian and Darwinian theories of language still underlies my own research. Donald recapitulates: "Our primal language was thought to be concrete—that is, lacking in abstract concepts—and based on facial expressions, gestures of the hand and body, and primitive vocalizations that had an imitative quality (Condillac, 1746; Rousseau, 1755). The earliest language-like sounds were either monosyllabic interjections (Vico, 1750) or sing-song modulation (Blackwell, 1735)." *Origins*, 25. Donald synthetizes Darwin's theory of the emergence of a specifically human articulate language—itself the second stage of development made possible by a first stage during which human speech must have evolved from a rudimentary form of song, *Origins*, 31–35. Of particular interest during my student years in Toulouse were Jean-Rémi Lapaire's thought-provoking lectures on metaphoric language and thinking and his reading recommendations on embodied thinking, such as *Philosophy in the Flesh: The Embodied Mind and its Challenge to Western Thought* (1999) by George Lakoff and Mark Johnson. Although this goes way back, I wish to express gratitude for how these mind-opening courses have had a lasting influence over my work in ecopoetics.

29. Abram, *Spell*, 76.
30. Ibid, 76.
31. Ibid, 76.
32. Rash, *Waterfall*, 136.
33. Ibid, 136.
34. Ibid, 137.
35. The [Kw] phoneme is held to have branched out into Latin and Germanic languages where it still marks many interrogative pronouns such as "who, what, why, when, where," "qui, que, quoi" in French, or "quien, que, quale, quando, quanto" in Spanish, and in the semantics of lexical words such as "quest," or "question." Following Vico's theory of language as evolved from primitive vocalizations, it can be hypothesized that [kw] corresponds to a monosyllabic interjection which potentially first arose spontaneously in primal forms of language as an immediate, affective and bodily response to the feeling of being faced with something quizzical. See Donald, *Origins*, 25.

Conclusion

Via my analyses of liminal realism and ecopoetic reading grid, this book has ventured new interpretations of texts penned by writers who had previously garnered profuse academic attention, such as Giono, Kingsolver, Powers, Proulx, or Silko, while casting light on work by comparatively less widely acclaimed writers such as Ann Pancake, Linda Hogan, Ron Rash, Susan Power, Starhawk, or Robin Wall Kimmerer. Having clarified what an ecopoetics of reenchantment may be, and how it can be useful in the light of the history of disenchantment, this monograph has explored important indigenous, ecofeminist, and postmodernist contributions triggering wonder in a way that, ultimately, brings us to ponder the possibility of a rational postmodern reenchantment. As Serenella Iovino and Serpil Oppermann argue: "Recognizing the vitality of things in all natural-cultural processes, and the co-extensivity of language and reality, ecological postmodernism perceives nature as being primarily constituted of interacting, interrelated phenomena. Its intention to 're-enchant' reality," they go on, "claiming that all material entities, even atoms and subatomic particles have some degree of sentient experience and that all living things have agency of their own, is essential in the making of the new materialist approaches, especially those that place emphasis on ontology and politics."[1] Throughout this monograph, the reenchantment of the world and its co-*responding ecopoiesis* have been tackled via approaches branching out from science as well as from the arts and humanities. In the end, this book has evidenced that an ecopoetics of reenchantment has a crucial part to play in building a postmodern onto-epistemology guided by new materialism, or, as David Abram might say, by "matter-realism."[2]

The ecofeminist and ecopoetic dimensions of this reenchantment project have been investigated throughout, specifically as I have examined the remystification of narrative occurring via the rehabilitation of the many faces of Gaia in the Anthrop-o(bs)cene. Focusing on the braiding of various ontologies within a Western context, close readings of the animistic and totemic dimensions of the stories at hand have furthermore led to the theorizing of "liminal realism"—a mode that shares much with magical realism, but that is approached here through an ecopoetic lens for the way that it specifically negotiates an interspecies kind of magic, situating us in-between human and other-than-human worlds. This book has demonstrated that by adopting a stance in-between scientific, mythical, and poetic worldviews, the interweaving of separate onto-epistemologies that is characteristic of liminal realism promotes a worldview based on relationships of reciprocity and symbiosis, while it restores and enhances our capacity for wonder and our sensitive intelligence. Focusing on the marked resurfacing of the song of the earth topos in my corpus, I have moreover demonstrated how the rematerializing of language and the sensuous *poiesis* at play in the texts under scrutiny work together to produce poetic echoes of the biophony and geophony composing the world's polyphonic soundscapes. Also paying a great deal of attention to the odorscapes and haptic dimensions of the world that are enmeshed in the earthly textures relayed in my corpus, I have revealed the initiatory function of an *ecopoiesis* that works like postmodern shamanism, mediating between human and other-than-human worlds. Starting with a thorough analysis of ecopoetic anamorphosis, which is central to the art of reenchantment, I have then tied the practice of ecopoetic anamorphosis to shapeshifting, to cross-species metaphors, and ultimately, to a kind of shamanic interspecies shifting of perspectives and multisensorial stereoscoping revealing some of the many visible and invisible dimensions of the world.

Throughout, this monograph offers an original cross-Atlantic take on ecopoetics as it straddles the two academic worlds and sparks a conversation between artworks, theory, and studies emerging from the English-speaking world as well as from Francophone contexts, while dealing with burning issues across the Environmental Humanities and Sciences. In addition, this book on reenchantment rehabilitates some of the pioneering work on ecofeminism and ecopsychology the better to show how early thinkers in those fields have propounded precious concepts and visions to solve today's many intermingled crises. It thus paves the way for a postmodern ecological way out of Modernity's disenchanted dead ends. Simultaneously, it gives evidence of the usefulness of transdisciplinary ecopoetic reading grids tapping into insights garnered from science while looking at ecoliterature exhibiting dense poetic qualities.

In this book, I have specifically demonstrated how an ecopoetics of reenchantment can sharpen our capacity to respond to the world in ways making room for senses other than sight. I have provided a methodology to analyze how the material texture of ecopoetic prose is concretely interwoven within the material textures of the world, spelling out the invisible soundscapes, odorscapes, and feelscapes that co-constitute the vibrant flesh of the world. This study thus establishes how environmental arts and literature can reinitiate us into a pluriverse that is inhabited by many different species and composed of many different elements, both biotic and abiotic, each possessing different means of agency and expression. It substantiates David Abram's claim that "[our] own chatter erupts in response to the abundant articulations of the world," as "human speech is simply our part of a much broader conversation."[3]

This book is unique in its ecopoetic take on the notion of reen-*chant*-ment as it demonstrates how ecopoets write as mediators between humans and other-than-human worlds, spelling a language ringing with many poetic echoes of the earth—in particular of its soundscapes—and reattuning us to the world's wild music. As Bernie Krause's work has shown, together with research caried out by bioacousticians and ethologists studying bird and cetacean populations in particular, our capacity to decipher the polyphony composing a multispecies soundscape can have very concrete effects. Indeed, we may thereby reveal those places where human encroachment and noise pollution threaten the health and reproductive capacity of certain species in need of harmonious territories to navigate the world, communicate, and mate. Such studies have percolated into the healthy regulation of anthropic activity in certain areas.

Meanwhile, ecologists and plant neurobiologists such as Monica Gagliano or Stefano Mancuso have illuminated the impact of music on plants—discoveries which shed new light on indigenous traditions associating religious singing and dancing with fertility rituals that indeed prove to enhance plant growth. As Gayil Nalls puts it, "music not only accelerates growth of plants but also significantly influences the concentration of various metabolites. Research [. . .] found that music not only increased oxygen uptake, 'the frequencies in these vibrations facilitate the physiological processes like nutrient absorption, photosynthesis, protein synthesis and an overall development of healthier plants with better yield.'" Relying on recent research that validates Hopi corn rituals, Nalls writes: "Life is sacred, with sensorial dimensions and connections that are sometimes beyond our immediate understandings. We must find a way of relating to the earth as a living organism and understand the inherent interconnected and interdependent natures of ecosystems."[4] Throughout, this study aligns with ecofeminism in its reclaiming of our corporeal intelligence and vitality, which moreover makes room for an ecopoetic spirituality grounded in earthly matter, energy, and creativity.

Having sifted my corpus for clues as to how an ecopoetics of reenchantment might pave the way for us to "[stay] with the trouble," as Donna Haraway puts it, or to keep dwelling ecopoet(h)ically in a damaged world, I have shown how reenchantment is about rehabilitating our sense of wonder, remedying today's exacerbated tendencies toward compassion fatigue and general hyposensitivity, and subsequently, restoring those parts of the world the future of which still hinges on our capacity to respond fast and appropriately to the current crisis. While modern theories of human language have long been fraught with cognitive biases influenced by the myth of human exceptionalism, making us believe in our capacity to wrench ourselves from nature, an ecopoetics of reenchantment illuminates the many ways in which language, as Knickerbocker eloquently puts it, is first and foremost "infused by the natural world."[5] *Ecopoiesis*, Knickerbocker has proved, "[performs] the complexity, mystery, and beauty of nature rather than merely [representing] it."[6] Throughout, this monograph corroborates Knickerbocker's definition of ecopoetics as "the foregrounding of poetic artifice as a manifestation of our interrelation with the rest of nature."[7] Moreover, in an epoch where many children form a worldview stymied by what Robert Pyle has called an "extinction of experience" of the more-than-human world,[8] this book impels an urgent reconsideration of what is most needed in our scientific, academic, and cultural curricula since, to quote Schiller's visionary understanding of the dangers inherent in the disenchantment of the world, and the subsequent need for an "aesthetic modulation of the psyche," "there is no other way of making sensuous man rational except by first making him aesthetic."[9] Indeed, as I have shown, by reviving our embodied intelligence and vitality, *ecopoiesis* possesses the great power to reverse the castration of the sensitive operating in most patriarchal, modern educational systems.

Having favored a cogent, comprehensive approach of a few novels over a more superficial catalogue approach, I have come to leave aside many of the reenchanting texts I had initially intended to explore via close readings. Yet, I presently have high hopes of seeing other researchers pick up the work many of us have started. There is much to be said for instance about the ecofeminist ecopoetics of reenchantment underlying Toni Morrison's *Song of Solomon* (1977), *Tar Baby* (1981), and more recently, Jean Hegland's *Into the Forest (1996)*, Delia Owens's *Where the Crawdads Sing* (2018), or Barbara Kingsolver's *Unsheltered* (2018). As I am nearing the completion of this book, I am simultaneously delighted to be working closely with PhD candidate Noémie Moutel, who has been hard at work in the past few years synthesizing Theodore Roszak's seminal contribution to ecopsychology and ecofeminism and trying to assess his influence on contemporary fiction. Noémie Moutel, having recently shown interest in adjoining some of the above novels to her corpus, we may look forward to exciting new developments in an ecopoetics of reenchantment.

Other novels that belong to postcolonial literature would benefit from being read in the light of ecopoetic liminal realism. This is the case for instance of *The Painted Drum* (2005), by Louise Erdrich, which magically ties the fate of a family to a totemic drum across several generations. The same goes for Alexis Wright's *The Swan Book* (2016) and the main character Oblivia's liminal experience at the heart of a eucalyptus tree early in the story. Lost in the forest—it is suggested that the child has landed there after surviving gang rape—she sojourns in a womb-like gum tree that teaches her a kind of sign language, moving her fingers "in slow swirls like music," "writing stanzas in ancient symbols" which, "dredged from the soup of primordial memory in these ancient lands," corresponds either to "the oldest language coming to birth again instinctively," or, to the "unconscious [child's] forming [of] words that [resemble] the twittering of birdsong speaking about the daylight [. . .], the old ghost language of warbling and chortling remembered by the ancient river gum."[10]

Being limited by length, I have also had to give up on including my analyses of French writer Anne Sibran's liminal realist ecopoetics in her autobiographical novel *Enfance d'un chamane* (2017). Nevertheless, Anne Sibran was the guest of honor at a one-day conference on an ecopoetics of biodiversity co-organized between the University of Perpignan Via Domitia and the Catholic University of Milan, with Davide Vago, in June 2021. As we are currently coordinating a collective book as a spinoff of the event, I am writing a chapter that takes up the work carried out in the present book to apply it to Anne Sibran's writing. Also written halfway between the ethnographic account and the novel, Nastassja Martin's *Croire aux fauves* (2019) is profoundly ecopoetic as it recounts the woman's negotiation of a liminal zone in-between bear and human realms. Another French prose writer whose ecopoetic feat deserves close readings along the lines delineated throughout this monograph—and like Sibran's, for both its liminal realism and the song of the earth relayed in the writing—is best-selling novelist Alain Damasio. I would like to encourage further exploration of his densely ecopoetic *La horde du contrevent* (2004), and his latest novel, *Les Furtifs* (2019). Beyond the uplifting thought-experiments packed into his fiction, the thick musicality of Damasio's wild prose takes up the world's unfurling creativity and polyphonic song.

Many recent novels that might fit the definition of the cli-fi genre equally call for more attention to disenchantment/reenchantment dialectics. This concerns for instance Margaret Atwood's *Oryx and Crake* trilogy, casting humming, genetically modified humans; or George R. R. Martin's *A Song of Ice and Fire* (1996), staging a character who, going against what the poetic title of the novel suggests, claims that "[life] is not a song,"[11] as opposed to the "forest people"—they who are called "the children of the forest" and who, in their own "True Tongue," call themselves "those who sing the song of the earth."[12] Ringing a bell with some of the points discussed throughout this monograph, these indigenous forest people exhibit many liminal traits and contend: "Before your old tongue was ever spoke, we had sung our songs ten thousand years."[13]

Of particular interest, and going down a very different historiographic and ontological route, is Claire Vaye Watkins's cli-fi novel *Gold Fame Citrus*

(2015), offering a highly dysphoric picture of both our disenchanted world and the pernicious reenchantment that cult leaders and other religious quacks can invest.[14] In Watkins's environmental fiction—inspired in great part by her father's biography, the latter having been Charles Manson's closest follower—the humming of the gigantic sand dune threatening to swallow up the landscape turns out to be no more than a myth perversely recycling the song of the earth motif the better to seduce gullible, adrift characters who are desperate for a sense of existential meaning and connection. Having benefited from the cautionary tale flagging the potential pitfalls of reenchantment encapsulated in the dark historical episode written by Charles Manson and his "Family," Watkins provides insight into various facets of disenchantment and reenchantment, both of which come with their own merits, booby traps, and dangers.

Finally, much great work is also waiting to be carried out in ecocinema in the light of reenchantment. While writing the last part of this book, I was brought in to supervise an application for a PhD scholarship integrated in my Perpignan research laboratory, the CRESEM, for a transdisciplinary study of ecocinema via a philosophical, ecopoetic, and ecofeminist lens. I am thrilled as I have started co-advising, with Elise Domenach, PhD candidate Julie Fortin who, in the next two years, will be conducting groundbreaking research on the way ecofeminist cinema can help reframe our environmental imagination.[15] From our collaboration so far, the thesis holds great promise of demonstrating how ecocinema promotes an ecopoetics of reenchantment and encourages biophilia, rather than the ecophobic responses to the world induced by more canonic Western movies, or, more recently, by many cli-fi movies and series. Indeed, as I hope to have made clear in this study, in forcing us to deny the many ways in which the more-than-human world is animated, sentient, and expressive, and in requiring that we repress our biophilic tendencies so as to embrace the myth of the autonomous reasonable self, held to be in control of both its own bodymind and its "environment," disenchantment plays a key role in the breeding of ecophobia. On the contrary, because it is fundamentally enchanting, ecofeminist, indigenous, and neopagan *ecopoieses* cultivate the biophilic tendencies inherent in humankind, allowing us to remember that we are partly animal, and, as Thoreau famously put it, that we are "leaves and vegetable mold" ourselves. Thus, may we come "to have," or to regain, "intelligence with the earth."[16]

I am hopeful that as we keep studying ecofeminist, indigenous, neopagan, and neo-indigenous works of art, we may be taking part in a much needed restor(y)ation and repoetization of the world. In doing so, we may promote immersive experiences, through fiction and other forms of *ecopoiesis*, that offer insight into enchanting worlds not "beyond our own," as James William Gibson had phrased it, but essentially *within* our own dwelling places.

Whether in our Covid-ridden times, or for obvious environmental reasons that should urge us to fly less to the faraway places many of us tend to be attracted to when we hanker for adventurous exploration, may we be enlightened by ecopoetic art revealing how we can in fact travel to amazing worlds right at the edge of the ones we inhabit, enfolded in our own, worlds at such a close distance that they are oftentimes just one sensitive or perceptive shift away.

NOTES

1. Iovino and Oppermann, "Material," 78.
2. Abram, *Becoming*, 10.
3. Ibid, 172.
4. I am indebted to Terry Harpold for drawing my attention to Nalls's paper on "The Power of Harmony."
5. Knickerbocker, *Ecopoetics*, 185
6. Ibid, 159.
7. Ibid, 159.
8. See Pyle, *Thunder*.
9. Schiller, *Aesthetic*, 161, qtd in Bennett, *Enchantment*, 203 and 142.
10. Wright, *Swan*, 6–7.
11. Martin, *Game*, 473.
12. Martin, *Dance,* 175. I am indebted to Raphaelle Moncomble, whose master's thesis has brought my attention to the ecopoetic dimension of Martin's work.
13. Martin, *Dance*, 175.
14. I am grateful to Claire Perrin, whose PhD on the representation of drought and climate change in cli-fi brought my attention to Clare Vaye Watkins's novel.
15. I am thankful for Pascale Amiot's precious support, as well as for Elise Domenach's enthusiastic response to the whole project. Not being competent to supervise research in film studies and philosophy, I have called onto specialist Elise Domenach to oversee those aspects of Julie Fortin's research.
16. Thoreau, *Walden*, 114.

Bibliography

Abram, David. *The Spell of the Sensuous*. New York: Vintage Books, 1996.
———. *Becoming Animal: An Earthly Cosmology*. New York: Vintage Books, 2010.
Adamson, Joni. "Why Bears Are Good to Think and Theory Doesn't Have to Be Murder: Transformation and Oral Tradition in Louise Erdrich's *Tracks*." *Studies in American Indian Literatures* 2, 4, no. 1 (Spring 1992): 28–48.
———. *American Indian Literature, Environmental Justice, and Ecocriticism: The Middle Place*. Tucson: University of Arizona Press, 2001.
———. "Whale as Cosmos: Multi-Species Ethnography and Contemporary Indigenous Cosmopolitics." *Revista Canaria De Estudios Ingleses*. 64 (April 2012): 29–45.
———. "Source of Life: *Avatar*, Amazonia, and an Ecology of Selves." In *Material Ecocriticism*, edited by Serenella Iovino and Serpil Oppermann, 253–68. Bloomington: University of Indiana Press, 2014.
Adamson Joni, and Salma Monani, "Cosmovisions, Ecocriticism, and Indigenous Studies." In *Ecocriticism and Indigenous Studies*, edited by Salma Monani and Joni Adamson, 1–19. New York: Routledge, 2016.
Alaimo, Stacy. *Bodily Natures: Science, Environment, and the Material Self*. Bloomington: University of Indiana Press, 2010.
Albee, Edward. *Who's Afraid of Virginia Woolf*. 1962. London: Penguin Books, 1983.
Albrecht, Glenn. "Psychoterratic Conditions in a Scientific and Technological World." In *Ecopsychology: Science, Totems, and the Technological Species*, edited by Peter H. Khan, Jr. and Patricia H. Hasbach, 241–64. Cambridge, MA: The MIT Press, 2012.
———. *Earth Emotions: New Words for a New World*. Ithaca, NY: Cornell University Press, 2019.
Allen, Paula Gunn. *The Sacred Hoop: Recovering the Feminine in American Indian Tradition*. 1986. Boston: Beacon Press, 1992.
———. "May it Be Beautiful All Around." In *Hozho: Walking in Beauty*, edited by Paula Gunn Allen and Carolyn Dunn Anderson, xi–xxiv. Chicago: Contemporary Books, 2001.
Ashcroft, Bill, Gareth Griffiths and Helen Tiffin. *Postcolonial Studies: The Key Concepts*. 1998. London: Routledge, 2013.

Asquith, Mark. *The Lost Frontier: Reading Annie Proulx's Wyoming Stories.* New York: Bloomsbury, 2014.
Atwood, Margaret. *Oryx and Crake.* London: Bloomsbury, 2003.
Bakhtin, Mikhail. *The Dialogic Imagination.* Translated by Emerson, Caryl and Michael Holquist. 1981. Austin: University of Texas Press, 1998.
Barad, Karen. *Meeting the Universe Halfway: Quantum Physics and the Entanglement of Matter and Meaning.* Durham, NC: Duke University Press, 2007.
Barthes, Roland. *Le plaisir du texte.* Paris: Editions du Seuil, 1973.
Bate, Jonathan. *The Song of the Earth.* Boston: Harvard University Press, 2000.
Bateson, Gregory. *Steps to an Ecology of Mind.* London: Paladin, 1973.
Baudelaire, Charles. *Les fleurs du Mal.* 1981. Paris: Classiques Larousse, 19e Edition.
Bennett, Jane. *The Enchantment of Modern Life: Attachments, Crossings, and Ethics.* Princeton, NJ: Princeton University Press, 2001.
Berendt, Joachim-Ernest. *The World is Sound, Nada Brahma: Music and the Landscape of Consciousness.* 1983. Rochester, VT: Destiny Books, 1991.
Bergé, Aline. "La geste forestière de Robert Marteau: une écopoétique du vivant." In *La Forêt Sonore*, edited by Jean Mottet, 117–35. Ceyzérieu: Editions Champ Vallon, 2017.
Berman, Morris. *The Reenchantment of the World.* Ithaca, NY: Cornell University Press, 1981.
Bernstein, Charles, ed. *Close Listening: Poetry and the Performed Word.* New York: Oxford University Press, 1998.
Berry, Thomas. *The Dream of the Earth.* 1988. Berkeley, CA: Counterpoint, 2015.
Bhabha, Homi K. *The Location of Culture.* London: Routledge, 1994.
Bohm, David. "Postmodern Science and a Postmodern World." In *The Reenchantment of Science,* edited by David Ray Griffin, 57–68. Albany: State of New York Press, 1988.
Bonabeau, Eric, Marco Dorigo and Guy Therulaz. *Swarm Intelligence: From Natural to Artificial Systems.* Oxford: Oxford University Press, 1999.
Bonneuil, Christophe and Jean-Baptiste Fressoz. *L'Evènement anthropocène: La Terre, l'histoire et nous.* Paris: Editions du Seuil, 2016.
Bowers, Maggie Ann. *Magic(al) Realism.* London: Routledge, 2004.
Brun, Frédéric, ed. *Habiter poétiquement le monde: Anthologie Manifeste.* Paris: Poesis Editions, 2016.
Buber, Martin. *Between Man and Man.* 1947. Translated by Ronal Gregor-Smith. London: Routledge, 2002.
Buell, Lawrence. *The Environmental Imagination: Thoreau, Nature Writing, and the Formation of American Culture.* Cambridge, MA: Harvard University Press, 1995.
———. *Writing for an Endangered World: Literature, Culture, and Environment in the U.S. and Beyond.* Cambridge, MA: Belknap Press of Harvard University Press, 2001.
Bushdid, C., M. O. Magnasco, L.B. Vosshall and A Keller. "Humans Can Discriminate More than 1 Trillion Olfactory Stimuli." *Science* 343, no. 6177 (March 21, 2014): 1370–72.

Callicot, J. Baird. *Beyond the Land Ethic: More Essays in Environmental Philosophy*. Albany, NY: State University of New York Press, 1999.
Cameron, Sharon. *Writing Nature: Henry David Thoreau's Journal*. Oxford: Oxford University Press, 1985.
Campagne, Armel. *Le Capitalocène: Aux racines historiques du dérèglement climatique*. Paris: Divergences, 2017.
Capra, "Foreword." In *The World is Sound, Nada Brahma: Music and the Landscape of Consciousness*, by Joachim-Ernest Berendt, xi–xiii. 1983. Rochester, VT: Destiny Books, 1991.
Capra, Fritjof, and Charlene Spretnak. *Green Politics*. New York: E.P. Dutton, 1984.
Carson, Rachel. *Silent Spring*. 1962. Boston: Mariner Books, 2002.
———. *The Sense of Wonder*. 1964. New York: HarperCollins Publishers, 1998.
Cavagna Andrea, Alessio Chiarelli, Irene Giardina, Giorgio Parisi, Raffaele Santagati, Fabio Stefanini, and Massimiliano Viale. "Scale-Free Correlations in Starling Flocks." *PNAS* 107 no. 26 (June 29, 2010): 11865–70.
Cazajous-Augé. "A Poetics of Traces in Rick Bass's Short Stories." In *Dwellings of Enchantment: Writing and Reenchanting the Earth*, edited by Bénédicte Meillon, 177–89. Lanham, MD: Lexington Books, 2021.
Chatwin, Bruce. *The Songlines*. New York: Penguin Books, 1987.
Choné, Aurélie, Isabelle Hajek, and Philippe Hamman. *Rethinking Nature: Challenging Disciplinary Boundaries*. London: Routledge, 2017.
Christ, Carol P. *Diving Deep and Surfacing: Women Writers on Spiritual Quest*. 1980. Boston: Beacon Press, 1995.
Clément, Gilles. *Le jardin planétaire: réconcilier l'homme et la nature*. Paris: Albin Michel, 1999.
Cochoy, Nathalie. "The Imprint of the 'Now' on the Skin of Discourse: Annie Dillard's Pilgrim at Tinker Creek." *Revue Française d'Etudes Américaines*, no. 106 (2005): 33–49.
———. "Le réalisme infime d'Annie Dillard dans *Teaching A Stone to Talk*." *Caliban*, 19 (2006): 291–98.
———. "Dillard Dancing: An American Childhood." In *Environmental Awareness and the Design of Literature*, edited by François Specq, 136–47. Brill, 2017.
Cohen, Robin. "Wild Indians: Kingsolver's Representation of Native America." In *Seeds of Change*, edited by Priscilla Leader, 145–56. Knoxville: University of Tennessee Press, 2010.
Collot, Michel. *Le chant du monde dans la poésie française contemporaine*. Clermont-Ferrand: Editions Corti, 2019.
Conn, "When the Earth Hurts, Who Responds?" In *Ecopsychology: Restoring the Earth, Healing the Mind*, edited by Theodore Roszak, Mary E. Gomes, and Allen D. Kanner, 156–71. San Francisco: Sierra Club Books, 1995.
Cooper, Lydia R. "'Woman Chasing Her God': Ritual, Renewal, and Violence in Linda Hogan's *Power*." *ISLE* 18, no. 1 (Winter, 2011): 143–59.
Cuddon, J. A. *Dictionary of Literary Terms and Literary Theory*. 1977. London: Penguin Books, 1991.
Cyrulnik, Boris. *Un merveilleux malheur*. Paris: Editions Odile Jacob, 1999.

Damasio, Alain. *La horde du contrevent*. Clamart: La Volte, 2004.

———. *Les Furtifs*. Clamart: La Volte, 2019.

Damasio, Antonio. *Descartes' Error: Emotion, Reason, and the Human Brain*. New York: Penguin Books, 1994.

Darwin, Charles. *On the Origin of Species*. 1859. Cambridge, MA: Harvard University Press, 1964.

Davis, Christina. "An Interview with Toni Morrison." In *Conversations with Toni Morrison*, edited by Taylor-Guthrie, 223–34. Jackson: University Press of Mississippi, 1994.

Davis, John. "Psychological Benefits of Nature Experiences: An Outline of Research and Theory with Special Reference to Transpersonal Psychology." 2004, Retrieved from: http://psichenatura.it/fileadmin/img/J._Davis_Psychological_benefits_of_Nature_experiences.pdf.

De la Cadena, Marisol. "Indigenous Cosmopolitics in the Andes: Conceptual Reflections beyond Politics." *Cultural Anthropology* 25, no. 2 (April 2010): 334–70.

Deleuze, Gilles and Felix Guattari. *Mille Plateaux*. Paris: Editions de Minuit, 1980.

DeLoughrey, Elizabeth, and George B. Handley, eds. *Postcolonial Ecologies: Literatures of the Environment*. Oxford: Oxford University Press, 2011.

Descartes, René. *Discours de la méthode*. 1637. Paris: Bibliothèque de la Pléiade, Gallimard, 1966.

Descola, Philippe. *Les lances du crépuscule*. Paris: Plon, 1993.

———. *Par-delà nature et culture*. Paris: Editions Gallimard, 2005.

Després, Olivier. "Mécanismes de localisation spatiale chez l'homme: interaction entre le système visuel et le système auditif." PhD Diss., Université de Louis Pasteur, 2004.

D'haen, Theo L. "Magical Realism and Postmodernism: Decentering Privileged Centers." In *Magical Realism: Theory, History, Community*, edited by Lois Parkinson Zamora and Wendy B. Faris, 191–208. Durham, NC: Duke University Press, 1995.

Diamond, Irene, and Gloria Feman Orenstein, eds. "Introduction." In *Reweaving the Wounds: The Emergence of Ecofeminism*, ix–xv. San Francisco: Sierra Club Books, 1990.

Diamond, Jared. *Collapse: How Societies Choose to Fail or Succeed*. 2005. New York: Penguin Books, 2006.

Dillard, Annie. *Teaching A Stone to Talk: Expeditions and Encounters*. 1982. New York: Harper Perennial, 1992.

Donald, Merlin. *Origins of the Modern Mind: Three Stages in the Evolution of Culture and Cognition*. Cambridge, MA: Harvard University Press, 1991.

Ducarme, Frédéric and Denis Couvet. "What does 'nature' mean?" *Palgrave Communications* 6, no.14 (2020). https://doi.org/10.1057/s41599-020-0390-y.

Durand-Rous, Caroline. "Le totem réinventé ou la transmission du mythe par la littérature amérindienne." *L'Atelier* 6, no 2. (2014): 103–20.

———. "Le totem réinventé: exploration de l'identité et redéfinition de soi dans la fiction amérindienne contemporaine." PhD diss., University of Perpignan Via Domitia, 2017.

Durix, Jean-Pierre. *Mimesis, Genre, and Post-Colonial Discourse: Deconstructing Magical Realism*. New York: St. Martin's, 1998.

Durkheim, Emile. *Les formes élémentaires de la vie religieuse*. 1985. Paris: Presses Universitaires de France, 1912.

Eisler, Riane. "The Gaia Tradition and Partnership Future: An Ecofeminist Manifesto." In *Reweaving the World: The Emergence of Ecofeminism*, edited by Irene Diamond and Gloria Feman Orenstein, 23–34. San Francisco: Sierra Club Books, 1990.

Elvey, Anne. "The Matter of Texts: A Material Intertextuality and Ecocritical Engagements with the Bible." In *Ecocritical Theory: New European Approaches*, edited by Axel Goodbody and Kate Rigby, 181–93. Charlottesville: University of Virginia Press, 2011.

Emerson, Ralph Waldo. "Nature." In *The Essential Writings of Ralph Waldo* Emerson, edited by Brooks Atkinson, 5–39. New York, The Modern Library, 2000.

Erdrich, Louise. *The Painted Drum*. New York: HarperCollins Publisher, 2005.

Estok, Simon. *The Ecophobia Hypothesis*. New York: Routledge, 2018.

Fancourt, Donna. "Accessing Utopia through Altered States of Consciousness: Three Feminist Utopian Novels." *Utopian Studies* 13, no.1 (2002): 94–113.

Faris, Wendy, B. "Scheherazade's Children." In *Magical Realism: Theory, History, Community*, edited by Lois Parkinson Zamora and Wendy B. Faris, 163–90. Durham: Duke University Press, 1995.

———. *Ordinary Enchantments: Magical Realism and the Remystification of Narrative*. Nashville: Vanderbilt University Press, 2004.

———. "'We, the Shamans, Eat Tobacco and Sing,' Figures of Shamanic Power in US and Latin American Magical Realism." In *Moments of Magical Realism in US Ethnic Literatures*, edited by Lyn Di Iorio Sandín, 155–74. New York: Palgrave Macmillan, 2012.

Flys-Junquera, Carmen. "Conversations with the Living World: Mutual Discovery and Enchantment." In *Dwellings of Enchantment: Writing and Reenchanting the Earth*, edited by Bénédicte Meillon, 193–206. Lanham, MD: Lexington Books, 2021.

Forbes, Jack D. *Columbus and Other Cannibals: The Wétiko Disease of Exploitation, Imperialism, and Terrorism*. 1979. Revised edition. New York: Seven Stories Press, 2008.

Frumkin, Howard. "Building the Science Base: Ecopsychology Meets Clinical Epidemiology." In *Ecopsychology: Science, Totems, and the Technological Species*, edited by Peter H. Khan, Jr. and Patricia H. Hasbach,141–72. Cambridge, MA: The MIT Press, 2012.

Gaard, Greta. "Mindful New Materialisms: Buddhist Roots for Material Ecocriticism's Flourishing." In *Material Ecocriticism*, edited by Serenella Iovino and Serpil Oppermann, 291–300. Bloomington: Indiana University Press, 2014.

Gagliano, Monica. *Thus Spoke the Plant*. Berkeley, CA: North Atlantic Books, 2018.

Gavillon, François. "Magical Realism, Spiritual Realism, and Ecological Awareness in Linda Hogan's *People of the Whale*." *ELOHI Indigenous Peoples and the Environment* 3 (2013): 41–56.

Genette, Gérard. *Figures III*. Paris: Seuil, 1972.

Gerth, H. H, and C. Mills Wright, eds. *From Max Weber: Essays in Sociology*. London: Routledge & Kegan Paul, 1948.

Gibson, James William. *A Reenchanted World: The Quest for a New Kinship with Nature*. New York: Metropolitan Books, 2009.

Ginzburg, Carlo. *Le sabbat des sorcières*. 1989. Translated by Monique Aymard. Paris: Gallimard, 1992.

Giono, Jean. *Que ma joie demeure*. Paris: Grasset, 1935.

———. *Joy of Man's Desiring*. Translated by Katherine Allen Clark. Berkeley, CA: Counterpoint, 1980.

Gipe, Robert. "Straddling Two Worlds: An Interview with Ann Pancake." *Appalachian Journal* 38, no. 2/3 (Winter/Spring 2011): 170–97.

Gleick, James. *La théorie du Chaos: Vers une nouvelle science*. 1987. Translated by Christian Jeanmougin. Paris: Champs Flammarion, 1999.

Glotfelty, Cheryll. "Literary Studies in an Age of Environmental Crisis." In *The Ecocriticism Reader*, edited by Cheryl Glotfelty and Harold Fromm, xv–xxxvii. Athens: The University of Georgia Press, 1996.

Gomes, Mary E. and Allen D. Kanner. "The Rape of the Well-Maidens." In *Ecopsychology: Restoring the Earth, Healing the Mind*, edited by Theodore Roszak, Mary E. Gomes, and Allen Kanner, 111–21. San Francisco: Sierra Books Club, 1995.

Grandjeat, Yves-Charles. "Retreating and Surrendering in Barry Lopez's *Resistance*." In *Le travail de la résistance dans les sociétés, les littératures et les arts en Amérique du Nord*, edited by Grandjeat. Pessac: Maison des Sciences de l'Homme d'Aquitaine, 2008.

———. "Poetic Co-operation in the Works of 'Nature Writers of Our Own Time." *Miranda* 11 (2015): https://doi.org/10.4000/miranda.7024.

———. "Under the Volcano: Gary Snyder's Ecopoetics of Reinhabitation." Leaves 1, (2015): 278–86. http://dx.doi.org/10.21412/leaves_0121.

———. "'I Turn Homeward, Still Wondering': Reasons for Enchantment." In *Dwellings of Enchantment: Writing and Reenchanting the Earth*, edited by Bénédicte Meillon, 33–48. Lanham, MD: Lexington Books, 2021.

Griffin, Susan. *Woman and Nature: The Roaring Inside Her*. San Francisco: Sierra Club Books, 1978.

Griffin, David Ray ed. *The Reenchantment of Science: Postmodern Proposals*. Albany: State University of New York Press, 1988.

Guéry François. « Maîtres et protecteurs de la nature. » *Le Cahier, Collège International de Philosophie* 5 (Avril 1988): 163–67.

Hansen, Margaret M., Reo Jones, and Kirsten Tocchini. "Shinrin-Yoku (Forest Bathing) and Nature Therapy: A State-of-the-Art Review." *International Journal of Environmental Research and Public Health* 14, no. 8. (Jul. 2017): 851; https://doi.org/10.3390/ijerph14080851

Haraway, Donna. *When Species Meet*. Minneapolis: University of Minnesota Press, 2008.

———. *Staying with the Trouble: Making Kin in the Chthulucene*. Durham, NC: Duke University Press, 2016.

Harding, Wendy. *The Myth of Emptiness and the New Literature of Place*. Iowa City: University of Iowa Press, 2014.

Hegland, Jean. *Into the Forest*. Corvallis: Calyx, 1996.

Hogan, Linda. *Dwellings: A Spiritual History of the Living World*. New York: W. W. Norton & Company, 1995.

———. *Solar Storms*. New York: Scribner Paperback Fiction, 1995.

———. *Power*. New York: W. W. Norton & Company, 1998.

———. *People of the Whale*. New York: W. W. Norton & Company, 2008.

Hölldobler, Bert, and Edward O. Wilson. *The Superorganism: The Beauty, Elegance, and Strangeness of Insect Societies*. New York: W. W. Norton & Company, 2009.

Hopkins, Gerard Manley. *The Poems of Gerard Manley Hopkins*. 4th edition, edited by W.H. Gardner and N. H. MacKenzie. New York: Oxford University Press, 1970.

Huggan Graham, and Helen Tiffin. *Postcolonial Ecocriticism: Literature, Animals, Environment*. 2010. London: Routledge, 2015.

Hutcheon, Linda. *A Poetics of Postmodernism*: *History, Theory, Fiction*. London: Routledge, 1988.

Ingold, Tim. *The Perception of the Environment: Essays in Livelihood, Dwelling and Skill*. London & New York: Routledge, 2000.

Intelligent Trees. Directed by Julia Dordel and Guido Tölke, Dorcon Film, 2017.

Iovino, Serenella & Opperman, Serpil. "Theorizing Material Ecocriticism: A Diptych." *ISLE* 19, no. 3 (Summer 2012): 448–75.

———. "Material Ecocriticism: Materiality, Agency and Models of Narrativity." *Ecozon@* 3, no.1 (2012): 75–91.

Jabr, Ferris. "The Social Life of Forests." *New York Times*, Dec. 6, 2020. https://www.nytimes.com/interactive/2020/12/02/magazine/tree-communication-mycorrhiza.html

Jacobs, Naomi. "Barbara Kingsolver's Anti-Western 'Unraveling the Myths' in *Animals Dreams*." *Americana: The Journal of American Popular Culture 1900 to Present* 2, no. 2, (2003).

Johnson, Nathanael. *Unseen City: The Majesty of Pigeons, The Discreet Charm of Snails and Other Wonders of the Urban Wilderness*. New York: Rodale, 2016.

Jones, Jeremy, B. "Ann Pancake's *Strange as This Weather Has Been*." *Iowa Review*. Jan. 8, 2011. https://iowareview.org/blog/ann-pancake%E2%80%99s-strange-weather-has-been.

Kahn, Peter H. and Patricia H. Hasbach, eds. "Introduction to Ecopsychology." In *Ecopsychology: Science, Totems, and the Technological Species*. Cambridge, MA: The MIT Press, 2012.

Keim, Brandon. "Never Underestimate the Intelligence of Trees: Plants Communicate, Nurture their Seedlings and Get Stressed," published online: http://nautil.us/issue/77/underworldsnbsp/never-underestimate-the-intelligence-of-trees?linkId=76593830.

Keller, Hellen. *Story of my Life*. New York: Doubleday Page & Co., 1903.
Killingsworth, M. Jimmie, and Jacqueline S. Palmer, "Millennial Ecology: The Apocalyptic Narrative from Silent Spring to Global Warming." In *Green Culture: Environmental Rhetoric in Contemporary America*, edited by Carl G. Herndl and Stuart C. Brown, 21–45. Madison: The University of Wisconsin Press, 1996.
Kimmerer, Robin Wall. *Braiding Sweetgrass: Indigenous Wisdom, Scientific Knowledge, and the Teachings of Plants*. Minneapolis: Milkweed Editions, 2013.
King, Ynestra. "Healing the Wounds: Feminism, Ecology, and the Nature/Culture Dualism." In *Reweaving the World: The Emergence of Ecofeminism*, edited by Irene Diamond and Gloria Feman Orenstein, 106–21. San Francisco: Sierra Club, 1990.
Kingsolver, Barbara. "Homeland." In *Homeland and Other Stories*, 1–22. 1989. New York: Harper Perennial, 1993.
———. "Jump-Up Day." In *Homeland and Other Stories*. 1989. New York: Harper Perennial, 1993.
———. *Animal Dreams*. 1990. New York: Harper Perennial, 1991.
———. *The Poisonwood Bible*. New York: Harper Collins, 1998.
———. *Prodigal Summer*. New York: Harper Collins, 2000.
———. *Flight Behavior*. New York: Harper Collins, 2012.
———. *Unsheltered*. New York: Harper Collins, 2019.
Kish, Daniel and Johanna Hook. *Echolocation and Flash Sonar*. Louisville: American Printing House for the Blind, 2016.
Knickerbocker, Scott. *Ecopoetics: The Language of Nature, The Nature of Language*. Amherst and Boston: University of Massachusetts Press, 2012.
Kohn, Eduardo. *How Forests Think: Toward an Anthropology Beyond the Human*. Berkeley: University of California Press, 2013.
Krause, Bernie. *Le grand orchestre des animaux: Célébrer la symphonie de la nature*. 2012. Translated by Thierry Piélat. Paris: Flammarion, 2018.
———. *Chansons animales et cacophonie humaine: Manifeste pour la sauvegarde des paysages sonores naturels*. 2015.Translated by Amanda Prat-Giral. Arles: Actes Sud, 2016.
Krech, Shepard, III. *The Ecological Indian: Myth and History*. New York: Norton, 1999.
Kristeva, Julia. *La révolution du langage poétique*. Paris: Editions du Seuil, 1974.
Lacroix, Alexandre. *Devant la beauté de la nature*. Paris: Allary Editions, 2018.
Lakoff, George. "The Contemporary Theory of Metaphor." In *Metaphor and Thought*, edited by Andrew Ortony, 202–51. Cambridge: Cambridge University Press, 1993.
Lakoff George, and Mark Johnson. *Metaphors We Live By*. Chicago: University of Chicago Press, 1980.
———. *Philosophy in the Flesh: The Embodied Mind and its Challenge to Western Thought*. New York: Basic Books,1999.
Lakoff George, and Mark Turner. *More than Cool Reason: A Field Guide to Poetic Metaphor*. Chicago: University of Chicago Press,1989.
Latour, Bruno. *Nous n'avons jamais été modernes*. Paris: La Découverte, 1991.
———. *Face a Gaïa*. Paris: La Découverte, 2015.

Lauwers, Margot. « Amazones de la plume: Les manifestations littéraires de l'écoféminisme contemporain. » PhD diss. University of Perpignan Via Domitia, 2015.
Le Clézio, Jean-Marie Gustave. 1973. *Les Géants*. Paris: Gallimard, 1997.
Leopold, Aldo. *A Sand County Almanach, with Essays on Conservation from Round River*. 1949, 1953. New York: Ballantine Books, 1966.
Lévi-Strauss, Claude. *La pensée Sauvage*. Paris: Plon, 1962.
———. "Le Bon Plaisir." Radio program on France Culture, directed by Colette Fellous, October 25, 1986.
Li, Qing, ed. *Forest Medicine*. Waltham: Nova Biomedical, 2012.
Louv, Richard. *Last Child in the Woods: Saving Our Children from Nature-Deficit Disorder*. Chapel Hill: Algonquin, 2005.
Lovelock, James. "Gaia: The World as Living Organism." *New Scientist* (December 18, 1986): 25–28.
———. *Gaia: A New Look at Life on Earth*. New York: Oxford University Press, 1979.
Macé, Marielle. « Styles animaux. » *Face aux bêtes. L'Esprit Créateur* 51, no.4 (Winter 2011): 97–105.
Machet, Laurence, Lionel Larré et Antoine Ventura, eds. *The Invention of the Ecological Indian. ELOHI Indigenous Peoples and the Environment* 4 (2013).
Macy, Joanna. "Working Through Environmental Despair." In *Ecopsychology: Science, Totems, and the Technological Species*, edited by Peter H. Khan, Jr. and Patricia H. Hasbach, 240–59. Cambridge, MA: The MIT Press, 2012.
Malm, Andreas. *L'Anthropocène contre l'histoire: le réchauffement climatique à l'ère du capital*. Paris: La Fabrique, 2017.
Mancuso, Stefano. *The Revolutionary Genius of Plants: A New Understanding of Plant Intelligence and Behavior*. 2017. Translated by Vanessa Di Stefano. New York: Astria Books, 2018.
Mancuso, Stefano and Alessandra Viola. *L'intelligence des plantes*. 2013. Translated by Renaud Temperini. Paris: Albin Michel, 2018.
Manes, Christopher. "Nature and Silence." In *The Ecocriticism Reader*, edited by Cheryll Glotfelty and Harold Fromm, 15–29. Athens: The University of Georgia Press, 1996.
Margulis, Lynn. *Symbiotic Planet: A New Look at Evolution*. Amherst: Sciencewriters, 1998.
Margulis, Lynn, and Dorian Sagan. *Microcosmos: Four Billion Years of Microbial Evolution*. New York: Summit Books, 1986.
Martin, George R. R. *A Song of Ice and Fire*. Book One: *A Game of Thrones*. 1996. New York: Bantam Books, 2011.
———. *A Song of Ice and Fire*. Book Five: *A Dance with Dragons*. 2011. New York: Bantam Books, 2013.
Martin, Nastassja. *Croire aux fauves*. Paris: Gallimard, 2019.
Marx, Leo. *The Machine in the Garden: Technology and the Pastoral Ideal in America*. 1964. Oxford: Oxford University Press, 2000.

Maufort, Jessica. "The Magic Realist Compost in the Anthropocene: Improbable Assemblages in Canadian and Australian Fiction." In *Dwellings of Enchantment: Writing and Reenchanting the Earth*, edited by Bénédicte Meillon, 239–56. Lanham,MD: Lexington Books, 2021.

McKay, Nellie Y. "An Interview with Toni Morrison." In *Conversations with Toni Morrison*, edited by Taylor-Guthrie, 138–55. Jackson: University Press of Mississippi, 1994.

McKinney, Dan. *Do Trees Communicate?*, directed by Dan McKinney, Julia Dordel, 2011.

Meillon, Benédicte. "Feminine Characters and Voices in Barbara Kingsolver's *The Bean Trees*. " Master's thesis, Université de Toulouse-Le Mirail, 1999.

———. "L'implicite dans 'Stone Dreams' de Barbara Kingsolver." In *L'implicite dans la nouvelle de langue anglaise*, edited by Laurent Lepaludier, 113–36. Rennes: Presses Universitaires de Rennes, 2005.

———. "Barbara Kingsolver's 'Homeland' and Other Stories about *Another America*." *Espaces et Terres d'Amérique/Mapping American Spaces*, edited by Nathalie Dessens and Wendy Harding *Anglophonia* 19, (2006): 261–69.

———. "Aimé Césaire's *A Season in Congo* and Barbara Kingsolver's *The Poisonwood Bible* in the light of Postcolonialism." *Divergences & Convergences. Anglophonia* 21 (2007): 197–212.

———. "Literary Resistance in Barbara Kingsolver's 'Homeland.'" In *Le travail de la Résistance*, edited by Yves-Charles Grandjeat, 85–98. Presses Universitaires de la Maison des Sciences de l'Homme d'Aquitaine et Pessac, 2008.

———. "Ecopoétique du regard et de la liminalité: Entrelacs des formes du vivant dans la fiction de Linda Hogan." *La planète en partage/Sharing the Planet. Caliban* 55 (2016): 129–48.

———. "Revisiting the Universal Significance of Mythologies: Barbara Kingsolver's Syncretic and Mythopoeic Short Story 'Jump-Up Day.'" *Representation and Rewriting of Myths in Southern Short Fiction. Journal of the Short Story in English* 66 (Autumn 2016): 173–86.

———. "Silent Nature as a Claw in the Gut: Shock Therapy Epiphanies in Annie Proulx's Wyoming Stories." *Natura Loquens/Natura Agens. Revista Canaria de Estudios Ingleses*. 77 (2018): 105–25.

———. "Le chant de la matière: vers une écopoétique du réenchantement à travers quelques auteurs des Appalaches." *Représentations de la nature à l'âge de l'Anthropocène*. Edited by Jean-Daniel Collomb et Pierre-Antoine Pellerin, *Transtexte(s) Transcultures* 13 (2018). https://doi.org/10.4000/transtexts.1202

———. "Writing and Crying with 'The Agony of Gaïa' in Ann Pancake's *Strange as this Weather Has Been*." *Cris et écrits de la nature/ Land's Sorrows and Furrows. Caliban* 61 (2019): 113–37.

———. "Measured Chaos: EcoPoet(h)ics of the Wild in Barbara Kingsolver's *Prodigal Summer.*" *Toward an Ecopoetics of Randomness and Design. Ecozon@* 10, no.1 (2019): 60–80.

———. "Toutes ces femmes qui vivent en moi." *Penser et agir l'écologie politique. Ecorev* 47 (Spring 2019): 212–15.

———. "Deconstructing Gender Roles and Digesting the Magic of Folktales and Fairy Tales in Annie Proulx's Omnivorous Wyoming Stories." *Resisting Norms and Confronting Backlash: Gender and Sexuality in the Americas. L'ordinaire des Amériques*, 224 (2019). https://doi.org/10.4000/orda.4868

———. "Writing a Way Home: Magical Realism, Liminality, and the Building of a Biotic Communitas in Linda Hogan's Solar Storms and People of the Whale." In *Dwellings of Enchantment: Writing and Reenchanting the Earth*, edited by Bénédicte Meillon, 207–38. Lanham, MD: Lexington Books, 2021.

Merchant, Carolyn. *The Death of Nature: Women, Ecology, and the Scientific Revolution*. 1980. San Francisco: Harper and Row, 1989.

Merleau-Ponty, Maurice. *Le Visible et l'invisible*. Paris: Editions Gallimard, 1964.

———. *La nature: Notes, cours du Collège de France*. Paris: Seuil, 1995.

Milcent-Lawson, Sophie. "Bruits et chant du monde chez Giono: Petite poétique de l'onomatopée." *Jean Giono* 66 (2006): 35–50.

———. "Point de vue et discours des animaux dans l'œuvre romanesque de Jean Giono." In *Mondes ruraux, mondes animaux. Le lien des hommes avec les bêtes dans les romans rustiques et animaliers de langue française (XXe–XXIe siècles)*, edited by Alain Romestaing and Alain Schaffner, 61–72. Dijon: Éditions universitaires de Dijon, 2014.

———. "Chant du monde et paroles d'animaux dans l'œuvre de Giono." *Revue Giono* 11 (2018): 211–33.

———. "Un tournant animal dans la fiction contemporaine française?" *Pratiques: Le récit en questions*, 181–82 (2019). https://doi.org/10.4000/pratiques.5835.

Moncomble, Raphaëlle. "Women of Power and Power of Women in *A Song of Ice and Fire* by G. R. R. Martin." Master's thesis, University of Perpignan Via Domitia, 2020.

Morizot, Baptiste. *Les diplomates: Cohabiter avec les loups sur une autre carte du vivant*. Marseille: Wildproject, 2016.

Morrison, Toni. *Song of Solomon*. 1977. New York: Vintage International, 2004.

———. *Tar Baby*. New York: Alfred A. Knopf, 1981.

Morton, Timothy. *Ecology without Nature: Rethinking Environmental Aesthetics*. Cambridge, MA: Harvard University Press, 2017.

———*Hyperobjects: Philosophy and Ecology after the End of the World*. Minneapolis: University of Minnesota Press, 2013.

Murphy, Patrick D. "Women Writers: Spiritual Realism, Ecological Responsibility, and Inhabitation." *Journal of Literature, Culture and Media Studies* 1 (June 2009): 5–11.

Nagel Thomas "What Is It Like to Be a Bat?" In *The Philosophical Review* 83, no. 4 (1974): 435–50.

Nalls, Gayil. "The Power of Harmony: Musical, Spiritual, and Environmental." *Plantings* 10 (April 2022).

Nash, Roderick Frazier. *Wilderness and the American Mind*. 1967. New Haven, CT: Yale University Press, 2014.

Nietzsche, Friedrich. "On Truth and Falsity in their Ultramoral Sense." 1873. In *The Complete Works of Friedrich Nietzsche*, edited by Oscar Levy, translated by Maximilian A. Magge, New York: Gordon Press, 1974.

Nixon, Rob. *Slow Violence and the Environmentalism of the Poor*. Cambridge, MA: Harvard University Press, 2011.

Noiray, Jacques. "Utopie et travail dans *Que ma joie demeure* de Jean Giono." *Travailler* 7, no. 1 (2002): 63–76.

Owens, Delia. *Where the Crawdads Sing*. New York: G.P. Puntnam's Sons, 2018.

Pancake, Ann. *Strange as this Weather Has Been*. Emeryville, CA: Shoemaker & Hoard, 2007.

———. "Creative Responses to Worlds Unraveling: The Artis in the 21st Century." *The Georgia Review* 67, no.3 (Fall 2013): 404–14.

Patoine, Pierre-Louis. *Corps/texte: Pour une théorie de la lecture empathique: Cooper, Danielewski, Frey, Palahniuk*. Lyon: ENS Editions, 2015.

Perrin, Claire. "La sécheresse et le changement climatique dans les romans étatsuniens, de John Steinbeck à la cli-fi." PhD diss., University of Perpignan Via Domitia, France, 2020.

Perry, Donna. "Barbara Kingsolver: Interview." In *Backtalk: Women Writers Speak Out*, 143–70. New Brunswick, NJ: Rutgers University Press, 1993.

Peterson, Brenda, and Linda Hogan. *Sightings: The Gray Whale's Mysterious Journey*. Washington, DC: National Geographic, 2002.

Pinson, Jean-Claude. *Habiter en poète: Essai sur la poésie contemporaine*. Ceyzérieu: Champ Vallon, 1995.

———. *Poéthique: Une autothéorie*. Ceyzérieu: Champ Vallon, 2013

Plant, Judith, ed. *Healing the Wounds: The Promise of Ecofeminism*. London: Green Print, 1989.

Plaskow, Judith and Carol P. Christ, eds. *Weaving the Visions: New Patterns in Feminist Spirituality*. New York: HarperCollins, 1989.

———. "Naming the Sacred." In *Weaving the Visions: New Patterns in Feminist Spirituality*, edited by Judith Plaskow and Carol P. Christ, 95–100. New York: HarperCollins, 1989.

Plumwood, Val. *Feminism and the Mastery of Nature*. London: Routledge, 1993.

Pope Francis. *Encyclical Letter Laudato Si,' On Care for our Common Home*. Vatican Press, May 2015. https://www.vatican.va/content/dam/francesco/pdf/encyclicals/documents/papa-francesco_20150524_enciclica-laudato-si_en.pdf.

Power, Susan. *The Grass Dancer*. London: Picador, 1994.

———. *Roofwalker*. Minneapolis: Milkweed, 2002.

———. *Sacred Wilderness*. East Lansing: Michigan State University Press, 2014.

Powers, Richard. *The Overstory*. London: William Heinemann, 2018.

Prigogine, Ilya, and Isabelle Stengers. *Order Out of Chaos: Man's New Dialogue with Nature*. New York: Bantam Books, 1984.

Proulx, Annie. *Close Range*. 1999. New York: Scribner, 2003.

———. *Bad Dirt*. 2004. Waterville, ME: Thorndike Press, 2005.

Pughe, Thomas. "Réinventer la nature: vers une éco-poétique." *Etudes anglaises* 58, no.1 (2005): 68–81.

Pyle, Robert. *The Thunder Tree: Lessons from an Urban Wildland.* Guilford, CT: Lyons Press, 1998.
Rash, Ron. *Above the Waterfall.* New York: Harper Collins, 2015.
Ricœur, Paul. *La métaphore vive.* Paris: Editions du Seuil, 1975.
Rigby, Kate. "Spirits that Matter: Pathways toward a Rematerialization of Religion and Spirituality." In *Material Ecocriticism*, edited by Serenella Iovino and Serpil Oppermann, 283–90. Bloomington: Indiana University Press, 2014.
Robin, Marie-Monique. *La fabrique des pandémies: Préserver la biodiversité, un impératif pour la santé planétaire.* Paris: La Découverte, 2021.
Roger Alain, and François Guéry, eds. *Maîtres et protecteurs de la* nature. Ceyzérieu: Champ Vallon, 1991
Roh, Franz. "Magic Realism: Post-Expressionism." In *Magical Realism: Theory, History, Community*, edited by Lois Parkinson Zamora and Wendy B. Faris, 15–32. Durham, NC: Duke University Press, 1995.
Romestaing, Alain. *Jean Giono: Le corps à l'œuvre.* Paris: Champion, 2009.
Roszak, Theodore. *The Voice of the Earth.* New York: Simon & Schuster, 1992.
———. "Where Psyche Meets Gaia." In *Ecopsychology: Restoring the Earth, Healing the Mind*, edited by Theodore Roszak, Mary E. Gomes, and Allen Kanner, 1–17. San Francisco: Sierra Books Club, 1995.
Rumpala, Yannick. *Hors des décombres du monde: Ecologie, science-fiction et éthique du futur.* Ceyzérieu: Champ Vallon, 2018.
Ryden, Kent. C. *Mapping the Invisible Landscape: Folklore, Writing and the Sense of Place.* Iowa City: University of Iowa Press, 1993.
Schafer, Murray. *Le Paysage sonore: Le monde comme musique.* 1977. Translated by Sylvette Gleize. Marseille: Wildproject, 2010.
Schiller, Friedrich. *On the Aesthetic Education of Man.* Edited and translated by Elizabeth Wilkinson and L.A. Willoughby. New York: Oxford University Press, 1967.
Schoentjes, Pierre. *Ce qui a lieu: Essai d'écopoétique.* Marseille: Wildproject, 2015.
Şenel, Neşe. "An Ecofeminist Reading on Margaret Atwood's *The Handmaid's Tale* and Starhawk's *The Fifth Sacred Thing.*" Master's thesis, Erciyes University, 2015.
Serres, Michel. *Le contrat naturel.* Paris: France Loisirs, 1990.
———. "Vents." In *Nouvelles du monde*, 57–70. Paris: Flammarion, 1997.
Servigne, Pablo and Raphaël Stevens. *Comment tout peut s'effondrer: Petit manuel de collapsologie à l'usage des générations présentes.* Paris: Editions du Seuil, 2015.
Servigne, Pablo, Raphaël Stevens and Gauthier Chapelle. *Une autre fin du monde est possible: Vivre l'effondrement (et pas seulement y survivre).* Paris: Editions du Seuil, 2018.
Sewall, Laura. "The Skill of Ecological Perception." In *Ecopsychology: Restoring the Earth, Healing the Mind,* edited by Theodore Roszak, Mary E. Gomes, and Allen D. Kanner, 201–15. San Francisco: Sierra Club Books, 1995.
———. "Beauty and the Brain." In *Ecopsychology: Science, Totems, and the Technological Species*, edited by Peter H. Khan, Jr. and Patricia H. Hasbach, 265–84. Cambridge, MA: The MIT Press, 2012.

Shakespeare, William. *As You Like It*. Edited by Alan Brissenden. Oxford: Oxford University Press, 1993.

Sheldrake, Rupert. *A New Science of Life: The Hypothesis of Formative Causation*. London: Blond & Briggs, 1981.

———. "The Laws of Nature as Habits: A Postmodern Basis for Science." In *The Reenchantment of Science*, edited by David Ray Griffin, 79–86. Albany: State of New York Press, 1988.

Shepard, Paul. "Ecology and Man: A Viewpoint." In *Ecocriticism: The Essential Reader*, edited by Ken Hiltner, 62–69. New York: Routledge, 2015.

Sibran, Anne. *Enfance d'un chamane*. Paris: Gallimard, 2017.

Silko, Leslie Marmon. *Ceremony*. New York: Penguin Books, 1977.

Simard, Suzanne. *Finding the Mother Tree: Discovering the Wisdom of the Forest*. New York: Alfred A. Knopf, 2021.

Simon, Anne. *Une bête entre les lignes: Essai de zoopoétique*. Marseille: Wildproject, 2021.

Slemon, Stephen. "Magic Realism as Postcolonial Discourse." In *Magical Realism: Theory, History, Community*, edited by Lois Parkinson Zamora and Wendy B. Faris, 407–26. Durham, NC: Duke University Press, 1995.

Slovic, Scott. *Seeking Awareness in American Nature Writing: Henry David Thoreau, Annie Dillard, Edward Abbey, Wendell Berry, Barry Lopez*. Salt Lake City: University of Utah Press, 1992.

———. "Epistemology and Politics in American Nature Writing: Embedded Rhetoric and Discrete Rhetoric." In *Green Culture: Environmental Rhetoric in Contemporary America*, edited by Carl G. Herndl and Stuart C. Brown, 82–110. Madison: University of Wisconsin Press, 1996.

———. *Going Away to Think: Engagement, Retreat, and Ecocritical Responsibility*. Reno: University of Nevada Press, 2008.

Slovic, Scott and Paul Slovic, eds. "The Psychophysics of Brightness and the Value of Life." In *Numbers and Nerves: Information, Emotion, and Meaning in a World of Data*, 1–22. Corvallis: Oregon State University Press, 2015.

Smith, Theresa S., and Jill M. Fiore. "Landscape as Narrative, Narrative as Landscape." *Studies in American Indian Literature* 22, no. 4 (2010): 58–80.

Snow, Charles Percy. *The Two Cultures and the Scientific Revolution*. 1959. London: Cambridge University Press, 2001.

Snyder, Gary. *A Place in Space: Ethics, Aesthetics, and Watersheds*. Berkeley, CA: Counterpoint, 1995.

Spill, Frédérique. *The Radiance of Small Things in Ron Rash's Writing*. Columbia: University of South Carolina Press, 2019.

———. "Ron Rash's Above the Waterfall, or the Square Root of Wonderful." In *Dwellings of Enchantment: Writing and Reenchanting the Earth*, edited by Bénédicte Meillon, 161–75. Lanham, MD: Lexington Books, 2021.

Sponsel, Leslie E. *Spiritual Ecology: A Quiet Revolution*. Santa Barbara, CA: Praeger, 2012.

Spretnak, Charlene. "Ecofeminism: Our Roots and Flowering." In *Reweaving the World: The Emergence of Ecofeminism*, edited by Irene Diamond and Gloria Feman Orenstein, 3–14. San Francisco: Sierra Club Books, 1990.
Stamm, Gina. "Post-pastoral and the Nonmodern: Jean Giono's Engagement with Nature." *Studies in 20th & 21st Century Literature 43, no. 1* (2018).
Starhawk. *The Spiral Dance: A Rebirth of The Ancient Religion of the Great Goddess.* 1979. New York: HarperCollins, 1999.
———. *Dreaming the Dark: Magic, Sex and Politics.* Boston: Beacon Press, 1982.
———. "Power." In *Reweaving the World: The Emergence of Ecofeminism*, edited by Irene Diamond and Gloria Feman Orenstein, 73–86. San Francisco: Sierra Club, 1990.
———. *The Fifth Sacred Thing.* 1993. New York: Bantam Books, 1994.
———. *The Earth Path: Grounding Your Spirit in the Rhythms of Nature.* New York: HarperCollins, 2004.
Stengers, Isabelle. « Faire avec Gaïa: pour une culture de la non-symétrie », *Ecopolitique Now! Multitudes*, 24, 2006, https://www.multitudes.net/wp-content/uploads/2006/04/24-stengers.pdf. Accessed July 25, 2019.
———. *Au temps des catastrophes: Résister à la barbarie qui vient.* Paris: La Découverte, 2009.
Stevenson, Sheryl. "Trauma and Memory in Kingsolver's *Animal Dreams*." In *Seeds of Change*, edited by Priscilla Leader, 87–108. Knoxville: University of Tennessee Press, 2010.
Stewart, Garrett. *Reading Voices: Literature and the Phonotext.* Berkeley: University of California Press, 1990.
Stone, Christopher D. *Should Trees Have Standing? Law, Morality, and the Environment.* 1972. Oxford: Oxford University Press, 2010.
Suhamy, Henri. *Stylistique anglaise.* Paris: Presses Universitaires de France, 1994.
Swimme, Brian. "The Cosmic Creation Story." In *The Reenchantment of Science*, edited by David Ray Griffin, 47–56. Albany: State of New York Press, 1988.
———. "How to Heal a Lobotomy." In *Reweaving the World: The Emergence of Ecofeminism*, edited by Irene Diamond and Gloria Feman Orenstein, 15–22. San Francisco: Sierra Club Books, 1990.
Taylor, Bron. *Dark Green Religion: Nature Spirituality and the Planetary Future.* Berkeley: University of California Press, 2010.
Taylor, Paul. *Respect for Nature: A Theory of Environmental Ethics.* Princeton, NJ: Princeton University Press, 1986.
Thaler, Lore and Melvyn A. Goodale. "Echolocation in Humans: An Overview." *Wiley Interdisciplinary Reviews. Cognitive Science* 7, no. 6 (November 2016): 382–93.
Thoreau, Henry David. *Walden and Other Writings.* New York: Barnes and Noble, 1993.
Tredinnick, Mark. *The Land's Wild Music: Encounters with Barry Lopez, Peter Matthiessen, Terry Tempest Williams & James Galvin.* San Antonio, TX: Trinity University Press, 2005.

Tudge, Colin. *The Secret Life of Trees: How They Live and Why They Matter*. London: Penguin Books, 2005.

Turner, Victor. "Betwixt and Between: The Liminal Period in Rites of Passage." In *Reader in Comparative Religion: An Anthropological Approach*, 1965, edited by Evon Z. Lessa and William A. Vogt, 234–43 1979.

———. *The Ritual Process: Structure and Anti-Structure*. 1969. Ithaca, NY: Cornell University Press, 1989.

Van Cauwelaert, Didier. *Les emotions cachées des plantes*. Paris: Plon, 2018.

Viveiros de Castro, Eduardo. "Cosmological Deixis and Amerindian Perspectivism." *Journal of the Royal Anthropological Institute* 4, no. 3 (September 1998): 469–88.

———. "Exchanging Perspectives: The Transformation of Objects into Subjects in Amerindian Ontologies." *Common Knowledge* 10, no. 3 (Fall 2004): 463–84.

Vizenor, Gerald. *Wordarrows: Indians and Whites in the New Fur Trade*. Minneapolis: University of Minnesota Press, 1978.

———. *Manifest Manners: Narratives on Postindian Survivance*. Lincoln: University of Nebraska Press, 1999.

von Mossner, Alexa Weik. *Affective Ecologies: Empathy, Emotion, and Environmental Narrative*. Columbus: Ohio State University Press, 2017.

von Uexküll, Jakob. *Milieu animal et milieu humain*. 1934. Translated by Charles Martin-Freville Paris: Payot & Rivages, 2010.

Watkins, Claire Vaye. *Gold Fame Citrus*. New York: Riverhead Books, 2015.

Watts, Alan. *The Book: On the Taboo Against Knowing Who You Are*. New York: Random House "Vintage," 1989.

Weber, Max. "Science as Vocation." In *From Max Weber: Essays in Sociology*. Translated and edited by Hans H. Gerth and C. Wright Mills. New York: Oxford University Press, 1946.

———. *The Protestant Ethic and the Spirit of Capitalism*. Translated by Talcott Parsons. New York: Charles Scribner's Sons, 1958.

Wheeler, Wendy. "The Biosemiotic Turn: Abduction, or, the Nature of Creative Reason in Nature and Culture." In *Ecocritical Theory: New European Approaches*, edited by Axel Goodbody and Kate Rigby, 270–82. Charlottesville: University of Virginia Press, 2011.

White, Jr., Lynn. "The Historical Roots of our Ecological Crises." 1967. In *The Ecocriticism Reader*, edited by Cheryll Glotfelty and Harold Fromm, 3–14. Athens: The University of Georgia Press, 1996.

Whitehead, Alfred North. *The Concept of Nature*. Cambridge: Cambridge University Press, 1920.

Williams, Terry Tempest. *Refuge: An Unnatural History of Family and Place*. 1991. New York: Vintage Books, 2018.

Wilson, Edward O. *On Human Nature*. Cambridge, MA: Harvard University Press, 1978.

———. *Biophilia: The Human Bond with Other Species*. Cambridge, MA: Harvard University Press, 1984.

———. *The Origins of Creativity*. New York: Penguin Books, 2017.

Woods, Angela. "The Voice-Hearer." *Journal of Mental Health* 22, no.3 (2013): doi: 10.3109/09638237.2013.799267
Wrede, Theda. "Barbara Kingsolver's *Animal Dreams*: Ecofeminist Subversion of Western Myth." In *Feminist Ecocriticism: Environment, Women, and Literature*, edited by Douglas A. Vakoch, 39–64. Lanham, MD: Lexington Books, 2012.
Wright, Alexis. *The Swan Book*. New York: Atria Books, 2016.
Ywahoo, Dhyani. "Renewing the Sacred Hoop." In *Weaving the Visions: New Patterns in Feminist Spirituality*, edited by Judith Plaskow and Carol P. Christ, 274–81. New York: HarperCollins, 1989.
Zamora, Lois Parkinson and Wendy B. Faris, eds. *Magical Realism: Theory, History, Community*. Edited by 163–90. Durham, NC: Duke University Press, 1995.
Zhong Mengual, Estelle and Baptiste Morizot. "L'illisibilité du paysage: Enquête sur la crise écologique comme crise de la sensibilité." *Nouvelle revue d'esthétique*, 22, no. 2 (2018): 87–96.

Index

Abram, David, 9, 35, 47, 95, 113–15, 129, 137–40, 148, 159, 172, 175–76, 225, 248, 265, 274, 280–81, 286–88, 327–28, 330, 333, 335
Adamson, Joni, 5–6, 16, 99, 140, 152
affect, 6, 60, 88, 122, 204, 228, 244, 252–53, 311
agency, 5, 9, 17, 31, 35, 47–48, 58, 60, 65, 67–68, 75, 78, 85, 102, 189, 200, 206, 208, 215, 219, 224, 232, 253, 275–76, 291, 303, 310, 311–13, 327, 333, 335
agential realism, 27, 47, 67, 217
agony, 87, 92, 186, 280;
    agony of gaia, 86, 89, 95
Alaimo, Stacy, 104, 228, 309
Albee, Edward, 229
Albretch, Glenn, 122
alienation, 4, 22–23, 26, 47, 62, 110, 120, 148, 204, 283, 287–88, 296, 299
Allen, Paula Gunn, 25, 27, 33–35, 44, 48, 77–78, 85, 104, 148, 234, 258, 300, 311, 327
amnesia, 65, 200, 292
anamorphosis, 52, 55–56, 58–60, 63, 67, 83, 92, 98, 203, 207, 233–35, 244–46, 248–59, 263–64, 278, 283, 291, 307, 309, 325–26, 330, 334

animal, 2–4, 7, 11–13, 15, 17, 24, 47, 54, 59, 63, 68, 70–71, 73–76, 83, 89, 91, 95, 113–15, 117, 130, 133, 139, 141, 143–60, 162–65, 169–73, 175–78, 183–87, 192, 199, 203–4, 209, 225, 247–49, 255–56, 263–68, 270–80, 282, 284, 286, 299, 301–2, 304, 306, 311, 313, 320, 323, 329, 338
animation, 4–6, 24, 42, 53, 59, 66, 84–85, 114, 119, 141, 152, 164, 171, 173, 176, 189, 207, 226, 248, 252, 260, 263, 276, 281, 287, 308, 313, 323, 325
animism, 5, 7–9, 17, 22–27, 37, 43, 53, 57, 59, 66–67, 130, 132–33, 140–41, 145, 150, 158, 162, 164–65, 169, 180, 189, 192, 196, 198, 203, 206–8, 215, 217, 219, 222, 232, 248, 333
anthropocene, 10, 18, 65, 67–68, 85, 101–2, 302
anthropocentric, 17, 23, 27, 31, 51, 59, 64, 66, 68, 87, 115, 157, 161, 181, 208, 209, 211, 231, 233, 244, 246, 247, 255, 260, 264, 296
anthropology, 5, 6, 10–11, 132, 133, 135, 137
anthropomorphic, 58, 61, 75, 76, 89, 99, 249, 252–54, 260, 272, 291;
    pathetic fallacy, 24, 84, 255, 260;

personification, 24, 57–59, 75, 80, 91;
prosopopoeia, 306
anthropophony, 3, 112, 114, 271, 284, 296
anxiety, 44, 94, 186, 190
apocalypse, 57, 77, 86, 121, 186, 195
Asquith, Mark, 104
Atwood, Margaret, 337

Bakhtin, Mikhail, 38
Barad, Karen, 16, 25, 27, 33–35, 38, 47, 67, 112
Barthes, Roland, 268, 279
Bate, Jonathan, 3, 14, 118, 122, 233, 239, 256, 258, 260–61, 280, 283, 292, 294, 302
Bateson, Gregory, 26–27, 37, 46
Baudelaire, Charles, 171
becoming, 9, 10, 36, 41, 70, 90, 174–76, 179–80, 224, 274, 281, 292, 302, 321, 326, 327;
    becoming animal, 176, 201, 278, 280, 282, 300, 321;
    becoming earth, 175;
    becoming plant, 307, 321;
    co-becoming, 8, 35, 47, 91, 112, 244, 248, 256, 286, 288, 299, 312–13
bee, 127, 201–4, 229, 235, 278, 327
Bennett, Jane, 16, 33, 36, 42, 110–11, 118, 127, 153, 154
Berendt, Joachim-Ernest, 2, 14–16, 111, 113, 118, 295
Bergé, Aline, 14–15, 321
Berman, Morris, 16, 22–23, 26–28, 30, 38, 46
Bernstein, Charles, 112, 121, 293
Berry, Thomas, 233
Bhabha, Homi K., 16
biocentric, 2, 12–14, 46, 51, 59, 97, 116, 135, 145, 160–62, 216, 231, 233–34, 244, 246, 264–65, 273, 291
biology, 5, 9, 12, 17, 31–32, 43, 45, 58, 61, 63, 66, 70, 74, 96, 99, 122, 132, 133, 155, 161, 179, 197, 205, 217, 219, 221, 238, 258, 268–69, 273, 277, 335
biomimesis, 64, 277–78
biomorphic, 249
biophilia, 43, 47, 227, 338
biophony, 3, 13, 112, 114, 208, 245, 257, 264, 274, 284, 286, 296, 301, 303, 319, 321, 334
biosemiotic, 5, 9, 48, 90, 154, 157, 172, 202, 208, 210–11, 226, 229, 272
biosphere, 5, 14, 45–46, 53, 56, 85, 101, 206, 284
biotic, 11, 66–67, 70, 75, 95, 139, 164, 214, 225, 267, 269, 272, 302, 334;
    abiotic, 302, 334
bodymind, 4, 8–13, 15, 22, 25–27, 31–32, 34–38, 42, 46–47, 52–54, 56–60, 62, 67–68, 70, 72–74, 77, 81, 86–89, 91, 93–95, 103–4, 111–12, 115, 119, 122, 129–30, 132, 135, 139, 141, 144–45, 147–49, 152–55, 158, 160, 162, 170–72, 174–76, 178, 180–86, 188–90, 196–205, 207, 210, 212–14, 218–20, 222–24, 226–28, 233, 243–45, 248, 250–52, 256–57, 259, 269–71, 273, 276–79, 281, 285–88, 290–95, 298, 301, 303, 305, 309–10, 313–15, 319–23, 326, 329–30, 338
Bohm, David, 30–31, 33–36, 38–39
Bonabeau, Eric, 330
Bonneuil, Christophe, 101
Bowers, Maggie Ann, 134–35
braiding, 6, 63, 73, 78, 117, 130, 141, 149, 164, 197–98, 206, 219, 243, 248; 273, 309, 333;
    intermingle, 83, 109, 175, 178, 214, 216, 264, 334;
    interweave: see weaving
Brun, Frédéric, 17
Buber, Martin, 160
Buell, Lawrence, 7, 77, 104, 121, 128, 243, 250, 259, 275, 281

Callicot, J. Baird, 44, 49, 87
Cameron, Sharon, 245
capitalism, 24, 29, 64, 68–69, 80–81, 101, 177, 179, 191, 195
Capra, Fritjof, 295, 305
Carson, Rachel, 10, 33, 66, 76–77, 109–10, 116–18, 121, 174–75, 211, 268, 319–20
Cavagna, Andrea, 330
Cazajous-Augé, Claire, 218
chaos, 4–5, 14–15, 23, 28, 33, 72, 76, 112, 118, 191, 224, 245, 248, 252, 256, 295, 321
Chatwin, Bruce, 291
Choné, Aurélie, 15
chora, 115–16, 118, 286–88, 304, 326
Christ, Carol P., 49, 54, 72, 79
Clément, Gilles, 107
cli-fi, 52, 86, 98, 140, 195, 197, 327, 337–39
Cochoy, Nathalie, 166
cognition, 5, 12, 32, 119–20, 154, 161, 205, 223, 231, 246, 251, 256, 258, 260, 288, 293, 313, 325, 336
Cohen, Robin, 100
Collot, Michel, 14
communitas, 10–12, 145, 161, 207, 223, 225, 228, 267
community, 8, 12, 41, 45–46, 53–54, 57, 60–61, 64, 67, 71, 76, 86, 95, 100, 101, 113, 133–34, 159–60, 163–64, 169, 178–79, 181, 195–96, 201, 203–4, 210–11, 214–16, 231, 233, 251, 255, 259, 269, 272–73, 291, 311–12
compassion fatigue, 335
Conn, Sarah A., 93, 106
conscious, 5, 11–13, 22–26, 30, 34, 36–37, 41, 43, 45–47, 49, 52–55, 58–59, 64, 72, 93, 97, 111, 113, 118, 129–30, 139–40, 145, 149–50, 154, 170, 180, 181, 185, 187, 189–90, 199, 201–2, 204, 212, 220, 222, 227, 233, 235, 244–45, 250, 256, 266, 270, 274, 276, 279, 287, 290, 324;

unconscious, 11–13, 26, 46–47, 138, 149, 155, 162, 177–78, 185, 187, 189–90, 196, 198, 200–201, 212, 216, 218, 220, 227, 233, 249–51, 256, 258, 274, 298, 337
Cooper, Lydia R., 162
Couvet, Denis, 15
Cuddon, J. A., 130–31, 260
Cyrulnik, Boris, 102

Damasio, Antonio, 12–13, 153, 337
dance, 3, 9, 41–42, 45, 64, 71, 75, 91, 95, 117, 146, 164, 173, 203, 208, 244, 248, 257, 265, 268, 275–76, 287, 295–300, 307–10, 312–13, 326–27, 335
Darwin, Charles, 62, 76–77, 330
Davis, John, 228
De la Cadena, Marisol, 6, 16
deep ecology, 14, 54
deer, 64, 84, 92, 155–57, 169–74, 177, 183–88, 192, 209, 263–66, 277, 286, 299, 301–2, 307–8
Deleuze, Gilles, 280, 281
DeLoughrey, Elizabeth, 16
Descartes, René, 5, 21, 26, 37, 65, 66
Descola, Philippe, 5, 24, 132–33, 140–41, 145, 159, 203, 218
despair, 94–95, 97, 107, 223, 338
Després, Olivier, 304
Diamond, Irene, 7, 16, 49, 79
Diamond, Jared, 140
diffraction, 13, 66, 87, 93, 107, 110, 211, 291, 295, 301, 315, 321, 325, 328–29
Dillard, Annie, 153–55, 157, 166
disenchantment, 4–5, 10, 15, 21, 22–24, 26, 31, 36, 42, 44, 68, 86, 88, 94, 96–97, 110, 113, 115, 117, 173, 195, 229, 232, 243, 273, 283, 333–34, 336–38
Donald, Merlin, 330–31;
dreaming, 42, 47, 58, 62, 96, 133, 146, 148–50, 155, 159,

162, 177–78, 185–88, 190,
199–201, 206, 220, 233, 257,
272, 304, 308, 310
Ducarme, Frédéric, 15
Durand-Rous, Caroline, 17, 141
Durix, Jean-Pierre, 129
Durkheim, Emile, 231–32
dwelling, 2, 6, 8, 14, 17–18, 32, 37,
52, 54, 61, 63–65, 76–77, 82–83,
92–93, 97–98, 100–101, 113, 115,
118, 131–32, 136, 144, 148, 150,
152, 154, 156, 158–60, 162, 164–65,
170–71, 173–74, 179, 181–82, 184,
189–91, 196, 201–2, 205, 218, 223,
225, 233, 243, 247–48, 254, 256–58,
260, 264–66, 268, 274, 276, 280,
283, 286–89, 292–93, 295, 299, 308,
312, 334–35, 338–39

earth, 2–3, 7, 11, 14–15, 17, 23, 28–29,
37, 42, 45–47, 52–56, 62–68, 72–75,
77, 80, 83–89, 91, 93, 96–98, 101–4,
107, 110–11, 114–16, 118–19, 122,
128, 134–35, 137, 148, 150–51, 159,
164, 169, 171–73, 175–79, 181, 187,
189, 191, 199–200, 204–5, 209, 214,
218, 220–21, 223, 226, 231, 233–34,
244–45, 248, 256, 260, 267–68, 281,
283–89, 291–94, 296–302, 307–8,
310–14, 319–23, 325–27, 330,
334–35, 337–38
Earth, 2–4, 7, 9–10, 14–17, 27–28, 41,
43–47, 49, 51–53, 55–63, 65, 68,
71, 73, 75–76, 79–83, 85, 87–91,
93–94, 107, 115–18, 122, 131, 135,
144, 169, 193, 206, 214, 216, 224,
233, 283–84, 287–88, 303, 319, 322,
325, 330, 359
earth goddess, 102
echo, 3, 27, 84–85, 95, 99, 114, 118,
148, 157, 171, 173, 185, 199, 200,
213, 239, 273, 291, 296–97, 301,
321, 323, 328;

poetic echoes, 2–3, 91, 110,
118, 122, 191, 226, 245, 291,
295, 297, 301,
308, 326–27, 334–35
ecocritic, 3, 6–7, 15, 33, 36, 43, 47–48,
103, 107, 116, 138, 141, 243, 245,
286, 312, 359
ecoeroticism, 289, 322, 329;
jouissance, 268, 279, 289, 298
ecofeminism, 7, 9–10, 15–17, 23–24,
28–32, 36, 38, 41–45, 47–49, 53–55,
63, 67–71, 73–74, 77, 79–80, 85,
87, 89, 97–98, 102–3, 117, 131, 134,
136, 140, 178–80, 195, 198, 204,
234, 268, 274, 288–90, 303, 305,
333, 335–36, 338, 359
ecology, 5–7, 9, 10, 14, 27–31, 44–49,
55–56, 62–66, 75–77, 85, 94, 102,
116, 120–21, 132–33, 135–37, 149,
152, 155, 160, 162, 164, 177, 185,
187, 189, 195, 198, 204–5, 210, 214,
217–18, 220, 224, 227, 229, 233,
243, 254, 268, 273, 277, 283, 288,
291, 298, 303, 305, 321, 333–35, 359
ecophobia, 43–44, 57, 72, 85,
102–3, 338
ecopoetics, 2–4, 6, 8–18, 24, 28, 41–43,
45–48, 51–54, 56, 58–62, 64, 66,
69–71, 73, 77–78, 80, 82, 84–86,
88–89, 91, 97–98, 105, 109–12,
114–20, 122, 127–33, 135, 138–42,
147–48, 150, 152, 154, 159, 161–62,
164, 166, 169–70, 172, 175, 177,
179–80, 189–90, 195, 197–99,
203–6, 212, 214–15, 217, 222–23,
225, 228–35, 239, 243–49, 251–52,
254–58, 260, 263, 265–66, 268, 274,
276, 280, 283–84, 286–93, 298–300,
302–3, 307–15, 321–26, 328–30,
333–39, 359;
ecopoiesis, 2, 6, 8, 14, 25, 47, 52,
58, 60, 71, 76, 85, 90–91, 96,
98, 110–12, 114–20, 122, 135,
137, 157, 161, 190–91, 200,
233, 244, 246–47, 251, 253,

256, 258, 260, 264, 283–84, 286–88, 291–96, 301, 307–9, 312, 320, 323, 325–27, 329, 333–36, 338
ecopsychology, 5, 43–46, 57, 93–94, 106, 116, 119, 122, 140, 324, 334, 336
effervescence, 33, 231–32, 298
Eisler, Riane, 29–30, 42–44, 62, 102
Elvey, Anne, 303–4
Emerson, Ralph Waldo, 189, 206
empathy, 11, 43, 92–93, 116, 120, 127, 134, 144, 160, 163, 174, 177, 186, 327
enchantment, 4–6, 8–10, 15, 22, 24, 28–29, 33, 36, 42, 48, 51–52, 58, 73, 79, 97, 110–11, 117–18, 120–21, 130, 137, 152–54, 172, 189, 201, 217, 225–26, 228, 230–33, 243, 247, 249, 252, 254–55, 257, 260, 266, 273, 276, 289, 294, 298–300, 307, 311–12, 319, 321–23, 325–27, 330, 333, 338
enlightenment, 5, 15, 139, 184, 213, 221, 226–28, 339
entanglement, 7–9, 13, 34–36, 38, 53, 59, 63, 75–76, 78–79, 89, 91, 95, 100, 112, 115, 137, 139, 141, 148, 152, 154, 174, 189, 196, 198, 228, 231, 233, 243, 247, 253, 257, 275, 286, 291, 293, 301, 303, 310, 315, 325, 330;
    enmesh, 3, 8, 35, 59, 63, 67, 76, 92, 97, 151, 172, 177, 200, 207, 216, 218, 232, 257, 264, 269, 279, 334;
        interlacing, 8, 13, 33, 35, 52, 77, 79, 82, 112, 118, 147, 149, 171, 180, 206, 218, 227, 247, 264–65, 275, 285, 308, 311, 315, 320, 327
environment, 2, 4, 6, 12–13, 17, 22, 28–29, 43–44, 46, 59, 62–63, 65, 77, 80, 87, 94–95, 97, 100–2, 104, 107, 111, 114, 116, 119, 128, 142, 149, 155, 157, 170, 205, 207–8, 216, 218, 224, 229–30, 231–32, 236, 247, 250–51, 255, 260, 284–85, 321, 328, 334, 338, 359
environmental literature, 7, 17, 47, 52, 127, 142, 209, 233, 243, 334
epiphany, 24, 71, 73, 85, 97, 111, 127, 131, 148, 150, 171, 178, 184–85, 188–89, 206–7, 217, 220, 222–23, 226–28, 230, 246, 298–99, 323
epistemology, 6–7, 9, 26, 31–33, 35, 42, 52, 78, 112, 128, 132, 136, 138–39, 142–43, 148, 162, 180, 189, 197, 206, 210, 225, 228, 230–33, 243, 246, 250, 296, 310, 312, 333–34
Erdrich, Louise, 134, 336
Estok, Simon, 43–44, 102
ethic, 7, 9, 28–30, 32, 35, 42, 46, 56, 61–62, 87, 110, 111, 116, 127–28, 162–63, 182, 214, 216, 225, 228, 230, 232, 246, 301, 311
ethnocentric, 138
ethology, 5, 9, 31, 130, 132, 136, 144, 157, 165, 192, 268–70, 335
extraordinary, 15, 45, 110, 111, 113, 144, 153, 178–79, 196–97, 214, 221, 223–24, 232, 244, 260, 294

Fancourt, Donna, 235
Faris, Wendy B., 18, 51, 98, 128, 129, 134–35, 137–40, 142–43, 158, 170, 183, 198, 220, 257, 266
fear, 15, 17, 24, 94, 96–97, 146, 185, 187, 197, 228, 253–54, 260, 266, 298, 309
feeling, 4, 8, 11–13, 15, 22–23, 30, 38, 43, 46, 53, 72, 85, 88, 90–91, 93–95, 97, 106–7, 111, 113, 116–19, 144–46, 152, 154–58, 162, 164, 171, 173, 176, 178, 181–88, 190, 196, 199–201, 204, 210, 212, 217–18, 220, 228, 243, 249–50, 256–57, 260, 264–66, 268, 271, 273, 276, 279, 284–86, 288–90, 297–99, 301, 311, 313, 322, 328, 331;

emotion, 7, 12, 17, 22–23, 54, 86, 96, 116, 127, 214, 231, 255, 260, 270–71, 288, 327;
emotional contagion, 327;
feelscape, 8, 264, 279, 301, 323, 334
flesh, 2, 10, 34–35, 67–68, 74, 88, 92, 111–12, 115, 117, 156, 171–73, 175–76, 211, 217–18, 231, 250, 264, 267, 289, 293, 323, 325, 334;
flesh of language, 112, 293;
flesh of the world, 2, 10, 34–35, 111–12, 117, 171–72, 218, 231, 264, 267, 293, 320, 323, 325, 327, 334
Flys-Junquera, Carmen, 281
focalization, 84, 87, 146, 162, 181, 244, 263, 265–66, 268, 270–71, 294, 301, 322;
defocalization, 266
Forbes, Jack D., 16, 148, 284, 303, 325
forest, 5–6, 14, 58–61, 69, 71, 75–76, 90, 98, 114, 130, 133, 144–45, 147, 153, 159, 169–74, 176, 178–79, 181, 189, 199, 205, 210, 212, 214, 216–17, 218–19, 223–24, 232, 248, 250, 253–54, 265–66, 268–70, 273, 290, 297, 308, 319–21, 336–37
Fressoz, Jean-Baptiste, 101
Frumkin, Howard, 238

Gaard, Greta, 44, 49
Gagliano, Monica, 213–14, 236, 335
Gaia, 5, 10, 18, 39, 44–48, 51–53, 55–60, 62, 74–77, 80–91, 97–99, 206, 225, 245, 290, 333
Gavillon, François, 136
Genette, Gérard, 265
genius loci, 23, 66, 150
geophony, 3, 13, 112, 114, 208, 245, 264, 284, 286, 296, 303, 314–15, 319, 334
Gibson, James William, 9, 49, 338
Ginzburg, Carlo, 218, 235

Giono, Jean, 2, 13–14, 71–74, 103, 115, 134–35, 140, 169–80, 191–93, 233, 248, 263–68, 275, 279, 297–302, 305–6, 309, 322, 333
Gipe, Robert, 105
Gleick, James, 15
Glotfelty, Cheryll, 36
god, 22–23, 38, 51, 77, 102, 310
goddess, 52, 57, 77, 235
Gomes, Mary E., 43
Goodale, Melvyn A., 304
Grandjeat, Yves-Charles, 7, 16, 119–21, 154, 293, 296, 309–10
Griffin, David Ray, 16, 21–22, 30–32, 36, 38, 44
Griffin, Susan, 36, 98, 102
Guattari, Felix, 280–81
Guéry, François, 101

Hajek, Isabelle, 15
Hamman, Philippe, 15
Hansen, Margaret M., 238
Haraway, Donna, 15, 90–91, 100–101, 106, 140–41, 258, 302, 335
Harding, Wendy, 105
harmonia mundi, 2, 4, 13, 16, 244, 295–96
harmony, 3–4, 16, 26, 53, 63, 87, 117, 148, 174, 191, 195, 201, 224, 295–96, 300–301, 306, 313–14, 320, 323
Hasbach, Patricia H., 239
hearing, 2–3, 5, 14–16, 46–47, 84, 110, 112–14, 116–18, 146–47, 156–57, 164, 172, 181, 183, 191, 196–97, 199, 207–8, 212–14, 218–23, 226, 231, 257, 263–64, 266, 285–88, 290, 294–97, 301, 304, 308, 313, 315, 319–21, 323–24, 326–27, 329, 335;
deaf, 47, 114, 157, 225, 286, 324
Hegland, Jean, 336
Hogan, Linda, 11, 13, 16, 72, 103, 127, 130, 135–36, 140, 143–45, 147–50, 152, 154–55, 157, 159–62, 164, 174, 224, 233, 248, 257–58, 307, 310–11, 313, 333, 359

Hölldobler, Bert, 235
Hook, Johanna, 304
Hopkins, Gerard Manley, 322, 325, 330
Huggan, Graham, 16
human, 2–8, 10–15, 17, 21–30, 32–36, 38, 41–48, 51–52, 54–55, 57–58, 61–74, 76, 78–86, 88, 90–92, 95, 97–98, 100–102, 104, 109–15, 119–20, 122, 127–28, 130, 132–33, 137–52, 154–60, 162–64, 169–76, 178, 181, 183, 185–91, 196–219, 221–29, 231–35, 238, 243–56, 260, 263–68, 270–81, 284–88, 291, 294–97, 299–304, 307, 309–13, 315, 320–21, 323–28, 330, 334–37
humility, 33, 43, 63, 68, 72, 102, 111, 159, 174, 176, 182, 189, 223, 273, 276, 291, 296, 311
Hutcheon, Linda, 38

immanence, 9, 27, 45, 54, 189, 280
incantation, 88, 111, 115, 191, 289, 294
indigenous, 4, 6–7, 9–10, 13, 17, 23–26, 29, 33, 41, 43–44, 47, 63–66, 68–69, 79, 100, 103, 113–14, 130, 132, 134–37, 139–40, 145, 147, 149–51, 158, 162, 164–65, 169, 179–80, 183, 192, 206, 218–19, 224–26, 234, 257, 284, 290–91, 307, 311, 333, 335, 337–38; neo-indigenous, 338
Ingold, Tim, 141, 165, 192
initiation, 10, 11, 18, 94, 115, 134, 143, 145, 149–50, 156, 158, 176, 180, 185, 199, 201, 208, 219, 222–23, 287, 324, 326–27
instrument, 6, 59, 117, 172, 200, 284, 292, 294, 296, 300, 315, 320
interaction, 3, 15, 35–36, 45, 58, 111, 204, 215, 217, 223, 244, 304, 333; intra-action, 4, 7, 33, 35–36, 48, 53, 90, 120, 133, 183, 189, 206, 209, 309, 311, 359
interconnection, 7, 27, 29, 43, 45–6, 51, 53, 55, 59, 69, 75–77, 94–96, 104, 107, 152, 188, 207, 228, 267, 273, 276, 309, 335
interdependence, 43–45, 68, 73, 76, 131, 149, 155, 164, 182, 225, 251, 311, 335
interspecies, 6, 25, 59, 133, 139, 142, 177, 184, 196, 201, 203, 206, 208, 211, 212, 224–25, 245, 254, 263, 271–72, 277–78, 281, 301, 334; interspecies communication, 6, 139, 177, 184, 196, 206, 208, 212, 224–25, 263, 272, 277–78, 301
intuition, 12–13, 22, 24, 38, 42, 47–48, 103, 117, 131, 149–50, 153, 155–56, 177, 187, 190, 196, 200, 210, 256, 264, 288, 295, 321
Iovino, Serenella, 16, 47, 49, 311, 333

Jabr, Ferris, 237
Jacobs, Naomi, 100
Johnson, Mark, 256, 330
Johnson, Nathanael, 237

Kahn, Peter H., 122
Kanner, Allen D., 43, 49
Keim, Brandon, 236
Keller, Hellen, 324
Killingsworth, M. Jimmie, 121
Kimmerer, Robin Wall, 13, 16, 76, 90, 98, 106–7, 135–36, 179–80, 182, 209, 224–25, 307, 312, 315–16, 333
King, Ynestra, 42
Kingsolver, Barbara, 13–14, 62–72, 75–77, 100, 102–4, 127, 129, 133, 135–36, 170, 248–51, 259, 263, 268–70, 273–77, 279–82, 307, 319–21, 333, 336, 359
Kish, Daniel, 304
Knickerbocker, Scott, 3, 110–11, 121, 284, 293, 304, 322, 336
Kohn, Eduardo, 5, 140, 160
Krause, Bernie, 2–3, 14, 113–14, 121–22, 264, 296, 303, 335

Kristeva, Julia, 103, 115, 122, 135, 286–87, 303–4

Lacroix, Alexandre, 243
Lakoff, George, 256, 330
land, 7–8, 17, 29, 46, 62–67, 70, 80–81, 86, 88–91, 93–98, 100, 116, 141, 143, 146, 149–50, 155–57, 163–65, 174–75, 179–83, 187, 199–200, 214, 216, 224–25, 248, 253, 258–59, 263–64, 267–69, 284–88, 291, 294, 296–97, 299–301, 312, 322–23, 326, 329
landscape, 8, 63–65, 68, 81, 84, 92–93, 95, 111, 113, 133, 144, 150, 152, 159, 171–73, 180, 199–200, 218, 226, 252, 259, 263–64, 271, 287, 301, 323, 325–26, 338
Latour, Bruno, 5, 16, 58, 84, 100
Lauwers, Margot, 359
Le Clézio, Jean-Marie Gustave, 191
Leopold, Aldo, 10, 33, 46, 97, 116–17, 121, 180, 206, 214, 269
Lévi-Strauss, Claude, 132, 161
lifeform, 7, 11, 15, 21, 47, 52, 59, 65–66, 70, 75, 77, 88, 91, 95, 128, 132, 147, 148, 161, 172, 189, 208, 215, 217, 249, 253, 273, 275, 291, 309
liminal realism, 9–13, 17, 26, 39, 46–48, 51–52, 71, 97–98, 120, 128–29, 131–32, 134–40, 142, 144, 148, 150–51, 154, 160–62, 164, 169–70, 180–81, 186, 189–90, 195–96, 203–4, 206, 212, 214, 219, 228, 230–35, 243–45, 257, 265–66, 268, 284, 309, 316, 333–34, 336–37, 359
liminality, 6, 10–13, 18, 24, 77, 131, 133, 135, 137, 139, 141, 144–45, 148, 150, 154, 159–62, 164, 170, 178, 184–85, 198–99, 201, 205, 210, 212, 214, 217–18, 220–23, 226, 228, 230–31, 263–64, 270, 307–8, 336–37;
in-between, 9, 11, 47, 51, 67, 120, 132–34, 138–40, 143, 145, 148–50, 159, 162, 174, 200, 208, 214, 243, 254, 257, 263–64, 268, 301, 312, 334, 337;
threshold, 10–11, 73, 89, 110, 131, 143–44, 149–50, 155–56, 206
logos, 21;
logocentrism, 68, 80, 113, 115, 119, 154, 228
Louv. Richard, 122
Lovelock, James, 32, 39, 46, 56–57, 59–60, 206
lyricism, 2, 4, 15, 24, 76, 91, 289, 292, 293, 296, 298, 314, 321, 323, 328

Macé, Marielle, 308
Machet, Laurence, 38
Macy, Joanna, 94, 106–7
magic, 4, 6–7, 9, 13, 21–22, 24, 27, 45–46, 49, 66, 69, 111, 121, 128–30, 136, 138–41, 145, 148, 155, 170–72, 176, 183, 189–90, 197–98, 200, 204, 214, 222–23, 225, 232–33, 235, 244, 252, 256–57, 266, 276, 293–94, 300, 308, 312, 324–26, 334
magical realism, 6, 9–11, 18, 51, 80, 98, 101, 128–30, 134–39, 141–42, 150, 158, 170, 180, 183, 190, 220, 231, 266, 268, 333, 359
Malm, Andreas, 101
Mancuso, Stefano, 236, 260, 281, 305, 335
Manes, Christopher, 14, 246
Margulis, Lynn, 32, 39, 46, 56–58, 60–62, 64, 85, 99
Martin, George R. R., 337, 339
Martin, Nastassja, 337
Marx, Leo, 88
material ecocriticism, 47–49, 138, 141
matter, 3–10, 15, 25, 27, 31–32, 34–36, 38, 42, 44, 46–49, 51, 53, 58–59, 63, 65–72, 75, 78–79, 81, 83, 85–86, 102, 104, 111–14, 129, 133–39, 141, 145, 147–48, 150, 159, 174, 179, 183, 185, 188–89, 196–98, 203–4,

206, 210, 214, 217–18, 222, 226–28,
230, 233, 244, 247–48, 255, 268,
272, 274–75, 277, 280–81, 284, 286–
88, 290, 292–93, 295–96, 308–11,
313, 323, 327, 329, 333–35
matter realism, 9
Maufort, Jessica, 141–42
McKinney, Dan, 236
mechanism, 5, 21–22, 28–29, 31, 34,
36, 43, 45, 57, 59–61, 93–94, 99,
115, 119–20, 131, 225, 230, 295
mediator, 3, 17, 64, 112, 116–17, 119,
148, 155, 159, 164, 179–80, 207–8,
215, 244–45, 255, 268, 284, 287,
292, 294, 308, 334–35
Meillon, Bénédicte, 15–16, 18, 100,
102, 104–5, 140–42, 165, 167,
280, 330, 359
memory, 31, 47, 65–66, 70, 93, 152,
165, 200, 227, 276, 290, 292, 337
Merchant, Carolyn, 15, 23–24, 28, 30,
36–37, 44, 56, 67, 80–81, 101–2
Merleau-Ponty, Maurice, 26, 34–35, 39,
47, 95, 103, 111–12, 292, 328
metaphor, 3–4, 27, 33, 48–49, 52–53,
58–60, 66, 69, 75–77, 79–81, 83,
89, 98–99, 104, 122, 128, 142, 186,
210, 224, 237, 244–46, 249, 251–56,
259–60, 264, 291, 330, 334;
    metonymy, 24, 68–69, 81, 84, 87,
    110, 185–86, 327;
    synecdoche, 83, 269, 291, 308
middle voice, 301, 309, 314
Milcent-Lawson, Sophie, 264, 279, 306
milieu, 15, 136, 165, 217, 247
mimesis, 304, 322, 328
mine, 23, 65, 67–68, 77, 81, 86, 88–89,
91, 93, 100, 112, 122, 154, 179,
181, 281, 329
modern, 4–6, 11, 13, 21–28, 31–32, 34,
41–43, 46, 63, 68, 80, 113, 119–20,
128, 130–31, 137, 139, 141, 144–45,
148, 150, 158–59, 162, 173, 180,
188, 196, 204, 207, 209, 211–12,
214, 220, 222–24, 226, 231, 233,

250, 255–56, 278, 283, 287, 290,
295, 298, 302, 307, 316, 335–36
Monani, Salma, 16, 99
Moncomble, Raphaelle, 339
Morizot, Baptiste, 243, 270, 281
Morrison, Toni, 137, 336
Morton, Timothy, 15, 85
mother earth, 27–28, 53, 55–58, 62–63,
65, 71, 75–76, 81, 85, 90–91,
104, 115–16, 118, 122, 135, 193,
284, 287–88
mountain, 23, 65, 67–68, 72, 75, 81–82,
84, 86–93, 95–97, 105, 133, 144,
150–53, 159, 171, 176, 179–86,
189, 199, 232, 248, 266–70, 272–73,
289–91, 321–23, 325
mountain lion, 151–53
multispecies, 3, 5, 8–9, 11, 48, 59–61,
70–72, 74, 76, 90, 95, 128, 139, 143,
145–47, 149, 158, 160, 174, 189,
204, 214, 218, 221–22, 228–29, 233,
244, 254–56, 264, 266–67, 273, 278,
286–87, 301–2, 312, 335
Murphy, Patrick D., 136, 141
music, 2–3, 7, 9, 14, 16–17, 41–42, 81,
85, 113, 116–18, 199–200, 202, 206,
222, 226, 236, 284–86, 292, 294–
301, 305, 308, 314–15, 320, 322–23,
326–27, 335, 337
muteness, 15, 66, 83, 85–86, 89, 105,
109–10, 121, 145, 154, 157, 171,
174–75, 208, 218, 225, 269, 273,
293–94, 299, 301–2, 312–13, 320,
323–25, 327–28
mystical, 4, 10, 41–42, 44, 60, 99, 134,
137–38, 142, 177, 182–84, 188–89,
205–6, 214, 218, 226–28, 298, 305,
322, 324, 327, 329, 336
myth, 2–8, 27–29, 33, 43, 45, 52,
56–59, 61, 65, 78, 85–86, 92,
99–100, 104–5, 133–35, 137–38,
142, 147, 149–50, 155–56, 158, 161,
179, 200, 216, 228, 230, 233, 243,
254–55, 258, 295, 304, 311, 327,
334, 336, 338

mythopoeia, 6, 44, 52, 69, 77, 80, 85, 129, 142, 216, 233, 254, 273, 359

Nagel, Thomas, 279
Nalls, Gayil, 335, 339
Nash, Roderick Frazier, 224
naturalism, 8, 24–25, 53, 66, 132, 139, 144–45, 149, 153, 164, 180, 206, 215, 218, 268
naturculture, 69, 72, 90, 97, 117, 162–63, 181, 224, 273, 291, 359
nature, 3–5, 7–9, 15, 17, 21–32, 34–37, 41–47, 51–55, 57–67, 69, 71–77, 80–85, 93, 95, 99–102, 104–7, 109–10, 116–20, 122, 128–29, 134–37, 139, 141–42, 146, 151–53, 157–58, 160–62, 169–72, 174, 176, 178, 180–84, 188–89, 203–6, 209, 213–14, 217–18, 223–25, 227–28, 232–34, 236, 238–39, 244–54, 256, 258–60, 265–66, 268, 271, 273–74, 276–78, 280–81, 284, 287–88, 291, 293–99, 309, 312, 319, 321–22, 328, 333, 335–36
network, 10, 46, 59, 76, 82, 119, 207, 211, 223, 228, 250–51
new materialism, 45, 47, 49, 333
Nietzsche, Friedrich, 252
Nixon, Rob, 100
Noiray, Jacques, 73, 170, 177–78
nonhuman, 3–5, 7, 9, 11–12, 15, 22–24, 30, 32, 35, 43–44, 48, 54, 58–60, 63, 68–69, 72, 88, 102, 110–12, 119–20, 122, 127–28, 130, 136–38, 141–42, 147–48, 152, 154, 159, 178, 184, 188, 199, 202, 204, 208, 213, 219, 225, 229, 232–33, 238, 246, 249, 255–56, 264, 266, 270–71, 274–75, 284, 293, 309, 311, 313, 325, 336, 338;
    greather-than-human, 69, 73, 174, 228, 245, 271, 273, 278, 300, 320;
    other-than-human, 4, 11–12, 15, 25, 38, 43, 59, 69, 75–76, 85, 90–91, 106, 110, 118, 120, 122, 128, 132, 139, 143–45, 152, 154–55, 160, 174, 188, 196, 199, 203–4, 206, 208, 222, 225–26, 243–44, 251, 255–56, 263–65, 267–68, 274–75, 280, 301–3, 308–9, 312, 321, 327, 334–35
numbness, 30, 93–95, 106–7, 116, 127, 171, 243, 247, 314

odorscape, 8, 157, 171, 211–12, 218, 263–64, 271, 276, 301, 334
oikos, 6, 41, 44, 59, 63–64, 90–91, 97, 104, 115, 118, 182, 209, 224, 228, 247, 254, 284, 287, 294, 311, 321, 325
onto-epistemology, 9, 33, 78, 142, 206, 230, 296, 310, 333–34
ontology, 4, 6–7, 9, 21, 24–27, 34–35, 42–43, 46, 112, 132, 134, 136, 138, 140–41, 143–44, 148, 162, 164, 185, 188–89, 192, 197, 205–6, 210, 225, 233–34, 243, 257, 268, 312, 333, 337
Oppermann, Serpil, 16, 47, 49, 311, 333
Orenstein, Gloria Feman, 7, 16, 49, 79
organicism, 32, 56, 61, 66
Owens, Delia, 336

pagan, 23, 29, 37, 42, 52, 68, 181, 187, 235;
    neopagan, 7, 44–45, 195–96, 338
pain, 26, 30, 41, 47, 70, 86, 88–89, 91, 92–94, 96–97, 102, 105–7, 146, 156, 163, 191, 197, 207, 222, 228, 280, 298, 323;
    pain of the Earth, 93
Palmer, Jacqueline S., 121
Pancake, Ann, 13, 86–97, 105–7, 135–36, 140, 169–70, 179–82, 185, 188–90, 192–93, 228, 233, 286, 289–91, 304, 319, 333, 359
panther, 109, 145–49, 151, 153–56, 158–60, 162–65, 221, 257, 269, 272, 311

paradigm, 5, 7, 21, 26–28, 31–32, 35, 44, 59, 61, 65, 68, 97, 131–32, 214
paronomasia, 110, 186, 298, 315
pathological, 23, 93
Patoine, Pierre-Louis, 127
patriarchy, 23, 31–32, 36, 43–45, 55, 62, 68, 79–82, 102, 105, 115, 195, 204, 250, 287–89, 303, 336
perception, 4–6, 8, 10, 12–13, 15–16, 24–26, 29, 34–35, 44, 47–48, 63–64, 66, 69, 90, 94, 106, 112–13, 117–20, 122, 128, 131, 137–39, 143–45, 148, 155–58, 161, 172, 174, 176, 180, 185, 189–90, 203–5, 207, 210, 220–22, 224, 228–29, 243–54, 256–60, 264, 266, 268, 270–71, 277, 283, 286, 301, 305, 308, 313, 321, 326, 333, 339
Perrin, Claire, 98, 195, 339
perspective, 2, 5, 9–10, 13–14, 22, 24, 43–45, 51–54, 59–60, 62, 64–68, 71, 73–74, 79–80, 84, 86–87, 97, 100, 111, 114, 127–28, 132–33, 135, 138, 144, 146, 148, 150, 153–54, 158–61, 172–74, 180, 186, 197, 200–203, 210–11, 216–17, 220, 223, 227, 232, 245–51, 256–57, 263–68, 270, 272–74, 276, 279–80, 284, 288, 296, 302–3, 307, 310–11, 321–22, 325–26, 334
Peterson, Brenda, 144
phenomenology, 26, 34–35, 47, 103, 111, 132, 175, 227, 253, 283
Pinson, Jean-Claude, 15, 17, 292, 293
place, 4–9, 17, 22–23, 28, 37, 48, 52–53, 56, 63–66, 71–72, 74, 81, 89–94, 97, 106, 117–18, 132, 141, 146, 148–49, 151–52, 154–57, 159, 162–63, 172, 175, 179–80, 183–85, 197, 199, 201–2, 204, 215, 218, 226, 230–31, 245, 251, 257, 259, 263–65, 267–69, 275, 277–78, 283, 288, 291, 294, 296, 298, 304, 307, 312–13, 324, 329, 333, 335, 338–39;

space, 35, 91, 119, 129, 139, 141, 149–50, 156, 159, 170, 177, 239, 257, 294
plant, 3–4, 7, 11–12, 15, 25, 47, 59, 68, 70, 75–76, 81, 91, 95, 98, 104, 113, 117, 133, 135, 139, 141, 147–49, 159–60, 164, 171, 174, 176, 178, 182–83, 199, 203, 207–9, 211–15, 219, 224–26, 229, 249–51, 253–54, 256, 260, 267, 269, 276, 278, 281, 289–91, 304–5, 307, 315–16, 335
Plant, Judith, 49
Plaskow, Judith, 49, 54, 79
Plumwood, Val, 15, 36
politics, 6, 17, 24, 41–42, 46, 48–49, 56–57, 80, 100–2, 109–10, 134–35, 170, 180, 201, 230–32, 234, 268, 333
Pope Francis, 44
postmodern, 3, 5, 8–10, 28, 30–32, 34, 38, 41, 60–62, 80, 129, 131, 136, 140–42, 185, 197, 198, 205–6, 209, 216–17, 219, 228, 230, 232–33, 243–44, 268, 333–34
postmodernism, 31, 52, 129, 134, 136, 333
post-pastoral, 140, 169–70, 179, 189, 268
Power, Susan, 13, 16, 307–8, 316, 333
Powers, Richard, 13, 56, 58, 61, 127, 130, 134, 135–36, 140, 170, 190, 205–6, 209, 212–14, 216–33, 235–37, 239, 243, 251–55, 276, 286, 333
Prigogine, Ilya, 15, 32
proprioception, 53, 59, 137, 257, 286, 294–95, 303, 324, 326
Proulx, Annie, 80–85, 99, 101, 104–5, 135–36, 333, 359
psyche, 10–12, 22–23, 38, 44, 57, 91, 93–94, 107, 115–16, 118, 129, 139, 143, 181, 185, 190, 196–97, 230–31, 248–49, 259, 296, 301, 336
psychology, 10, 43, 93–94, 107, 132, 135, 205, 228, 231, 233, 259, 311
Pughe, Thomas, 15, 232, 244
Pyle, Robert, 120, 122, 336

quantum physics, 5, 26, 28, 30–31, 33–36, 38, 44–45, 48

rape, 67, 81, 96, 337
Rash, Ron, 13, 322–30, 333, 359
rational, 4, 6, 12, 21, 29, 42, 52, 72, 121, 128, 130–32, 134, 136, 139, 145, 181, 190, 202, 206, 214–15, 220, 224–26, 244, 298, 333, 336;
    irrational, 7, 12, 21, 72, 172, 177, 220, 223–24
reason, 7, 12–13, 17, 24, 36, 42, 74, 94, 163, 175, 177, 187, 189, 208, 210, 211, 231, 246, 264, 269, 298, 338
reciprocity, 43, 64, 76, 78, 85, 152, 160, 164, 170, 182, 201, 219, 225, 300, 309, 311, 334
reductionism, 4, 21, 61, 115, 119, 131, 137
reenchantment, 2, 5, 7, 9–10, 14–16, 26–28, 30–33, 42–44, 46–47, 51, 54, 60–61, 83–86, 94, 96–98, 109, 111, 114, 117, 119–20, 140–41, 173, 175–76, 179, 195–96, 205–6, 212, 228, 230, 232–33, 243, 256, 258, 268, 274, 283–84, 291, 296, 305, 310, 312, 323, 325, 330, 333–38, 359
religion, 10, 21, 23, 30, 38, 44–45, 49, 53, 64–65, 217, 268
remystification, 51, 333
restor(y)ation, 338
restoration, 44, 94, 97, 114, 119–20, 151, 196, 269, 312
reverberation, 112, 118
rhizome, 216, 223, 229, 232, 247, 267, 281
Ricœur, Paul, 245–46, 259–60, 283, 292
Rigby, Kate, 49
ritual, 6, 8, 10–11, 17–18, 42, 56, 78, 111, 133–34, 137, 152, 158, 161, 164, 201, 215, 219, 222, 224, 276, 300, 307–8, 312–13, 335
river, 22–23, 65–68, 86, 88, 101, 116–17, 159, 201–2, 232, 259, 269, 310, 315, 337

Robin, Marie-Monique, 13
Roger, Alain, 101
Roh, Franz, 138
Romestaing, Alain, 103
Roszak, Theodore, 16, 46–47, 60–61, 87–88, 93, 99, 106, 140, 336
Rumpala, Yannick, 195
Ryden, Kent, 8, 92

sacred, 10, 27, 42, 44–45, 48, 52–56, 65–66, 68, 78, 150, 152, 189–90, 258, 284, 301, 307–8, 316, 335
Schafer, Murray, 2, 13–14, 113, 285
Schiller, Friedrich, 22, 336
Schoentjes, Pierre, 15
semiotic, 115, 135, 286–87, 290, 294, 303–4, 320, 326
Şenel, Neşe, 98
sensitive, 2–4, 8, 11–12, 17, 24, 29, 33–34, 38, 42, 52, 54, 98, 106, 111–14, 116–18, 120, 127, 131, 135, 137, 139, 145, 147, 150, 154–56, 171–76, 196, 208, 210, 212, 218, 244, 247, 252–53, 258, 263, 268, 276–77, 285–86, 288–89, 293–94, 308, 323, 325, 334–36, 339
sensuous poiesis, 110–11, 120, 245, 290, 293, 304, 321–22, 325, 334
sentience, 5, 25, 31–32, 56, 59, 77, 119, 133, 145, 173, 189, 206, 208, 211, 216, 219, 224, 229–30, 249, 255–56, 263, 265, 274, 280, 293, 301, 305, 333, 338
Serres, Michel, 83, 86, 105, 116, 247, 314
Servigne, Pablo, 140
Sewall, Laura, 107, 119, 122, 243
Shakespeare, William, 224
shamanism, 4, 6, 11, 25, 73, 95, 111, 130, 133, 137, 142, 145, 148–49, 155–56, 158–60, 162, 164, 169–70, 174, 178, 185, 187, 190, 196, 198–201, 203, 207–9, 213–14, 216, 219, 222, 235, 263, 268, 276, 334

shapeshifting, 6, 11, 150, 174, 203, 209, 245, 248, 254, 291, 334
Sheldrake, Rupert, 31, 38, 196, 205, 234
Shepard, Paul, 204
Sibran, Anne, 337
sight, 8, 17, 27, 35, 38, 56, 58, 60, 112–13, 118, 130, 137–38, 144–45, 147, 157, 172, 179, 183–84, 186, 190, 196–97, 202, 207, 223, 231, 244, 247, 250–51, 263, 269, 273, 276, 278, 285, 307, 312, 314, 323, 325, 334;
> blind, 2, 6, 8, 21, 25, 29, 38, 66–67, 76, 92–93, 113, 119, 137–38, 147, 176, 180, 190, 196, 209–10, 215, 218–19, 223, 226, 247, 251, 252, 254, 256, 265, 271, 286, 292, 293–94, 304, 310–11, 314, 324, 334;
> ocular hypertrophy, 113, 118, 285

Silko, Leslie Marmon, 13, 27, 77–78, 135–36, 140, 143, 150–52, 154, 233–34, 333
Simard, Suzanne, 60, 205, 210–11, 236, 237
Simon, Anne, 279
Slemon, Stephen, 231
Slovic, Paul, 107, 127, 243
Slovic, Scott, 7, 17, 107, 109, 121, 127, 243
smell, 8, 72, 87, 90, 113–14, 118, 120, 138, 145–46, 151, 155–57, 171, 173, 197, 200, 202–4, 211–14, 219, 226–27, 263, 269–72, 275–78, 281, 285–86, 289–90, 308, 323;
> anosmia, 157, 247

Snow, Charles Percy, 32
Snyder, Gary, 46, 217, 225, 260
soil, 41, 56, 70–71, 76, 86, 101, 116–17, 148, 177, 179, 200, 206–7, 228, 267, 288, 293, 305
solastalgia, 94, 96
song, 2–3, 7, 9–10, 14–15, 41, 73, 77–78, 87, 90–91, 98, 110–12, 114, 117–18, 128, 130, 137, 143, 164, 172–74, 191, 200–201, 208, 213, 221, 225, 239, 244, 254, 257, 263, 268, 284–86, 290–97, 300–302, 304, 308–10, 312–13, 319–20, 322–23, 326–28, 330, 335, 337;
> song of the earth, 2–3, 14, 48, 91, 97, 109–10, 114–16, 118, 169, 172, 191, 202, 233–34, 244–45, 264, 283–88, 292, 294–96, 299, 308, 313, 321, 330, 334, 337–38

sound, 3, 7–8, 15, 71, 78, 83, 87, 109–11, 113–14, 117–18, 120–21, 145, 146, 156–57, 171–73, 175, 183–84, 186–87, 196–97, 200, 202, 211, 220, 226, 257, 264–65, 267, 274, 284–86, 291, 293–301, 305, 311, 313–15, 319–20, 322–23, 327–28, 330;
> soundscape, 2, 3, 8, 13–15, 110, 112, 114, 146, 157, 171–72, 174, 218, 264, 284, 293, 295, 301, 323, 334–35

spider, 6, 28, 66, 75–79, 104, 328–29
Spill, Frédérique, 329
spirit, 6, 9, 22–24, 26–27, 29, 34, 42, 45, 51, 54–55, 64, 72, 77, 103, 134–35, 139, 145, 147, 150, 153, 156, 158, 162, 197–201, 216, 219, 225, 231, 248, 258, 287–88, 307, 309–13
spirituality, 4, 7, 10, 15, 25, 27–31, 38, 41–45, 49, 53, 55, 57, 68, 72, 74, 84–85, 91, 93, 129, 136, 138–39, 152, 164, 173, 177, 179, 182, 185, 189, 195, 206, 216–18, 221, 226–29, 236, 248, 268, 288–89, 305, 309–12, 335
Sponsel, Leslie E., 49
Spretnak, Charlene, 44, 79, 102, 305
Stamm, Gina, 169, 170
Starhawk, 5, 13, 15–16, 44–46, 49, 52–55, 85, 95–96, 104, 127, 129, 135–36, 138, 140, 148, 170, 190, 195–201, 203–4, 212, 229, 232–35, 284, 292–94, 312, 327, 333

Stengers, Isabelle, 5, 15, 32
Stevenson, Sheryl, 259
Stewart, Garrett, 293
Stone, Christopher D., 230, 239
storytelling, 7, 33, 36, 41, 47–48, 78, 129, 158, 164, 179, 198–99, 229, 238, 243–44, 254, 307, 309–13
subliminal, 11, 13, 47, 90, 196, 211, 218, 227, 286, 303
Suhamy, Henri, 131, 177
supernatural, 6, 12, 25–27, 36, 128–30, 133, 137, 139, 158, 172, 174, 190, 198, 223, 235
superstition, 4, 7, 137, 158
Swimme, Brian, 30, 32–33, 41, 79, 113, 118
symbiosis, 51, 56, 59, 61–62, 76, 82–83, 90–91, 98, 100, 157, 278–79, 294, 334
sympoiesis, 3, 15, 258, 302, 307, 309, 314

Taylor, Bron, 49, 53
Taylor, Paul, 14
technology, 30, 42, 80, 81, 172, 175, 196, 283, 302
Thaler, Lore, 304
thanksgiving, 64, 219
Thoreau, Henry David, 17, 24, 206, 214, 218, 280, 338
thought experiment, 204, 213, 219, 235, 263
Tiffin, Helen, 16
Tölke, Guido, 236
totemism, 7–9, 11, 17–18, 23, 43, 130, 132–33, 140–41, 145, 150–52, 155, 162–65, 169–70, 178, 180, 183–84, 189–90, 192, 196, 200, 204, 206, 215–16, 229, 304, 309–10, 316, 333, 336
touch, 35, 74, 91, 112, 137, 146, 176, 182, 184, 247, 250, 270, 285, 289–90, 299, 308, 315, 319, 321, 323–25, 334

transcendence, 8–9, 24, 27, 31, 43, 54, 57, 70, 91, 138, 189, 206, 218, 231, 299, 309
translate, 3, 9, 13, 25, 34, 56, 176–77, 198, 208, 210, 212, 217, 222, 225, 232, 260, 272, 285, 288, 290, 306, 308, 313, 319
translation, 21, 27, 48, 114, 117, 130, 136–37, 146–47, 152, 158, 164, 170, 188, 190–92, 196–97, 208, 211, 213, 215, 220, 222, 233, 244–45, 255, 271, 286–87, 290, 297, 301, 305–8, 314, 320, 322, 328–29
trauma, 27, 78, 93, 102, 181, 259, 323
Tredinnick, Mark, 7, 14, 16, 286, 296, 309
tree, 2, 17, 22–23, 27, 56, 58–61, 65, 71–72, 74–76, 80, 89, 95, 114–15, 127, 130, 134, 145, 147–48, 151, 153, 156, 160, 170–72, 178, 183, 185–86, 199–200, 205–30, 232, 236–37, 239, 243, 248–56, 259–60, 265–67, 275–76, 278, 284, 286, 290, 297, 299, 312, 315, 336–37
Tudge, Colin, 62, 205, 211
Turner, Victor, 10–12, 18, 160–61, 201, 259, 261

Van Cauwelaert, Didier, 260
vibrancy, 3, 8, 35, 113–14, 131, 138, 147, 158, 189, 196–97, 201, 203–4, 213, 228, 259, 264–65, 276, 284–86, 291, 293, 295, 297–98, 305, 311, 313, 323, 325, 329, 333–36
Viola, Alessandra, 236, 260, 281, 305
violence, 65, 68–69, 71–72, 80–81, 83–84, 87–88, 92, 94–95, 105, 162, 191, 196, 231, 266, 297, 303, 323
vision, 4, 24–25, 27, 32, 43, 45–46, 48, 52–55, 60, 63–69, 71–73, 76–77, 81–83, 89, 92, 99, 109–10, 117, 133, 137–38, 144–45, 147–48, 157–58, 161, 175, 177–80, 187, 189–90, 195 96, 199–201, 204, 207, 209–10, 215,

219, 230, 232, 246–54, 256–58, 263, 274–75, 295, 302, 310–12, 325, 327
Viveiros de Castro, Eduardo, 5–6, 16
Vizenor, Gerald, 224–25, 304
voice, 2, 10, 15, 46, 47, 83, 86, 87, 88, 97, 109, 110, 114, 115, 118, 127, 159, 164, 173, 180, 187, 188, 189, 199, 205, 206, 212, 213, 214, 219, 220, 221, 222, 226, 229, 231, 232, 233, 238, 252, 257, 258, 264, 266, 268, 284, 287, 288, 290, 291, 292, 294, 296, 301, 302, 308, 309, 312, 313, 314, 315, 319, 320;
    voice of the earth, 15, 4, 47, 98, 110, 118, 164, 187, 189, 288, 291, 292, 301, 319, 327
von Mossner, Alexa Weik, 127
von Uexküll, Jakob, 165

water, 36–37, 53–56, 65–67, 92–93, 100–101, 110, 116, 144, 146–47, 149, 155–56, 164, 171, 175–76, 206, 217, 255, 257, 265, 277, 300, 307–8, 310, 313, 315, 324–25, 329
Watkins, Claire Vaye, 337–39
Watts, Alan, 46, 61, 148
weaving, 17, 43, 45, 49, 76–80, 87, 107, 110–11, 118, 132, 136, 141, 152, 162, 189, 200, 224, 228, 230, 244, 275, 284, 287, 300–301, 307, 311, 313, 328, 329;
    interweaving, 5, 7, 27, 35, 41, 59, 71–73, 76, 86, 112, 118, 133–34, 149, 152, 164, 204–5, 218, 230, 232–33, 247, 258, 266, 273, 311, 328, 334;
    spinning, 77–78, 328–29.
    See also braiding
web, 5, 17, 28, 75–79, 104, 107, 172, 218, 273, 328
Weber, Max, 4, 21–22, 29, 30
whale, 79, 127, 130, 143–44, 149, 152
Wheeler, Wendy, 172
White, Jr., Lynn, 4, 23, 37, 49, 64, 68
Whitehead, Alfred North, 32, 39
wild, 3, 7–8, 14, 36, 73, 83–85, 116, 146–47, 151–57, 162, 171–72, 178, 182–83, 185, 189, 201–2, 210, 224, 227–28, 232, 238, 248, 250, 255, 267, 270, 284, 290, 294, 296–99, 305, 319, 321–22, 326–27, 335–37
Williams, Terry Tempest, 13, 287–88, 329
Wilson, Edward O., 12, 43, 47, 155, 157, 235, 238, 246
witch, 4, 11, 24–25, 45, 52, 195–201, 220, 284, 312
wolf, 103, 268–70, 272
wonder, 3, 15, 22, 30, 33, 42, 45, 51, 62, 71, 96, 109–11, 119–21, 130, 153, 172–73, 178, 190, 205, 210, 214–15, 217, 223, 268, 273, 275, 281, 283, 322, 325–29, 333–35
Woods, Angela, 221
world, 2–15, 17–18, 21–36, 38, 42–48, 51–54, 56–57, 59–61, 63, 66, 68–73, 76–81, 85–88, 92–97, 101, 103–4, 106–7, 110–16, 118–22, 127–38, 141, 143, 144–63, 169–79, 182–85, 187–91, 195–98, 200–204, 206–10, 212, 214–15, 218–19, 221–25, 228, 230–34, 239, 243–49, 251–60, 263–65, 268–69, 271–74, 276–78, 283–85, 287–88, 290, 292–93, 295–304, 307–13, 316, 320, 322, 324–30, 333–38
Wrede, Theda, 100, 102
Wright, Alexis, 141, 336

Ywahoo, Dhyani, 93, 300

Zamora, Lois Parkinson, 134, 139, 140
Zhong Mengual, Estelle, 243
zoocentric, 225, 246
zoopoetics, 264, 279, 306

About the Author

Bénédicte Meillon is associate professor at the University of Perpignan and currently affiliated with the LARCA-CNRS at the Université de Paris via a one-year fellowship. She codirects the UPVD ecopoetics research workshop, has created a network and Internet platform dedicated to ecopoetics and ecocriticism (https://ecopoetique.hypotheses.org), and sits on the EASLCE Advisory Board.

Her research explores ecopoetics of reenchantment, focusing on magical realism and "liminal realism," mythopoeia, and ecofeminism, and paying close attention to the intra-actions between naturcultures and the texture of language itself. She has published papers dealing with ecopoetic readings of environmental fiction by Barbara Kingsolver, Annie Proulx, Linda Hogan, Ann Pancake, and Ron Rash. She has also written papers on fiction by Russell Banks, Roald Dahl, and Paul Auster. She has co-edited, with Dr. Margot Lauwers, a special issue of *Crossways Journal: Lieux d'enchantement: approches écocritiques et écopoét(h)iques des liens entre humains et non-humains* (2018). She has directed a volume on *Dwellings of Enchantment: Writing and Reenchanting the Earth* (Lexington Books, 2021), and has also co-edited two transdisciplinary issues of *Textes & Contextes* on the reenchantment of urban wildness: the first, with Rachel Bouvet and Marie-Pierre Ramouche, dealing with eco- and geo-poetic approaches (June 2021) and the second, codirected with Sylvain Rode and Hélène Schmutz, with contributions in the Environmental Humanities (Nov. 2021). She is engaged in multimedia creative projects mobilizing arts and performance as a way to restore ecological attention. She successfully completed her Habilitation à Diriger des Recherches in 2021.

www.ingramcontent.com/pod-product-compliance
Lightning Source LLC
Chambersburg PA
CBHW021338300426
44114CB00012B/989